Chemisorption of Gases on Metals

Chemisorption of Gases on Metals

F. C. TOMPKINS

Department of Chemistry
Imperial College of Science and Technology
London

1978

ACADEMIC PRESS

London New York San Francisco

A Subsidiary of Harcourt Brace Jovanovich, Publishers

ACADEMIC PRESS INC. (LONDON) LTD.
24/28 Oval Road
London NW1

United States edition published by
ACADEMIC PRESS INC.
111 Fifth Avenue
New York, New York 10003

Library of Congress Catalog Card Number: 77-93207
ISBN: 0-12-694650-7

Printed in Great Britain by Page Bros (Norwich) Ltd, Mile Cross Lane, Norwich

Preface

Some valuable monographs on specific aspects of chemisorption phenomena on metals have been published during the last decade, but the most recent comprehensive survey in the English language appears to be the second edition (1964) of "Chemisorption" by D. O. Hayward and B. M. W. Trapnell. Since that time, a considerable advance in experimental techniques and in the theoretical treatment of the nature of the chemisorption bond have taken place. The present volume is an attempt to bring the subject up to date and to cater for the needs of the younger post-graduate students. Attention has been confined to metal adsorbents, since the acquirement of detailed knowledge of gas–metal surface interactions has recently been particularly informative; inclusion of oxides and similar adsorbents would have involved other concepts.

The present treatment is an attempt to provide a fairly comprehensive survey of chemisorption on metals and, by the inclusion of selective references to the more important reports in the original literature, to form a basis for further detailed studies. The earlier chapters deal with the more classical aspects of the subject and include an introduction to the thermodynamics and statistical thermodynamics of chemisorption; those on the kinetics of adsorption and desorption processes, however, place emphasis on the more recent investigations of sticking probabilities, precursor and multiplet states, energy accommodation and molecular beam scattering.

A brief introductory outline of the free-electron theory of metals is followed by a fuller treatment of the electronic properties of the surface metals. The perturbation of these properties by a chemisorbed layer and its effect on the electrical conductance and magnetic susceptibility of the adsorbent are included to place these earlier studies in their correct perspective. Work functions are, however, considered in some detail because they have now been placed on a firm quantitative theoretical basis and also they provide an essential parameter in chemisorption theory. The geometric structures of the metal surface and chemisorbed layer as revealed by LEED are then discussed.

The remaining chapters are primarily concerned with the most recent experimental techniques. The theoretical foundations of the methods, together with some essential experimental details, are outlined, and their contributions to our understanding of chemisorption phenomena are discussed. These techniques include transmission and absorption–reflectance infra-red spectrometry, field-emission and field-ion microscopy, particularly in relation to the surface mobility of adspecies, and the many forms of electron spectroscopy (Auger, ultra-violet and X-ray photoelectron, ion neutralization, electron energy loss, appearance potential, elastic resonance and inelastic tunnelling field-emission) that are now providing information about local density of states, bonding orbitals and binding energies. There follows a brief account of modern theories of chemisorption and of their relevance to electron-spectroscopic results. In the final chapter an attempt is made to assess the present status of knowledge by reference to some simple gas/metal systems.

F.C.T.

Contents

1

Introduction

1.1. TERMINOLOGY AND DEFINITIONS

Atoms at the surface of a solid exert an attractive force normal to the surface plane. Consequently, at a gas/solid interface, the concentration of gas exceeds that in the gas phase. The excess concentration at the surface is termed adsorption; the solid is the adsorbent and the gas the adsorbate. The amount adsorbed is the difference between the number of adsorbate molecules in the adsorbed layer and that in an equivalent layer in the absence of the adsorbent. Its magnitude depends on the physical and chemical properties of both adsorbent and adsorbate, the temperature of the gas/solid system and the ambient pressure of the gaseous adsorbate.

The term adsorption is restricted to the accumulation of adsorbate at the gas/solid interface. For porous adsorbents, gas molecules can diffuse down capillaries and channels and be adsorbed on their walls. Gas molecules can also penetrate the lattice of the adsorbent by a process akin to dissolution. This internal uptake is termed absorption. A process involving both adsorption and adsorption is called sorption. Adsorption may also be followed by a chemical process which is continued into the lattice, finally resulting in a three-dimensional compound, such as an oxide when oxygen is in contact with some transition metals. This process is termed incorporation, and the surface is said to have undergone reconstruction or reconstitution (Fig. 1.1).

Adsorption of a gas at the surface of a solid proceeds spontaneously and the free energy G of the gas/solid system is decreased. Moreover, the adsorbed state is more ordered than the gaseous state and the entropy change ΔS is negative. Since

$$\Delta G = \Delta H - T\Delta S,$$

it follows that the enthalpy change ΔH is negative, i.e. adsorption is an exothermic process, the heat liberated being termed the heat of adsorption q.

In general, the directly-measured physical quantities are the amount sorbed (in cm^3 s.t.p.) and the weight of the adsorbent (in g). For a physically uniform form of the adsorbent, its surface area is directly related to its weight; for a uniform (or homogeneous) surface, the amount adsorbed, at a particular

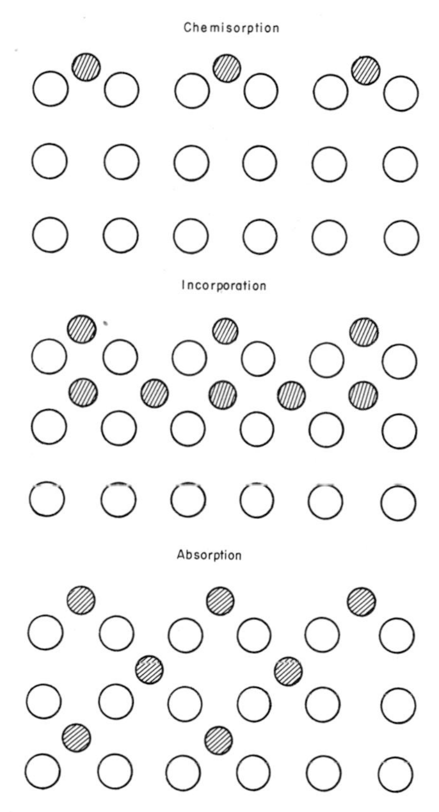

FIG. 1.1. Diagrammatic representations of chemisorption, incorporation and absorption.

temperature and a specific ambient pressure of gaseous adsorbate, is directly proportional to the surface area. Despite the fact that the experimental accuracy of surface-area determinations is usually not better than 20%, the amount adsorbed is often expressed in terms of the monolayer capacity v_m, i.e. the amount to cover the whole surface with an adsorbed layer of one molecule thickness. The ratio of the amount adsorbed v to that necessary to form a complete monolayer is the fractional coverage $\theta = v/v_m$. At higher pressures and lower temperatures, the monolayer capacity can be exceeded by formation of a multilayer of many molecules thickness and the amount adsorbed is then expressed as a multiple of the monolayer content.

1.2. PHYSICAL ADSORPTION (OR PHYSISORPTION) AND CHEMISORPTION

1.2.1. Physisorption

In physisorption, the adsorbate is held to the surface by the van der Waals forces that arise primarily from the interaction of fluctuating dipoles. The enthalpy of physisorption, or the binding energy of the adsorbate, is, therefore, of similar magnitude to that of the corresponding heat of condensation of the gaseous adsorbate (i.e. ~ 20 kJ mol^{-1} or less). Since metals contain a large number of mobile conduction electrons, they might be expected to have unique adsorption properties not possessed by dielectrics. However, for many gas/metal systems, e.g. inert gases and transition metals, the resonance frequencies of the adsorbate atoms are so high that the conduction electrons are unable to adjust to the dipole fluctuations of the adsorbed atoms. Consequently, the metals are incompletely polarizable and their adsorptive properties are not greatly different from those of dielectrics.

Nevertheless, the physisorption of inert gases effects a substantial reduction in the effective work function of the metal. This change in the surface distribution of conduction electrons indicates that an energy term, additional to the dispersion contributions, should be included in the binding energy. The simplest concept is that the adsorbed atoms are polarized by the large electrical field gradient at the metal surface. The energy to produce these field-induced dipoles can be approximated to $\frac{1}{2}F\alpha^2$, where F is the field strength and α the volume polarizability of the adsorbed atom; the additional energy is, therefore, greater the larger the atomic weight of the adsorbate. As an example, for argon, krypton and xenon adsorbed on a tungsten surface, the heats of adsorption at small fractional coverages [1] are 8, 18 and 35 kJ mol^{-1}, compared with condensation heats of 6·7, 9·0 and 12·7 kJ mol^{-1}, respectively.

Since physisorption heats are small, the amount adsorbed only becomes substantial near or below the boiling point of the adsorbate; multilayers are easily formed at higher pressures. Similarly, the extent of adsorption on the surfaces of different solids under the same conditions of temperature and pressure do not greatly vary.

1.2.2. Chemisorption

Chemisorption may be considered to be a chemical reaction between an adsorbate molecule and the surface array of metal atoms. The greatest interaction is with metal atoms in immediate contact with the adsorbed molecules, but the weaker interactions with all the other surface atoms are not negligible. Chemisorption binding energies have a similar magnitude to those of primary chemical bonds in free molecules. Thus, heats of chemisorption for many gas/metal systems range from 50 to 400 kJ mol^{-1}. For carbon

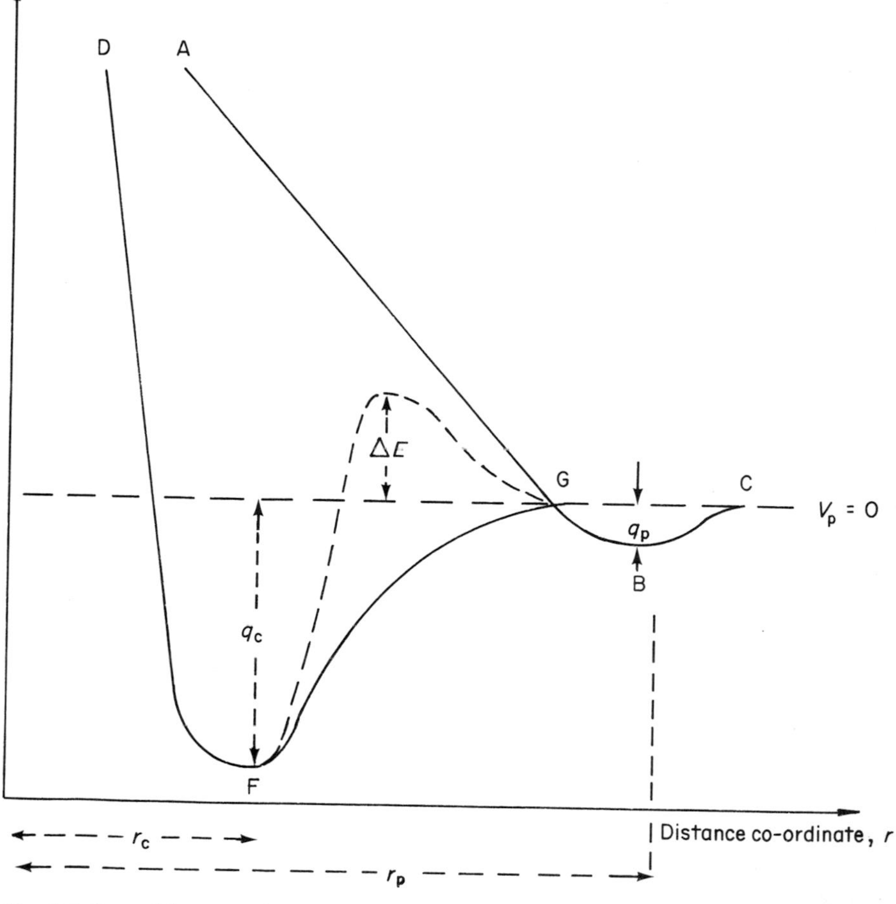

FIG. 1.2. Potential energy of a system as a function of the distance of the adsorbate molecule from the metal surface, with respect to energy zero for the adsorbate molecule at infinite distance from the surface.

Curve ABC represents physisorption with adsorption heat q_p at the equilibrium distance r_p. Curve DFG represents chemisorption with adsorption heat $q_c > q_p$ at the equilibrium distance $r_c < r_p$, corresponding to a short-range chemical interaction. The cross-over point G at $V_p \sim 0$ indicates that the chemisorption is a non-activated process. The dotted line corresponds to a system for which an activation energy ΔE is required for passage into the chemisorbed state.

monoxide on transition metals, as an example, values between about 170 and 350 kJ mol^{-1} are obtained. This wide variation indicates marked chemical specificity for different gas/metal systems. Figure 1.2 shows the potential energy curves for physisorption and chemisorption as the adsorbate molecule approaches the metal surface.

Since chemisorption involves short-range chemical forces, it is normally limited to the monolayer. Nevertheless, at higher gas pressures and at moderately low temperatures, the formation of a physisorbed or/and a weak chemisorbed layer of a different character may, in some systems, proceed over the primary chemisorbed layer.

1.2.2.1. *Dissociative chemisorption*

The specificity is exemplified by the fact that a gaseous adsorbate molecule may be dissociated into its component atoms and/or radicals. The adsorbed species (or adspecies) then comprise adsorbed atoms (or adatoms) and/or

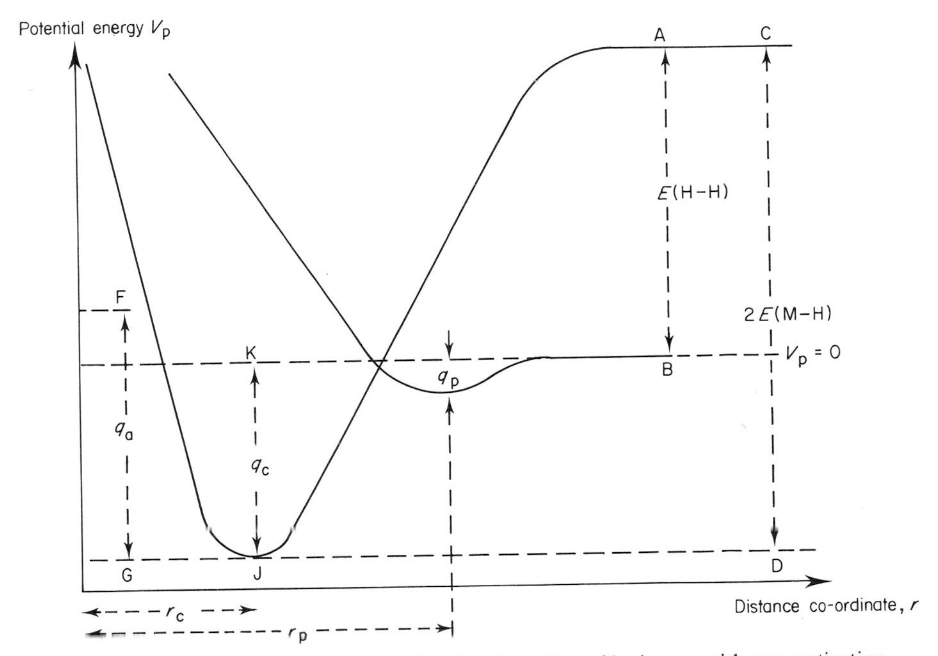

FIG. 1.3. Potential energy plot for dissociative chemisorption of hydrogen with zero activation energy. AB represents the dissociation energy of the free hydrogen molecule; CD is twice the bonding energy of the hydrogen atom to the metal surface and FG represents E(M—H) (i.e. $\frac{1}{2}$CD). The desorption energy to give a free hydrogen molecule is KJ, and FG is that to desorb a hydrogen atom with $q_a > q_c$, where the heat of chemisorption of the hydrogen molecule q_c equals $2E(\text{M—H}) - E(\text{H—H})$.

adsorbed radicals (or adradicals). This process is termed dissociative chemisorption. When the adspecies has the same form as the molecule of the gaseous adsorbate, the term associative chemisorption is used. Some adsorbates may undergo both dissociative or associative chemisorption on the same metal, the former being favoured at higher temperatures.

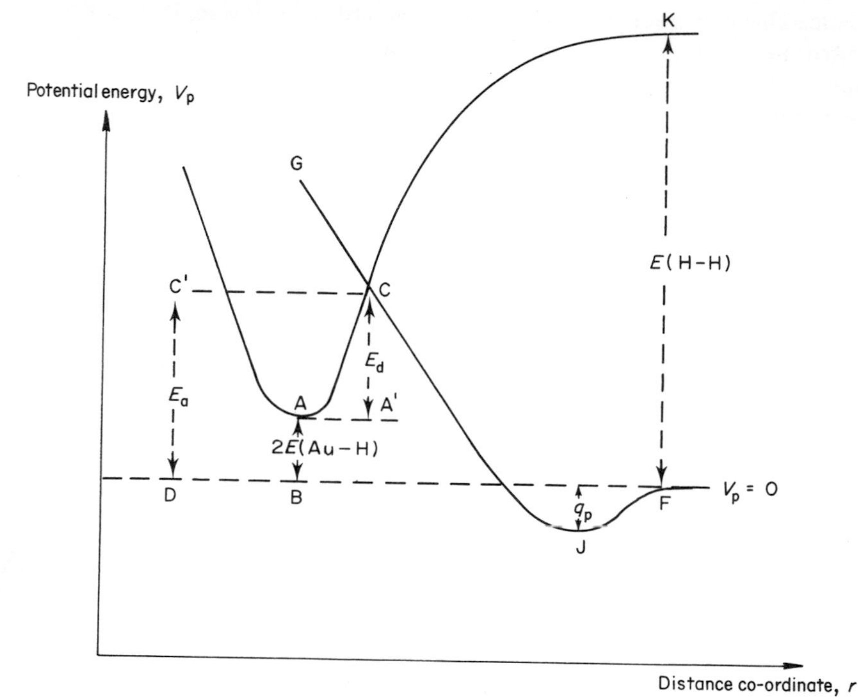

FIG. 1.4. Potential energy plot for the adsorption of hydrogen atoms on gold. AB is the (endothermic) heat of chemisorption of two hydrogen atoms relative to $V_p = 0$ for gaseous molecular hydrogen; KF is the dissociation energy of free hydrogen molecules; CA′ is the energy of desorption E_d as hydrogen molecules, and C′D the activation energy for chemisorption E_a as hydrogen atoms from gaseous hydrogen molecules. GJ is the plot for physisorption of hydrogen molecules.

The occurrence of dissociative chemisorption may be accounted for by the presence of unsaturated metal bonds at the surface. With hydrogen as adsorbate, two adjacent metal atoms react with the hydrogen molecule. When the energy gained in forming two metal atom/hydrogen adatom bonds exceeds the dissociation energy of the free hydrogen molecule, i.e. $2E(\text{M—H}) > E(\text{H—H})$, dissociative chemisorption is thermodynamically possible (Fig. 1.3). The chemisorption of most gases on transition metals is a non-activated process, consequently dissociative chemisorption is controlled by the thermodynamics of the process and not by the kinetics. However, there are exceptions; for example, a moderately large activation energy is necessary for the chemisorption of nitrogen by iron.

Transition metals are particularly active in chemisorption and their magnetic and electrical properties are modified in the process. This activity has, therefore, been correlated with the presence of unpaired *d* electrons in the metal and their participation in the formation of a strong chemisorption bond. Metals such as silver and gold have an *sp* character and the (M—H) bond energy is weaker and insufficiently large to effect dissociative chemisorption. Nevertheless, hydrogen atoms are strongly attached to these metals at low temperatures with a binding energy of 200 to 220 kJ mol^{-1}, i.e. not substantially less than that with transition metals (270–310 kJ mol^{-1}).

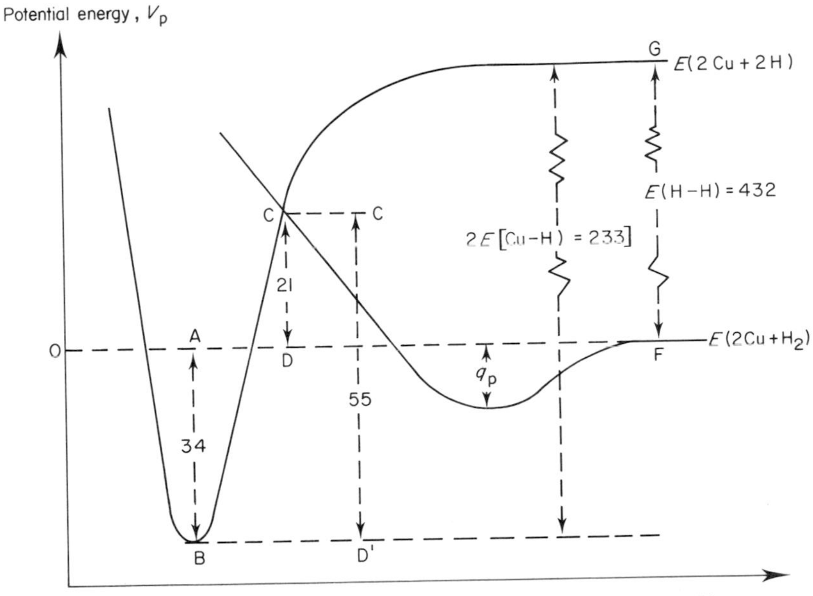

FIG. 1.5. Potential energy plot for the dissociative chemisorption of hydrogen on copper. GF is the dissociation energy of the free hydrogen molecules; CD is the activation energy for chemisorption; AB is the heat of chemisorption and CD′ the heat of desorption of hydrogen molecules.

Figure 1.4 shows the form of the potential energy plot for hydrogen atoms on gold. The potential energy of the two adsorbed hydrogen atoms is higher than that of a free hydrogen molecule, but lower than that of two gaseous hydrogen atoms. Chemisorption of atoms, therefore, proceeds when the hydrogen gas is pre-atomized in the presence of a hot tungsten filament and the adsorbent is at a sufficiently low temperature. With increase of temperature, the adsorbed atoms become mobile over the surface and, on collision, two atoms are evolved as a hydrogen molecule; the desorption energy is, therefore, roughly equal to the activation energy for migration. But, since the activation energy for adsorption of gaseous hydrogen molecules is greater than the heat of desorption, only physisorption can take place.

However, hydrogen is chemisorbed on the *sp* metal copper [2] when its gas pressure exceeds 5×10^{-1} Pa. The activation energy for adsorption is 21 kJ mol^{-1} and the heat of adsorption of 34 kJ mol^{-1} corresponds to a binding energy of hydrogen atoms [E(Cu—H)] of 233 kJ (see Fig. 1.5).

1.3. LOCALIZED, MOBILE AND IMMOBILE ADSORPTION

The potential energy over a uniform metal surface varies roughly sinusoidally. The binding energy of the adspecies is a maximum at the minima of these potential energy troughs; such localities are referred to as adsorption sites. The height of the potential energy barrier that restricts site-to-site movement of the adspecies is small compared with its binding energy to the site; consequently, the adspecies are mobile at temperatures insufficiently high for desorption to be significantly (Fig. 1.6). Nevertheless, most of the lifetime of the migrating adspecies is spent at the adsorption site and the adsorption is,

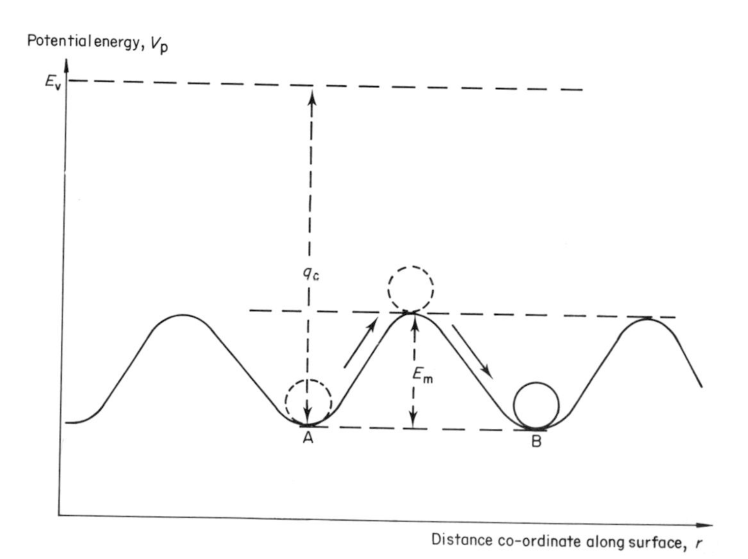

FIG. 1.6. Variation of potential energy across an ideal surface. E_m is the activation energy for an adsorbed molecule initially at A to migrate to an adjacent, orginally unoccupied, site B, with $E_m \ll q_c$.

therefore, localized. Even at high temperatures, at which they are "freely" mobile, chemisorption is still substantially localized. At sufficiently low temperatures, migration can be almost completely inhibited; an immobile adlayer is then obtained.

REFERENCES

1. G. Ehrlich, *in* "Transactions of the 8th Vacuum Symposium", Pergamon Press, Oxford, 1961, p. 126 and Table I, p. 127.
2. J. Pritchard, *Trans. Faraday Soc.*, 1963, **59**, 437.

2

Preparation and Maintenance of Clean Metal Surfaces

2.1. INTRODUCTION

Up to about 40 years ago, most of the investigations on the interaction of gases with metal surfaces were performed on powders, the surfaces of which had been freed of superficial oxide by reduction with hydrogen at moderately high temperatures, followed by outgassing of the absorbed hydrogen at elevated temperatures while maintaining a residual gas pressure of around 10^{-4} Pa.† Measurements, by Roberts [1] in 1935, of the heats of absorption of gases on a tungsten filament which had been subjected to high-temperature flashing in the best obtainable vacuum, however, clearly demonstrated that much, if not all, of the previous work had been conducted on metal surfaces that were seriously contaminated. This conclusion was confirmed [2, 3] by results obtained from the adsorption of gases on evaporated tungsten films from 1940 onwards by many workers. In this work, the surface was generated by the condensation of metal vapour on a cold substrate in a vacuum apparatus. Despite the insufficiently low pressures ($\sim 10^{-4}$ Pa) of residual gases present during this operation, the gettering action of the metal vapour and the large surface area of the film resulted in the production of surfaces that were initially free of serious contamination. Absorption measurements clearly demonstrated that the previous results had contributed little in providing information about the

† The Torr (or mmHg) is commonly used as a unit of presssure in adsorption studies. The S.I. unit is the Pascal (or N m^{-2}) and equals $7{\cdot}50 \times 10^{-3}$ Torr.

absorptive properties of clean metal surfaces. In all modern work, particular attention is directed to the production of clean surfaces.

2.2. PREPARATION OF A CLEAN SURFACE

A clean surface may be defined as one which contains less than a few per cent of a single monolayer of foreign atoms either absorbed on, or replacing, surface atoms of the parent lattice. Such surfaces can be prepared only in an environment in which the ambient pressures of residual gases are less than 10^{-7} to 10^{-9} Pa; the metal should be of the highest purity available because impurity atoms tend to segregate at the surface in much higher concentrations than in the bulk material. The four most effective methods of generating clean surfaces are: (i) high-temperature thermal volatilization of surface contaminants; (ii) ion-bombardment by inert gas ions, followed by outgassing and annealing; (iii) high-intensity field desorption; and (iv) evaporation and condensation of a metal vapour. Method (i) is limited to filaments and wires; method (ii) is largely confined to the surfaces of single crystal planes; method (iii) is specific to tips in field-emission and -ion microscopy; method (iv) is the most general, being applicable to all metals.

2.2.1. High-temperature thermal desorption [4–6]

Heating is effected by passage of an electric current through the metal filament. Contaminants are first minimized by heating to a temperature lower than that of the melting or sublimation point of the filament; however, the temperature must be high enough to ensure that contaminant removal proceeds at a sufficiently fast rate—with negligible evaporation of the metal—in a background pressure of less than 10^{-7} Pa over a period of hours. This heating should not be continued over a too lengthy period of time, since diffusion of impurities, such as C, Si, etc., to the surface from the bulk often occurs. A preliminary treatment of alternative exposures to oxygen and hydrogen is often effective in reducing the extent of such contamination [7]. The final treatment is to flash the wire at a temperature approaching its melting point for a few seconds in ultra-high vacuum ($<10^{-8}$ Pa). Polycrystalline filaments are more frequently used, but monocrystalline ribbons of well-defined morphology are available commercially. Unfortunately, thermal etching at the high temperatures employed during cleaning may partially destroy its monocrystallinity. The absorbent surface area is only 1 to 2 cm^2; consequently, the amount of gas absorbed is very small, but flash desorption into a known volume containing a sensitive ionization gauge allows its accurate determination. A great advantage is that the clean surface can be regenerated by raising the filament to a high temperature for short periods. The technique has been successfully applied to W, Mo, Re, Pt, Ni and a few other metals.

2.2.2. Ion bombardment by inert gas ions [6, 8, 9]

The metal single crystal, mounted in a holder which had been previously outgassed, is first subjected to electron bombardment at an elevated temperature (800 to 1400 K) in high vacuum (see Fig. 2.1). Highly purified inert gas (usually Ar, but He, Ne, Xe have been used) is then admitted at a pressure of 1 to 10^{-1} Pa at room temperature. Application of a potential difference (200 to 600 V) between a secondary electrode and the crystal (the cathode) produces a glow discharge (or ionization may be effected by a heated filament) and Ar ions are accelerated towards, and collide with, the crystal face. Normally, conditions are controlled such that a current density of 100 μA cm^{-2} is generated and maintained over a period of a few minutes. These mild conditions minimize pitting and creation of atomic defects on the surface. The crystal is then annealed at a high temperature *in vacuo*

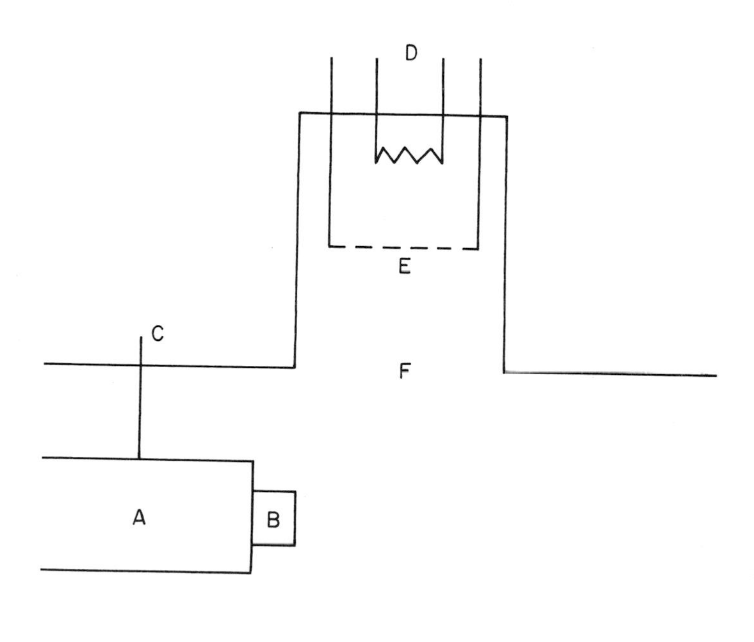

FIG. 2.1. One form of an ion-bombardment system. A, sample holder; B, single metal crystal; C, anode at 200 to 600 V; D, heated filament; E, grid; F, ionization chamber.

to eliminate lattice imperfections and remove occluded inert gas, the bombardment–annealing cycle being repeated several times. Careful design of the apparatus and control of conditions of the discharge minimize desorption of gas from the walls and crystal holder, and prevent sputtering of metal on the walls. Contamination that may be brought about by bombardment of ions of residual gaseous impurities (usually at pressure of $<10^{-7}$ Pa) can be prevented by increasing the overall ion current so that the surface impurities are removed as rapidly as they are adsorbed. Single crystals of Ni, W and Ti, etc., have been successfully cleaned using this method. A difficulty is the subsequent accurate assessment of the small amount of gas adsorbed on the crystal plane. A procedure commonly used is to pass the gaseous adsorbate into the reaction cell at an accurately controlled rate and to calculate the amounts adsorbed from the ambient pressures and an estimate of its sticking probability derived from the literature, for the same gas/metal system under comparable conditions.

2.2.3. Field desorption [10–12]

The potential of the field emission tip† is made positive and a voltage difference of 10^4 V is applied between it and an auxilliary electrode. Since the radius of curvature of the tip is 10^{-4} cm, field intensities of 10^8 V cm^{-1} are set up, with the result that the adspecies are desorbed in an ionic state. The surface comprises a large number of ideally-regular low- and high-index planes, but the area available for adsorption is extremely small, being of order 1×10^{-10} cm^2. Moreover, the chemisorption studies can only be carried out within the microscope itself. However, with the tip at liquid-He temperatures, and by use of a directed beam of adsorbate molecules (which have a sticking probability of unity at these low temperatures), absorption can be confined to specific localities at the surface of the tip. Surface diffusion phenomena can then be investigated in considerable detail.‡ Any metal can be cleaned by this technique, provided tips of sufficiently small dimensions of the metal can be fabricated.

2.2.4. Deposition of films by thermal evaporation and condensation on a cold surface [13]§

This method differs in principle from the other methods in that an attempt is made to prepare a clean surface directly; subsequent additional cleaning

† See Chapter 17.
‡ See Chapters 5 and 19.
§ See also Roberts [6].

is not possible. A metal source is heated to a sufficiently high temperature, usually electrically, to effect rapid evaporation, the vapour being condensed on a cold glass surface. A temperature at which the vapour pressure is 1 to 10^{-1} Pa gives a convenient rate of evaporation. The range of temperatures extends from ~3000 K (W, Ta, Nb etc.) down to around 1700 K for the more volatile metals (Ni, Pd, Fe, etc.). The evaporation source may be a filament in the form of a hairpin or coil attached to thick W rods through which a current of 5 to 15 A is continuously passed until a film of sufficient thickness is obtained; or the metal may be contained in a conical basket constructed from a filament or foil of another metal of higher melting point which does not alloy with the metal to be vaporized; or the basket may be replaced by an alumina crucible provided with an embedded W heater or with some external source of heat [13].

To obtain the film surface free from contamination requires stringent control of conditions before, during and after deposition [14]. Prolonged preliminary outgassing below the evaporation temperature of the metal with a dynamic pressure of 10^{-6} Pa or less is essential; during this period, the supporting leads and glass substrate are maintained at elevated temperatures. Even so, with the glass vessel in an ice-bath, the temperature of the inner condensation surface can increase to 400 K or higher, particularly for continuous evaporation of metals of high m.p.; at these temperatures, evolution of gas from Pyrex walls that have been inefficiently outgassed can be significant. An increase in the background pressure during deposition indicates that the rate of removal of gaseous contaminants by the extremely rapid rate of gettering by the film is less than their rate of evolution. Consequently, the film should be rejected should any significant pressure rise be detected during deposition; and, furthermore, the absence of such an increase does not necessarily mean that the film is uncontaminated. Extreme care should, therefore, be taken during all preliminary operations preceding deposition.

The considerable disadvantage of films is that they comprise an agglomerate of very small crystals with a high concentration of both surface and bulk defects [15]. Films are usually deposited on a borosilicate glass surface at around room temperature or below. The rate of surface diffusion of the metal atoms is small at these temperatures and only becomes appreciable around 0·1 T_m, where T_m is the m.p. in kelvins. Consequently, the metal nuclei that are initially formed are closely spaced and grow laterally by accumulation of atoms adsorbed on the surface and perpendicularly to the surface by capture of metal atoms from the vapour phase. The crystals are, therefore, randomly oriented and columnar in form. The film is porous and has a large roughness factor, defined as the ratio of the actual surface area to the geometric area of the substrate. Except for very thin films, the total surface area accessible to gaseous adsorbate molecules is usually a linear function of its weight. The outer surface is atomically rough and parallel to the substrate, so that the thickness of the film is also directly proportional to its weight. Such films are thermodynamically unstable and are usually thermally annealed at a temperature higher than that to be used in the subsequent chemisorption investigations.

At temperatures in the range 0·1 to 0·3 T_m, surface diffusion is rapid and a continuous impervious film of columnar crystals having normal grain boundaries is formed with a surface area close to the geometric area. Above 0·3 T_m, recrystallization occurs and the mean grain size increases, but the surface remains geometrically rough. During annealing, considerable equilibration of surface energy takes place and there is a decrease of the number of lattice defects (which originally are in concentrations in excess of their equilibrium value due to the rapid quenching during the condensation process). Mechanical strains arising from the misfit of the film on the substrate are eliminated, and the original finely-crystalline structure is transformed to one that is more representative of the bulk metal. The porosity, total surface area and roughness factor are substantially decreased, and planes of low surface energy are preferentially formed to give a high proportion of low-index planes. The surface properties of annealed films are closely representative of those of the bulk metal, although a variety of different planes, largely of low-index, is present in different (unknown) proportions. However, by stringent control of conditions of evaporation and subsequent annealing, films of essentially the same surface area and of the same mean surface properties can be reproduced.

2.2.5. Chemical and other methods

Metals, dispersed on an inert oxide support, are used in transmission infra-red spectrometric investigations of adsorbates on metals. The final process, in an attempt to obtain a clean surface, is the reduction of the metallic oxide by hydrogen, followed by high-temperature outgassing to remove the chemisorbed hydrogen. It is, however, highly unlikely that a clean metal surface is generated by this method. Thus, after 100 h treatment at 800 K with hydrogen at 1 atm pressure, the surface of nickel has been found to be 20% contaminated [16]. The main impurity is probably unreduced oxide, but trace impurities in the hydrogen which can accumulate at the surface may also contribute.

Such methods as cleavage [6] or crushing of single crystals in high vacuum are limited to materials which are brittle and easily fractured; they are, therefore, not applicable to metals in general.

2.3. SURFACE PROPERTIES OF FILMS, FILAMENTS AND CRYSTAL PLANES

2.3.1. Films and filaments

The most general method of obtaining surfaces of all metals of interest in chemisorption investigations is to obtain a film by evaporation from a metal source of ultra-high purity that has been thoroughly outgassed under ultra-high vacuum conditions and, subsequently, condensed on an efficiently outgassed borosilicate-glass substrate. This method can provide surfaces that are almost completely free of contamination. The disadvantages of films—that their surface properties may be different from those of the bulk metal, their porosity and high concentration of micro- and macro-defects, and their excess surface free energy—can be substantially eliminated by a careful annealing procedure conducted in ultra-high vacuum conditions. Investigations by various electron diffraction and electron microscopy techniques indicate a high degree of crystal perfection, but also the presence of stacking faults, twins, grain boundaries, etc., and a variety of low-index and, less numerously, higher-index planes. This situation can be improved by condensing the film on crystal planes obtained by cleavage of a solid substrate in high vacuum. For example, films of nickel and palladium are predominantly oriented with the (100) face parallel to a sodium chloride substrate and with the (111) face parallel to mica [17]. Orientation is favoured by a slow rate of deposition, and substrate temperatures of 600 K or higher. Considerable information about the crystal planes, crystal size and orientation characteristics can be extracted from the results of diffraction techniques, but this is time-consuming and is not a particularly effective operation.

Nevertheless, the use of films will undoubtedly continue, particularly when the main interest is the "average" adsorptive properties of clean metals—indeed, no other technique has, as yet, provided clean surfaces of so many different metals. Certainly, the results that can now be obtained on reasonably well-characterized films are not inferior to those on many polycrystalline filaments.

2.3.2. Single crystal planes

Detailed information about chemisorption on surfaces of specific geometry necessitates the use of single crystal planes, which are becoming increasingly available. Just as the cleaning procedure for filaments, i.e. flashing to a high temperature to volatize or thermally decompose surface contaminants, may provide a surface less free of impurities than that of a film, so that adopted in single crystal work does not, with any certainty, provide an ideally clean surface. Moreover, the complete removal, by annealing, of surface defects and, in particular, of steps arising from the emergence of dislocations at the surface, is rarely accomplished. Furthermore, the area of the crystal plane selected to examine its degree of cleanliness is extremely small, the accuracy of measurement of the amount absorbed is not high, and most investigations have so far been confined to the low-index planes.

2.4. ULTRA-HIGH VACUUM TECHNIQUES†

2.4.1. Attainment of low pressures

A clean metal surface can be contaminated in a few seconds by a monolayer of absorbed gas when the ambient pressure of the gas is around 10^{-4} Pa.‡ Since the collision rate of gas molecules with the surface is proportional to their pressure, this time can be extended to about an hour if the residual gas pressure is decreased to 10^{-7} Pa or less. Such low pressures can be easily attained by a multi-state diffusion pump. Mercury pumps are usually preferred, since mercury vapour can be efficiently removed by interposition of a series of traps cooled in liquid nitrogen between the pump and the adsorption system. Such pumps operate up to ~ 1 Pa and are used in conjunction with a rotary backing pump. Sputter-ion pumps have little or no advantage and indeed they may be a source of contamination in certain circumstances.

Tubing of about 2 to 3 cm in diameter is employed in the pumping system, since the conductance of a cylindrical tube increases as the cube of the diameter and inversely as its length. Since gas evolution from the walls increases directly as the surface area, the ratio of tube diameter to length should be as large as conveniently possible. Systems are usually constructed of borosilicate glass, although stainless steel is commonly used. Arrangements must be made for baking-out the entire apparatus with ovens (~ 750 K) and tapes (~ 600 K) for a period of hours to ensure thorough outgassing.

2.4.2. Measurement of low pressures

The attainment and maintenance of pressures of 10^{-7} to 10^{-8} Pa can now be regarded as a routine operation. The general application of ultra-high vacuum techniques followed closely upon the introduction of the inverted ionization gauge of Bayer and Alpert [19–21]. Its essential features are the emission of electrons from a hot cathode and their acceleration by a potential to a cylindrical anode. Within this anode volume gas molecules are ionized by collision with electrons and the ions are collected by a fine wire probe.

† An excellent detailed account of these techniques is provided by Ehrlich [18].
‡ See Chapter 4.

The gauge measures gas density and requires calibration against a McLeod gauge. The pressure is given by the ratio of the ion current to the electron current multiplied by an empirical factor, termed the sensitivity, and is obtained by the above calibration. Pressures are linearly related to this ratio between 5×10^{-3} and 10^{-8} Pa. A complication is the pumping effect of the gauge itself; this may be as high as 0·1 $dm^3 s^{-1} mA^{-1}$; it arises from the trapping of ions and/or of free radicals produced by dissociation of the gas at the hot cathode. Reduction of the electron emission to 10 μA decreases this rate of pumping by a factor of 10^{-3} and an ion current of 10^{-14} A at the lowest pressures can still be maintained. Dissociation of the gas is also substantially reduced by lowering the temperature of the cathode, but this lowering is accompanied by a simultaneous decrease of the electron emission current unless the work function of the emitter is reduced. To this end, rhenium coated with lanthanum boride, and thoriated tungsten and iridium filaments, have been employed with some success. Another device to minimize the errors of gauge pumping is the employment of a nude gauge inserted in the region in which the pressure is being recorded; the grid potential is applied only during the short period necessary to measure the ion current while running the hot cathode continuously.

All ionization gauges measure the total gas density, and, because of the presence of impurities in the gas itself, or produced from or by the gas during the operation of the gauge, mass spectrometers are being more frequently employed. Their range of operation varies from 10^{-6} to 10^{-12} Pa up to 10^{-2} Pa; the lowest pressures are measurable when shielded collectors are distant from the ion source, so that the generation of X-rays is made negligible. Various types of resonance spectrometers are available [22]. Frequent use is made of the omegatron and quadrupole spectrometers—both of which operate with crossed steady magnetic fields and an oscillating electrostatic field of radio frequency—and of time-of-flight mass spectrometers.

2.4.3. Gas purity

Gases of a purity of 10^3 p.p.m. are commercially available, but, particularly for flow methods extending over long period of time, minute amounts of impurity can introduce serious errors, as in some sticking probability measurements [23]. Hydrogen, by diffusion through palladium, and argon, by diffusion through quartz, may be highly purified to 10^{-3} p.p.m. and 10 p.p.m., respectively, but these are exceptions. Other methods involve low-temperature low-pressure distillations, and the selective gettering of impurities. For example, tungsten or molybdenum films remove most contaminant gases effectively from samples of inert gases, and nickel films reduce the concentrations of H_2, CO, O_2 and CO_2 in N_2, probably down to 10 p.p.m. Another procedure comprises the thermal decomposition of an appropriate solid compound of high purity; for example, good quality N_2 can be prepared by decomposition of alkali-metal or alkaline-earth azides, CO from molybdenum hexacarbonyl, and O_2 from Ag_2O or CuO. In general, however, the problem of preparing many gases of ultra-pure quality still remains.

REFERENCES

1. J. K. Roberts, *Proc. Roy. Soc. A*, 1935, **152**, 445 and 464.
2. O. Beeck, A. E. Smith and A. Wheeler, *Proc. Roy. Soc. A*, 1940, **177**, 62.
3. J. A. Allen, *Rev. Pure Appl. Chem.* (*Australia*), 1954, **4**, 133.
4. J. K. Roberts, *Proc. Roy. Soc. A*, 1935, **152**, 445.
5. R. J. Gomer, *J. Chem. Phys.*, 1953, **21**, 293.
6. R. W. Roberts, *Brit. J. Appl. Phys.*, 1963, **14**, 537.
7. E. W. Müller, *Ergebn. exact Naturwiss.*, 1953, **27**, 290.
8. H. E. Farnsworth and J. Tuul, *J. Phys. Chem. Solids*, 1958, **9**, 48.
9. R. E. Schlier and H. E. Farnsworth, *J. Chem. Phys.*, 1959, **30**, 917.
10. E. W. Müller, *Adv. Electron. Electron Phys.*, 1960, **13**, 83.
11. R. Gomer, "Field Emission and Field Ion Microscopy", Harvard University Press, Cambridge, Mass., 1961.
12. E. W. Müller and T. Tsong, "Field-ion Microscopy", Elsevier, New York, 1969.
13. L. Holland, "Vacuum Deposition of Thin Films", Wiley and Sons, Inc., New York, 1960.
14. T. W. Hickmott and G. Ehrlich, *J. Phys. Chem. Solids*, 1958, **5**, 47.
15. J. V. Sanders, *in* "Chemisorption and Reactions on Metallic Films" (J. R. Anderson, ed.), Academic Press, London and New York, 1971, Vol. 1, p. 1.
16. M. W. Roberts and K. W. Sykes, *Trans. Faraday Soc.*, 1958, **54**, 548.
17. D. W. Pashley, *Adv. Phys.*, 1956, **5**, 173.
18. G. Erlich, *Adv. Catalysis*, 1963, **14**, 391.
19. R. T. Bayard and D. Alpert, *Rev. Scient. Instrum.*, 1950, **21**, 571.
20. D. Alpert, *J. Appl. Phys.*, 1953, **24**, 960.
21. P. A. Redhead, *Can. J. Phys.*, 1958, **36**, 255.
22. W. G. Brombacher, "National Bureau of Standards Monograph No. 35", National Bureau of Standards, Washington, D.C., 1961.
23. D. O. Hayward, D. A. King and F. C. Tompkins, *Proc. Roy. Soc. A*, 1967, **297**, 305 and 321.

3

The Adsorption Process

3.1. THE CONDENSATION COEFFICIENT

The initial step involves the collision of the gaseous adsorbate molecule with the surface. The rate Z of impingement, usually expressed as molecules s^{-1} cm^{-2}, is given by the Herz–Knudsen equation,†

$$Z = P/(2\pi mkT)^{\frac{1}{2}},$$

where P is the gas pressure and m the mass of the gaseous molecule. As the molecule approaches the surface, it is subjected to an attractive potential with a resulting acceleration of its velocity component normal to the surface into a region of low potential energy. The higher the subsequent binding energy of the molecule to the surface, the greater the attractive force and the deeper the potential well. However, unless the incoming molecule can transfer to the lattice an amount of energy equal to or greater than its translation energy perpendicular to the surface (i.e. kT_g, where T_g is the absolute temperature of the gas molecules), condensation can not occur. The efficiency of this energy transfer is greater the higher the binding energy of the molecule to the surface, since the collision time is longer. As a result of the collision, a compressional wave is propagated through the lattice, i.e. energy is absorbed by phonon excitation of the lattice vibrations.†

The capture efficiency, i.e. the condensation coefficient, obtained experimentally (even for weakly bound molecules) on metals, approaches unity. However, capture merely indicates that a loss of energy kT_g perpendicular

† See Chapter 4 for a more detailed account and relevant references.

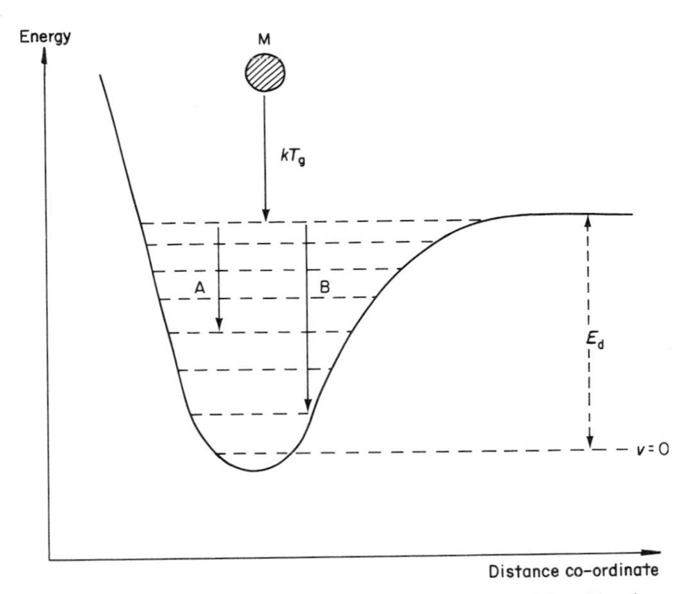

FIG. 3.1. A, B, transition to bound states: M, incoming molecule with a kinetic energy component of kT_g perpendicular to the surface; E_d, heat of desorption; v, vibration quantum number.

to the surface has occurred, but not that thermal equilibrium with the surface has been achieved. The captured molecule may still be in an excited state, and it will subsequently make transitions to lower vibrational states in order that a Boltzmann distribution over the vibrational levels characteristic of the surface temperature may be finally attained (Fig. 3.1). The molecule may also retain kinetic energy components along the surface, and will execute lateral movements until its migration energy is lost to the lattice when it becomes localized at an unoccupied adsorption site (Fig. 3.2).

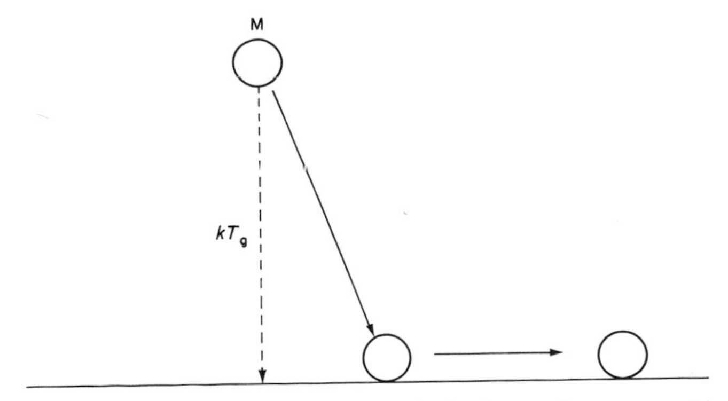

FIG. 3.2. Incoming molecule transfers energy kT_g to lattice, but retains an excess kinetic energy component parallel to surfaces.

The condensation coefficient, therefore, is concerned with the capture of a molecule in a physisorbed state, i.e. in the van der Waals potential well. This process precedes the final act of chemisorption for which the capture efficiency is referred to as the sticking probability.

3.2. THE STICKING PROBABILITY†

The sticking probability is defined as the ratio of the rate of capture of a molecule in the chemisorbed state to the rate of collision of the gaseous molecule with the surface. This probability is, almost invariably, less than the condensation coefficient. One reason is that the condensation can occur with quite high efficiency on an absorbed molecule, whereas the formation of a chemisorbed molecule requires the presence of an unoccupied adsorption site. Transformation from the weakly-adsorbed physisorbed state to the final strongly-bound chemisorbed state may take place rapidly at the site of condensation (provided it is not already occupied by a chemisorbed species), before the physisorbed molecule has time to desorb. Collision of an incoming molecule with a chemisorbed adspecies may involve the transient formation of a weak physisorbed state of extremely short life, and the molecule may be returned to the gas phase with some loss of kinetic energy. In this circumstance, the sticking probability decreases approximately linearly with increase of coverage.

However, if adsorption takes place on sites occupied by a chemisorbed species with a larger (though still weak) binding energy, a precursor state can be formed.† This mobile precursor is, subsequently, either desorbed or transformed to the chemisorbed state before desorption takes place when it encounters an unoccupied site. In this situation, the sticking probability is substantially independent over a wide range of coverages.

In dissociation chemisorption (for example, of hydrogen on a tungsten surface), the interatomic distance between the hydrogen atoms in the free molecule is significantly smaller than the distance between adjacent troughs of the potential energy surface. The increase of potential energy resulting from the initial bonding of the adsorbate molecule to the surface is, therefore, less than that in the final equilibrium chemisorbed state in which the two adatoms occupy the minima of two adjacent troughs. However, because the height of the potential barrier separating adjacent troughs is considerably smaller than the energy of the metal atom/hydrogen atom bond, the initial sticking probability is high, i.e. the energy transfer is still very efficient. In the chemisorption of gases on metals, the problems associated with energy transfer need not be considered in most circumstances.

† See Chapter 4 for a more detailed account and relevant references.

3.3. ADSORPTION EQUILIBRIUM

An adsorption isotherm depicts the equilibrium relationship between the number of molecules adsorbed on the surface of a solid adsorbent and the pressure of gaseous adsorbate molecules in contact with that surface, the entire gas/solid system being maintained at a constant known temperature. In an ideal system, the surface is planar, all adsorption sites have identical adsorptive properties, the adsorption process is non-activated and is restricted to a monolayer. Should there be no interaction between the adsorbed molecules, a random occupation of the uniform sites corresponds to the equilibrium state of the adsorbed layer. However, in practice, most surfaces are heterogeneous, i.e. the adsorption potential varies over an array of non-uniform sites. Since the sticking probability of gas molecules on collision with a metal surface is generally of order 10^{-1}, the initial distribution of the adspecies is essentially uniform, because sites of different adsorption potentials have the same (or nearly the same) capture efficiency. A non-equilibrium configuration is, therefore, set up; this is, subsequently, transformed to the equilibrium state, in which the adspecies preferentially occupy sites of highest adsorption potential according to a Boltzmann distribution of their binding energies. Equilibrium is attained by: (i) the preferential desorption of the adspecies from sites of low adsorption potential and their subsequent random re-adsorption at unoccupied sites having a wide range of binding energies; and (ii) the migration of adspecies from lower energy sites and subsequent capture at an unoccupied site having the highest adsorption potential (Fig. 3.3).

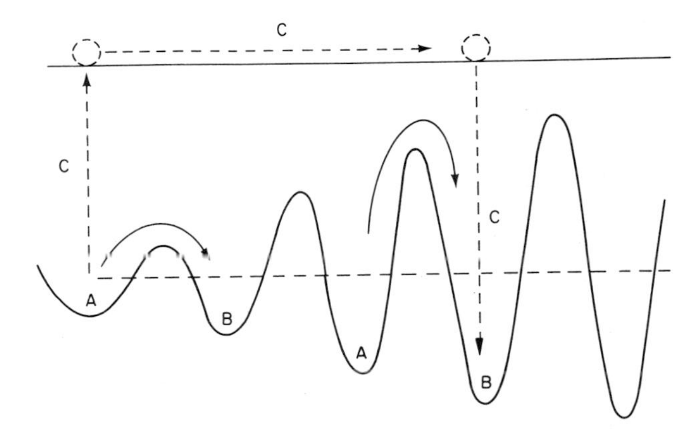

FIG.3.3. Establishment of adsorption equilibrium on a heterogeneous surface. A, Initial random occupation; B, migration of an admolecule to sites of higher adsorption; C, Desorption from a lower energy site and subsequent random adsorption at an unoccupied higher energy site.

For the determination of thermodynamic parameters, e.g. of heats of adsorption by calorimetry, it is essential that adsorption equilibrium should be rapidly attained within the experimental time available for measurements. Consequently, an independent assessment of the rate of attainment of the equilibrium state is advisable. In an extreme case, e.g. for a heterogeneous surface and a completely immobile layer, a constant heat of adsorption independent of coverage results since adsorption sites are uniformly occupied. Only average values can be evaluated and these depend on the particular distribution of adsorption potential over the total number of sites.

3.4. SURFACE DIFFUSION† AND ATTAINMENT OF EQUILIBRIUM

The surface of a metal ideally comprises a geometric plane. This plane is populated with the ion-cores of the metal in the form of a regular two-dimensional network which reflects the lattice structure of the metal and the particular crystallographic plane that is exposed. The minima of the potential energy oscillation at the surface are the adsorption sites at which the binding energy of the adsorbate is a maximum. The number of such sites per unit area can be evaluated from the lattice geometry of the plane. At these sites, the admolecule vibrates in the surface plane and perpendicular to it. Migration from an occupied to an unoccupied site can take place when the admolecule acquires, from lattice vibrations of the adsorbent, sufficient thermal energy for it to surmount the sinusoidal potential energy barrier separating adjacent sites. Surface diffusion, therefore, takes place by an activated hop-site mechanism, the energy of activation E_m approximating to the height of the barrier. This energy is considerably smaller than that to effect desorption. Consequently, surface mobility is the dominant factor in the rapid attainment of adsorption equilibrium on a non-uniform surface and becomes highly effective at a temperature approximating to $E_m/10RT$. Usually, E_m is between one fifth and one seventh of the desorption energy, E_D. Thus, for hydrogen on tungsten at low coverages, E_D is about 190 kJ mol^{-1}; E_m is, therefore, *ca* 34 kJ mol^{-1}, so that the hydrogen adatoms have considerable mobility at *ca* 400 K. The lifetime of the adatom at a site is, however, about 10^4 times longer than its time in flight, so that the layer still comprises a localized array.

At much lower temperature, movement of adatoms is highly restricted and the rate of attainment of adsorption equilibrium may be so slow that it is never attained within the time interval over which measurements

† See Chapter 9.

are made. The temperature of the system must, therefore, be raised to accelerate this rate, and later returned to the temperature of measurement. The temperature cycling is continued until a constant reproducible pressure is obtained.

Since chemisorption heats are high, the equilibrium pressures are extremely low at small fractional coverage. However, with the advent of the ionization gauge, which extends accurate pressure measurements down to 10^{-8} Pa, the uncertainty as to whether or not completely reversible conditions were attained in earlier work no longer obtains. When manometers having a lower limit of 10^{-4} Pa are used, equilibrium can only be rigorously tested at coverages approaching a full monolayer.

4

Rates of Adsorption, Thermal Accommodation and Molecular Beam Scattering

4.1. RATES OF ADSORPTION

The *maximum* rate of adsorption at a gas/solid interface is the collision rate of the gas molecules with the adsorbent surface. From kinetic theory, the number, Z, of molecular collisions (cm surface)$^{-2}$ s^{-1} is given by the Herz–Knudsen equation,

$$Z = n\bar{v}/4 \text{ cm}^{-2} \text{ s}^{-1}, \tag{4.1}$$

in which the average velocity $\bar{v}$ of molecules of molecular weight M at a temperature of T K is

$$\bar{v} = (8RT/\pi M)^{\frac{1}{2}} = 14\,500(T/M)^{\frac{1}{2}} \text{ cm s}^{-1}. \tag{4.2}$$

The number of molecules n in 1 cm^3 of an ideal gas is

$$n = P/kT = 7{\cdot}24 \times 10^{16}\, P/T \tag{4.3}$$

in which k is the Boltzmann constant ($= 1{\cdot}38 \times 10^{-23}$ J K^{-1}), T is the absolute temperature, and P is given in Pa. Hence,

$$Z = 2{\cdot}64 \times 10^{20}\, P/(MT)^{\frac{1}{2}} \text{ cm}^2 \text{ s}^{-1}. \tag{4.4}$$

For nitrogen ($M \sim 28$) and $T = 273$ K with $P = 1$ Pa,

$$Z = 3 \times 10^{18}\ \mathrm{cm^2\,s^{-1}}. \tag{4.5}$$

The average number of adsorption sites on a crystal plane of a transition metal is *ca* 10^{15} cm^{-2}; a site would, therefore, experience $\sim 3 \times 10^3$ collision s^{-1}. For simple gaseous adsorbates, chemisorption is, essentially, a non-activated process and the sticking probability is usually 0·3 to 0·6; hence, the initial rate of chemisorption would be $\sim 10^{-3}$ to 2×10^3 molecules cm^{-2} s^{-1} at $P = 1$ Pa. Before the development of ultra-high vacuum techniques, the residual gas pressure after evacuation was usually about 5×10^{-4} Pa, corresponding to an initial rate of adsorption of $\sim 1{\cdot}0$ molecule s^{-1} per site and decreasing as the free sites became occupied. The time Δt that elapsed before the surface was substantially contaminated can, therefore, be crudely estimated from the approximate relationship

$$P\,\Delta t \sim 5 \times 10^{-4}\ \mathrm{Pa\,s}. \tag{4.6}$$

Thus, for $P \sim 5 \times 10^{-4}$ Pa, a period of only about one second is available. Ultra-high vacua of $P \sim 10^{-7}$ Pa are now routinely obtainable, so that the initial rate per site is reduced to between 10^{-4} and 2×10^{-4} s^{-1}. Only about 1 % of the surface is, therefore, contaminated in about 2 min. Measurements can be initiated well within this period of time, following completion of the cleaning procedure; and, since changes of pressure of $< 10^{-8}$ Pa can be recorded using modern ionization gauges, extremely rapid initial rates of adsorption can be accurately measured.

4.2. EXPERIMENTAL DETERMINATION OF RATE CONSTANTS

4.2.1. Closed systems

Experimental rates of adsorption have been obtained using both closed and flow systems. The closed system at a temperature T comprises a constant volume V containing the adsorbent in the form of a metal filament and a calibrated ionization gauge of high sensitivity within an ultra-high vacuum apparatus. (See, for example, Ehrlich [1].) Following orthodox procedure for generating a clean surface, a measured amount of adsorbate is rapidly injected into the evacuated volume V ($< 10^{-8}$ Pa) to give a calculated initial pressure P_i. The decreasing pressure P_t is then recorded as a function of time t as adsorption takes place.

Conservation of mass requires that the rate of loss of the number of molecules from the gas phase $[(-V/kT)(\mathrm{d}P_t/\mathrm{d}t)]$ equals the *net* rate of adsorption by the filament, provided that all other surfaces within the system are effectively inert. The net rate of adsorption is the difference between the actual rate of adsorption and that of desorption. Consequently,

$$-\frac{V}{kT}\frac{\mathrm{d}P_t}{\mathrm{d}t} = A[P_t k_a(1 - \theta_t) - n_m \theta_t k_d], \tag{4.7}$$

where V = the volume of the systems (cm^3), A = the surface area of adsorbent (cm^2), θ_t = the fractional coverage, and P_t = the pressure (Pa) at time t (s); k_a = the rate constant of adsorption (molecules cm^{-2} Pa^{-1} s^{-1}) by the clean surface ($\theta = 0$) at unit pressure. The product $k_d n_m$ = the rate of desorption (molecules cm^{-2} s^{-1}) from a full monolayer ($\theta = 1$), and n_m = the number of molecules in the monolayer per cm^2 surface.

Equation (4.7) is valid when: (i) the surface is uniform; (ii) there is negligible interaction between adsorbate molecules; and (iii) the capture efficiency of an unoccupied site is unity. In these circumstances, k_a is independent of θ.

At low coverages ($\theta \ll 1$), the desorption rate is neglected and $(1 - \theta)$ is approximated to unity; Eq. (4.7) may then be written as

$$Ak_a P_t = \frac{-V}{kT}\frac{\mathrm{d}P_t}{\mathrm{d}t}, \tag{4.8}$$

or

$$\ln P_t = -(k_a AkT/V)t \ln P_i. \tag{4.9}$$

The plot of $\ln P_t$ against t is linear, of slope $-k_a AkT/V$, from which k_a can be evaluated (Fig. 4.1).

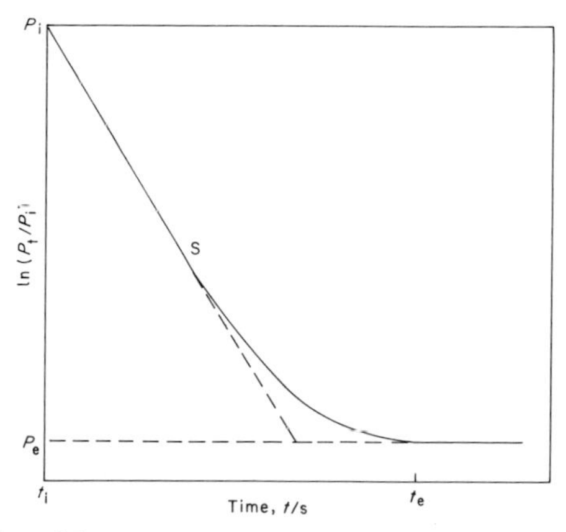

FIG. 4.1. Determination of the net rate of adsorption in a closed system of volume V at a temperature T. The rate constant of adsorption k_a is derived from the slope, $-k_a AkT/V$, where A is the surface area of the filament. After a sufficient interval of time, $\mathrm{d}P_t/\mathrm{d}t = 0$ and a steady-state pressure P_e is set up such that $k_a P_e(1 - \theta_e) = k_d n_m \theta_e$ and the number of molecules adsorbed $= An_m(\theta_t) = V(P_i - P_e)/kT$.

With increase of θ_t, the rate of adsorption decreases and that of desorption increases until $dP_t/dt = 0$; the two rates are then equal and a steady-state pressure P_e is set up (Fig. 4.1). As soon as the ln P_t versus t plot departs from linearity, the desorption rate can no longer be neglected. It is then necessary to perform a series of experiments in which the initial pressure P_i is varied, and, from the different P_t versus t plots, values of dP_t/dt at a constant amount adsorbed are extracted. Since the amount adsorbed at any time equals the decrease in the number of molecules in the gas phase, then

$$An_m\theta_t = V\Delta P/kT = V(P_i - P_t)/kT, \tag{4.10}$$

i.e. in each experiment of the series, a constant amount adsorbed is synonymous with constant value of ΔP (Fig. 4.2). Hence, from Eq. (4.7), the linear plot of

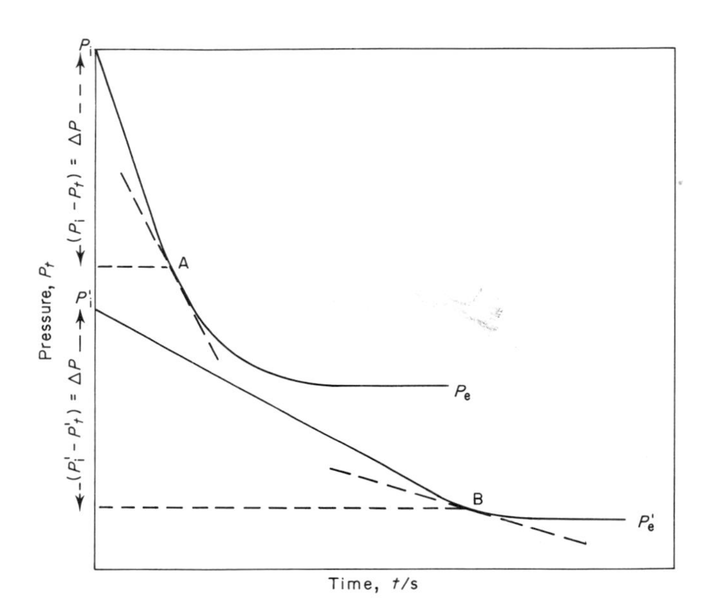

FIG. 4.2. Plots of P_t as a function of time for initial pressures P_i and P_i'. $\Delta P = P_i - P_t = P_i' - P_t'$ corresponds to the same surface coverage, $\theta = \Delta PV/AkTn_m$. Tangents at A and B give dP_t/dt and dP_t'/dt, respectively, for the same value of θ.

$(-V/kT)(dP_t/dt)$ against P_t at constant $\Delta PV/kT$ has a slope of $Ak_a(1 - \theta)$ and an intercept of $Ak_dn_m\theta$; $\theta = V\Delta P/An_mkT$ from Eq. (4.10). Hence, both k_a and k_d may be evaluated.

4.2.2. Flow method

In this method [2, 3], the gaseous adsorbate is admitted, at a measured fixed rate r_i, to the constant-volume adsorption system maintained at a fixed temperature T; simultaneously, gas is removed by pumping at a constant speed, either through a capillary of known conductance or through a controlled leak, to give an outflow rate r_0 for unit pressure from the reaction cell (Fig. 4.3). When there is no adsorption by the filament, a condition that can be attained by maintaining it at a sufficiently high temperature, a steady-state pressure P_o is set up, indicating that the inflow and outflow rates are equal, i.e. $r_i = P_o r_o$, in which the rates r_i and r_o are in units of number of molecules per second. The magnitude of P_o can be varied by changing the inflow rate r_o.

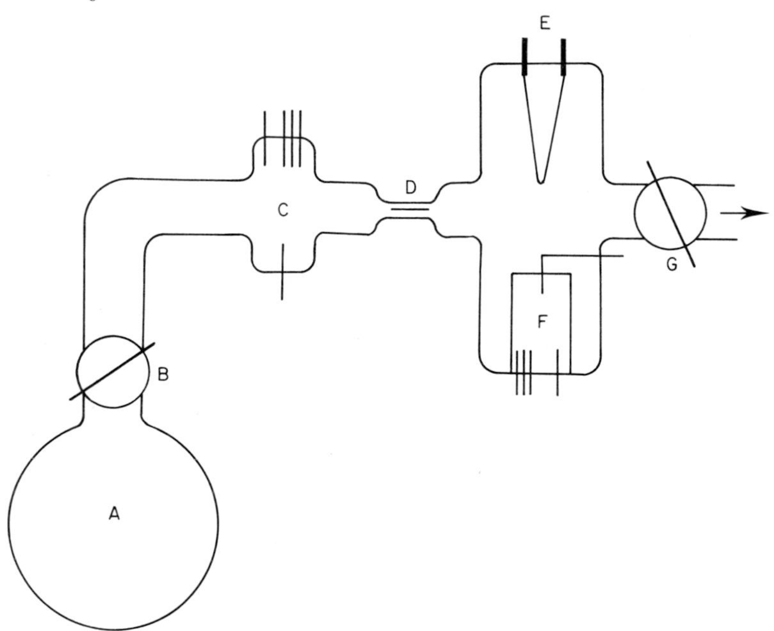

FIG. 4. 3. Flow system for adsorption studies. A, gas reservoir; B, adjustable valve for controlling inflow; C, ionization gauge for measuring inlet pressure to calibrated capillary D; E, metal filament adsorbent; F, ionization gauge to measure pressure in reaction cell; G, adjustable valve to regulate gas outflow to pumps.

When the steady-state pressure is obtained, the temperature of the filament is rapidly decreased and held constant at the experimental temperature T_f. Gas is then adsorbed by the filament with a consequent decrease of pressure such that at time t,

$$-V\,dP/kT\,dt = A[P_t(1-\theta_t)k_a - n_m\theta_t k_d] + (P_o - P_t)r_o, \qquad (4.11)$$

the final term being the *net* inflow rate expressed as a number of molecules per second.

A lower steady-state pressure P_s is finally attained when the net inflow rate of gas is equal to its net rate of adsorption, i.e.

$$(P_o - P_s)r_o = A[P_s(1 - \theta_s)k_a - n_m\theta_s k_a], \tag{4.12}$$

and the amount adsorbed $An_m\theta_s$ is given by the relationship,

$$An_m\theta_s = \frac{V}{kT}(P_o - P_s) + r_o\int_0^{t_s}(P_o - P_t)\,dt, \tag{4.13}$$

in which the right-hand side represents the difference in the amounts that have entered the volume V and have been removed by the pumps during the time t_s. The integral is evaluated graphically from the $(P_o - P_t)$ versus t plot, the upper limit t_s being the time to attain the steady-rate pressure P_s. It is then only necessary to determine the value of $(P_o - P_s)$ at the specific value of r_o in order to calculate the net adsorption rate at the coverage θ_s. Different values of P_o, P_s and θ_s are then obtained by varying r_o and, from any pair of values, k_a and k_d can be separately evaluated.

The amount adsorbed at the steady-state pressure P_s at time t_s can also be obtained by flashing the filament at a sufficiently high temperature to effect complete desorption [3] in such a short interval of time that the amount of gas lost from the cell can be neglected. The instantaneous pressure P_m has to be corrected for gas that entered the cell during the time interval $(t_s - t_o)$ of adsorption to obtain the amount desorbed. Alternatively, a lower constant temperature of desorption T_d may be employed. After sufficient time t_f, the original steady-state pressure P_o is attained. At this stage, the net amount of gas added to the cell during the adsorption time interval $(t_s - t_o)$ must equal the net loss during the desorption time interval $(t_f - t_s)$ and the number of molecules adsorbed at t_s equals the number desorbed at T_d in the interval $(t_f - t_s)$.

4.3. DETERMINATION OF THE STICKING PROBABILITY

4.3.1. Filaments

The sticking probability s has been defined as the ratio of the rate of chemisorption by a surface to the rate of collision of gaseous molecules with that surface. The term k_a denotes the number of molecules adsorbed per second by 1 cm^2 of a clean surface $(\theta \to 0)$ at unit pressure of the gaseous adsorbate.

The collision rate Z, expressed in the same units, is given by the Herz–Knudsen equation as $Z = (2\pi mkT)^{-\frac{1}{2}}$, where m is the mass of the gaseous molecule and T the absolute temperature. The sticking probability is, therefore,

$$s = k_a/Z \tag{4.14}$$

and is independent of pressure.

In many investigations, the net adsorption rate has been recorded; i.e. no account has been taken of the concurrent desorption rate. However, since chemisorption heats are normally large, the activation energy for desorption is high, and, particularly at low coverages, the rate of desorption is negligibly small. Nevertheless, chemisorption heats may decrease substantially as the coverage is increased and, consequently, a sticking probability derived from the experimental rate of uptake can be significantly lower than its true value at higher coverages. The accurate determination of k_a using a filament as adsorbent has been given above. The rate of non-dissociative adsorption is $AP_t(1 - \theta_t)k_a$ and the collision rate is AP_tZ; hence, $s = k_a(1 - \theta_t)/Z$ at a coverage θ_t. For dissociative chemisorption, $s = k_a(1 - \theta_t)^2/Z$.

4.3.2. Evaporated films

A flow method can also be used for the determination of the sticking probabilities on evaporated films. The main difficulty is the accurate evaluation of the collision rate of molecules with the surface, in particular: (i) the gas must be introduced symmetrically to the film in order to obtain a uniform collision rate over the whole surface; (ii) the ionization gauge should be adjacent to the film, and conduction effects in the entry tubing must be minimized; (iii) no deposition of the film should occur within the gauge assembly; and (iv) gauge pumping must be eliminated, e.g. by flushing with the gaseous adsorbate before film deposition. A particular apparatus design is shown in Fig. 4.4 [4].

The method of operation differs in some respects from that employed in the filament work. The conductance F of the gas-inlet tubing is accurately calibrated and the inflow rate r_i is controlled by a constant low pressure P_S in a storage vessel containing an ionization gauge. The reaction cell is closed to the pumps, so that the outflow rate is zero. Because of the very high surface area ($\sim 10^3$ cm^2) of the film, the rate of uptake of gas is very much greater than by a filament of area between 1 and 2 cm^2; consequently, the inflow rate is much higher and greatly exceeds the rate of change of the number of molecules in the gas phase. The term $(-V/kT)(dP_t/dt)$ appearing in Eq. (4.9) is, therefore, negligibly small and the net rate of adsorption equals the inflow rate

$r_i = P_S F$. The pressure P_r of gas close to the film surface is recorded on a second ionization gauge to give the collision rate $P_r(2\pi mkT)^{-\frac{1}{2}}$, or $P_r Z$. At low coverages, the rate of desorption from the film is negligible, and $s = r_i/P_r Z$. The validity of this assumption can be confirmed by terminating the inflow rate ($r_i = 0$) at some particular time t_t; P_r should then decrease rapidly (<5 s) to the background pressure ($P_b < 10^{-8}$ Pa).

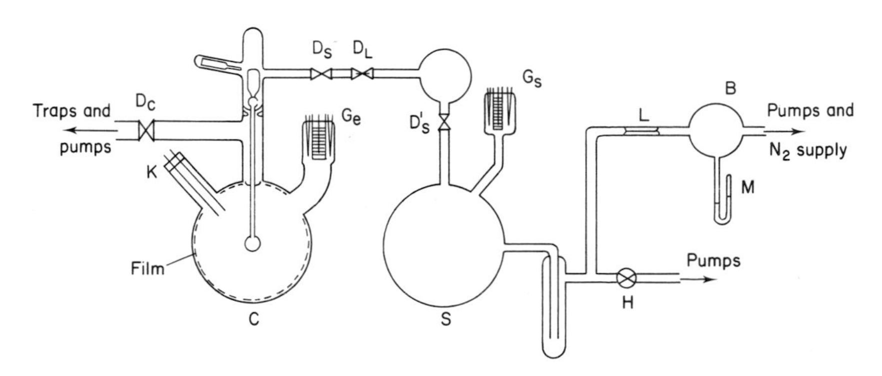

FIG. 4.4. The apparatus for the determination of sticking probabilities on evaporated films. C, reaction cell; S, storage vessel; D_c, D_s, D'_s, Decker valves; G_s, G_e, ionization gauges; D_L, Decker valve with a capillary leak; H, graduated mercury cutoff; L, fine capillary leak; M, manometer; B, nitrogen supply bulb.

A constant dynamic pressure in S is maintained by the inflow of gas through the capillary leak L and simultaneous pumping through the cut-off H. The diffuser is at the centre of C for uniform radial distribution. The film is evaporated from the hairpin attached to K with the diffuser withdrawn and the mouth of the gauge closed by a magnetically-operated nickel disc.

At higher coverages, at which the desorption rate is significant, the pressure decrease is slower and a steady-state pressure ($P_e > P_b$) is finally set up in the reaction cell. At this stage,

$$AP_e k_a(1 - \theta_e) = An_m \theta_e k_d, \tag{4.15}$$

and the amount adsorbed is

$$An_m \theta_e = r_i t_t - P_e V/kT. \tag{4.16}$$

The inflow of gas is then continued for a further short period before being terminated for a second time. Two different values of P_e and of θ_e are, thereby, obtained; the above equations are solved for k_a and the sticking probabilities evaluated as before.

4.4. THE MAGNITUDE OF THE STICKING PROBABILITY

Sticking probabilities are almost invariably less than unity. Possible reasons are: (i) the act of chemisorption may require an activation energy. For small molecules and clean surfaces of transition metals, however, the process is normally non-activated. (ii) The translational component of the kinetic energy of the gas molecules perpendicular to the surface may not be completely transferred to the adsorbent on impact, and some desorption within the period of vibration in the "pre-chemisorbed" state takes place. (iii) The incoming gas molecule collides with an occupied site and is back-scattered to the gas phase; *s* is then proportional to the fraction of the sites that are still unoccupied.

Such elastic collisions are assumed in the derivation of the Langmuir isotherm. Nevertheless, Taylor and Langmuir [5] in 1933, demonstrated that the sticking probability of caesium molecules at a tungsten surface remained constant and approximately equal to unity until an almost complete monolayer of Cs adatoms had been deposited. Constant values of *s*,

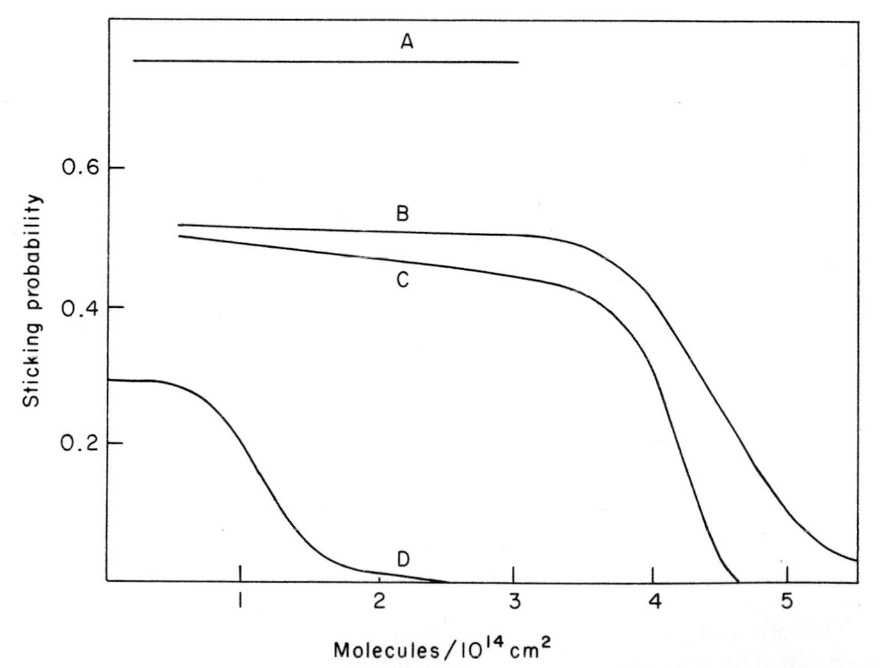

FIG. 4.5. Sticking probabilities as function of amount adsorbed for different adsorbates. Curve A, caesium: curve B, nitrogen at surface temperature $T_s = 113$ K; curve C, carbon monoxide, $T_s = 340$ K; curve D, nitrogen at $T_s = 298$ K.

independent of coverage at low temperatures, have since been reported by many workers, and Langmuir's original explanation, that adsorption occurs on top of the primary chemisorbed layer, is still accepted. A few examples of the variation of s with coverage are given in Fig. 4.5.

4.4.1. The precursor state and the variation of the sticking probabilty with coverage

On collision with the surface, the gas molecule is first physisorbed to form a precursor state. It may then transform to the chemisorbed state at an unoccupied site of the adsorbent or may acquire sufficient thermal energy from the metal lattice to be desorbed [6]. In Figure 4.6, possible potential energy plots for the process of chemisorption (ABC) and that of physisorption (DFF) to form the precursor state at E are shown. Although the cross-over point at G lies below V_p, the potential energy zero, an activation energy E_p is necessary for formation of the transition-state complex at G (with $E_p < q_p$) for transition into a chemisorbed state.

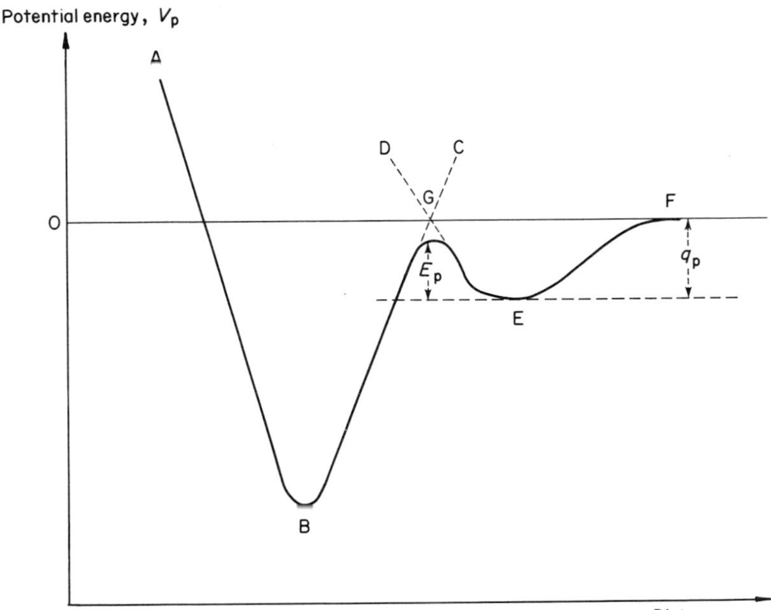

FIG. 4.6. Potential energy as a function of distance co-ordinate for precursor state.
ABC is the chemisorption potential energy as a function of the distance of the chemisorbed species from the metal surface, and DEF that corresponding to a physisorbed precursor state with heat of adsorption q_p. E represents a precursor state which has to acquire an activation energy $E_p < q_p$ before transition to the final chemisorbed state at B. The cross-over point G is below the potential energy zero $V_p = 0$ and the activation energy for chemisorption is zero.

The magnitude of s is, therefore, dependent on the values of E_p, q_p and the probability of transition κ from the precursor to the (dissociative) chemisorbed state. The sequence may be represented as

$$A_2(g) \underset{k_d}{\overset{\alpha Z}{\rightleftharpoons}} A_2(p) \xrightarrow{k_c f(\theta)} 2A(c),$$

where α is the condensation coefficient into the precursor state $A_2(p)$, and Z the collision flux. The rate constants for desorption and conversion from the precursor state are k_d and k_c, where k_c $(=k'_c/\kappa)$ includes the transmission factor. The fractional coverage available for chemisorption is $f(\theta)$ and equals unity for a clean surface. In general, $s_\theta = \alpha Z/(1 + k_d/k_c f(\theta))$, where $k_d/k_c = [\exp(E_p - q_p)/RT]/\kappa$. The value of κ is less than unity when the activated complex has fewer degrees of freedom than the initial mobile precursor state; e.g. for a diatomic molecule forming an immobile complex, it may be as low as 10^{-3} from transition-state theory.†

For the chemisorption of nitrogen on W(100), the probable precursor is the γ-state for which q_p is *ca* 43 kJ mol^{-1}. The experimental value [7] of s_θ is ~0·4 between 190 and 600 K, but decreases to 0·08 at 1100 K. The s_θ versus $f(\theta)$ plot is well fitted by the above equation. The plot of $\log(k_d/k_c)$ against $1/T$ gives activation energies of 3 kJ mol^{-1} for $T < 600$ K and 12 kJ mol^{-1} for $T > 600$ K; hence, E_p is 39 and 30 kJ mol^{-1}, respectively. The change in activation energy at 600 K is in accord with the onset of mobility of the chemisorbed adatoms at this temperature.

An alternative approach involving the migration of the mobile precursor to unoccupied chemisorption sites has also been proposed [8, 9]. The different events that may occur on collision of a gas molecule with a surface are envisaged [9] as: (i) little, or none, of the kinetic energy of the incident molecule is transferred to the surface and it is elastically scattered back into the gas phase. (ii) Most, or all, of the translational kinetic energy of the component perpendicular to the adsorbent surface is transferred to the adsorbent. For zero angle of incidence (i.e. for impacts perpendicular to the surface), the gaseous molecule is either temporarily immobilized at the collision site in a physisorbed state, which is associated with a small potential well, before passing into the much deeper well corresponding to the chemisorbed state at the same site; or, it may acquire the necessary thermal energy from the lattice vibrations of the solid for desorption and is inelastically scattered back into the gas phase. At other angles of incidence, most of the kinetic energy of the perpendicular component is lost on impact; the molecule enters the physisorbed state, but still retains kinetic energy in the components parallel to the surface. It, therefore, migrates from the collision site; it may

† See Chapter 5.

then be physisorbed at a site distant from the site of impact following dissipation of its mobility energy to the solid lattice; or it may acquire momentum from lattice vibrations in a direction outwards from the surface and be desorbed; or it may pass into the chemisorbed state at an unoccupied site at some distance from the original collision site.

Thus, an incoming molecule may be either elastically scattered or transformed to a physisorbed state, i.e. to a precursor state; it may then pass into a chemisorbed state at an unoccupied metal site, or undergo desorption before it can locate such a site. The lifetime of a molecule in the precursor state is very short, since its binding energy to the surface is small; in contrast, it remains in the chemisorbed state for a very long time because a high activation energy is necessary for desorption. A distinction is, therefore, made between the condensation coefficient α, i.e. the probability of capture in the precursor state on collision of the gas molecule with the surface, and the sticking probability s_0, which denotes the capture probability into the final chemisorbed state. If all molecules in the precursor state become chemisorbed, then $\alpha = s_0$; in general, however, some are desorbed from the precursor state, and $s_0 < \alpha \approx 1$. A quantitative treatment of these events has been formulated by Kisliuk [8] and, later, by King and Wells [9] in the following, somewhat simpler, terms.

The probabilities of desorption, migration or chemisorption of a precursor-state molecule are expressed in terms of the rates of desorption (r_d), migration (r_m) and chemisorption (r_c) at both empty and occupied sites. Thus, at empty sites,

$$P_d = r_d/(r_d + r_m + r_c), \qquad P_m = r_m/(r_d + r_m + r_c) \tag{4.17}$$

and

$$P_c = r_c/(r_d + r_m + r_c). \tag{4.18}$$

At occupied sites,

$$P'_d = r'_d/(r'_d + r'_m), \qquad P'_m = r'_m/(r'_d + r'_m). \tag{4.19}$$

Equality of rates of desorption and migration from both empty and filled sites is assumed; i.e. $r'_m = r_m$, $r_d = r'_d$.

At a coverage θ, a fraction α of the incident molecules enter the precursor state and $(1 - \alpha)$ are elastically scattered back into the gas phase. The total desorption probability from this state for a series of successive migratory jumps over the surface is given by:

$$\{\theta(1 - P'_m) + (1 - \theta)P_d\} + \{\theta P'_m + (1 - \theta)P_m\}\{\theta(1 - P'_m) + (1 - \theta)P_d\} + \{\theta P'_m + (1 - \theta)P_m\}^2\{\theta(1 - P'_m) + (1 - \theta)P_d\} \ldots = 1 - s_\theta/\alpha. \tag{4.20}$$

Summation of this geometric progression for an infinite number of jumps leads to

$$\frac{\theta(1 - P'_{m}) + (1 - \theta)P_{d}}{1 - \theta P'_{m} - (1 - \theta)P_{m}} = 1 - \frac{s_{\theta}}{\alpha}. \tag{4.21}$$

Substitution of the probabilities in terms of the appropriate rate quotients then gives

$$s_{\theta} = \frac{\alpha(1 - \theta)}{(1 + r_{d}/r_{a}) - \theta(1 + r_{d}/r_{m})^{-1}}, \tag{4.22†}$$

which reduces at zero coverage to

$$s_{0} = \alpha/(1 + r_{d}/r_{a}). \tag{4.23}$$

With the assumption that the pre-exponential factors for desorption and transformation to the chemisorbed state are the same,

$$r_{d}/r_{a} = \exp[-(q_{p} - E_{p})/RT]. \tag{4.24}$$

In an investigation of the sticking probabilities of nitrogen on a polycrystalline tungsten foil, s_0 was found [9] to be 0·89 and independent of both coverage and temperature below 150 K. The roughness factor of the surface was probably 1·9 (see Section 4.4.3 below), thereby reducing the value of s_0 to 0·8 for a planar surface. It is highly probable that all precursor molecules are transformed to the chemisorbed state at these low temperatures; the value of the condensation coefficient was, therefore, taken as 0·8. Insertion of this value into Eq. (4.23) gives $r_d/r_a = 0{\cdot}30$. The value of s_θ was then calculated as a function of θ by adjustment of two parameters, viz. $(1 + r_d)/r_m = 0{\cdot}91$ and a saturation coverage of $\sim 3 \times 10^{14}$ molecules cm^{-2}. Figure 4.7 gives an example of the good agreement between the theoretical values of s_θ up to the saturation value.

4.4.2. Effect of the temperature

For s to be constant and independent of coverage, a molecule in the precursor state must migrate to an unoccupied metal site at which it is chemisorbed

† For dissociative of a homonuclear diatomic molecule,

$$s_{\theta} = \frac{\alpha(1 - \theta)^2}{(1 + r_{d}/r_{a})\{1 - \theta(2 - \theta)(1 + r_{d}/r_{m})^{-1}\}}.$$

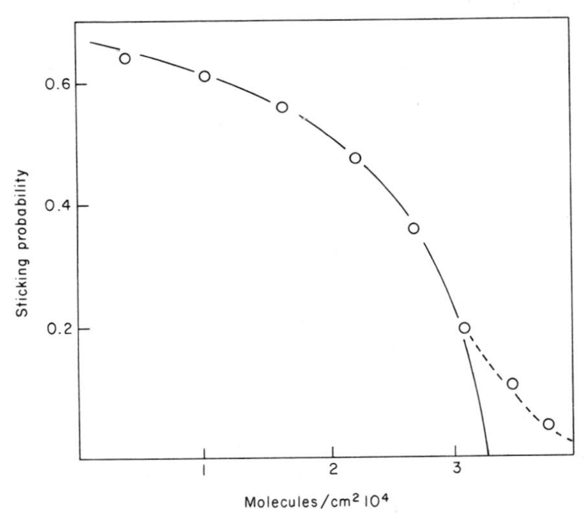

FIG. 4.7. Sticking probabilities for nitrogen on polycrystalline tungsten foil at a surface temperature of 300 K, O, Experimental points from molecular beam measurements; —, theoretical best fit by adjustment of the parameters, r_d/r_a, n_m (and $\alpha = s_m$ at 77 K = 0·8), in Eq. (4.22).

before it is desorbed. The mean distance $\bar{x}$ travelled by the precursor molecule in a time t is

$$\bar{x} = (Dt)^{\frac{1}{2}}, \tag{4.25}$$

where D is the diffusion coefficient, given by

$$D = a^2 \nu \exp(-E_m/RT). \tag{4.26}$$

The quantity ν is the jump frequency ($\sim 10^{12}$ s^{-1}), a is the jump length and E_m the activation energy for diffusion. The time interval t before desorption occurs, or the lifetime in the precursor state, is the reciprocal of the probability P for desorption in units of s^{-1}, i.e.

$$P = 1/t = \nu \exp(-E_d/RT), \tag{4.27}$$

where E_d is the activation energy for desorption.

The maximum value of $\bar{x}$ is obtained by combining Eqs. (4.25) and (4.26) to give

$$\bar{x}^2/t = a^2 \nu \exp(-E_m/RT). \tag{4.28}$$

Substitution for $1/t$ from Eq. (4.27) then gives the following relationship:

$$E_d = E_m + 4{\cdot}6\, RT \log(\bar{x}/a), \tag{4.29}$$

and the number of jumps n is given by

$$\bar{x} = an^{\frac{1}{2}}. \tag{4.30}$$

From field-emission studies for N_2 on W, $E_d - E_m = 5{\cdot}9\ \mathrm{kJ\,mol^{-1}}$ and $a \sim 3$ Å; hence, at 78 K, $\bar{x} \sim 270$ Å decreasing to ~ 10 Å at 290 K. At the lower temperature, some 800 sites are samples before the precursor molecule is desorbed; consequently, there is a high probability that it will pass into the chemisorbed state even when the surface coverage is substantial. However, as the temperature is raised, the number of jumps rapidly decreases to about 10 at 290 K; the probability of desorption, therefore, rapidly increases with increasing coverage. Hence, s is independent of temperature at a sufficiently low temperature, but decreases as the temperature is raised (provided that a precursor state that can be transformed to the chemisorbed state exists).

4.4.3. Effect of the non-planar surface

Another factor affecting the magnitude of s, particularly for films, is the roughness of the surface. On an ideally smooth surface, for a sticking probability of s_s, a fraction $(1 - s_s)$ of molecules impinging on the surface is reflected back into the gas phase. On a rough surface, a fraction f of the reflected molecules undergoes a second collision; and a fraction f of these twice-reflected molecules experiences a third collision, and so on. The measured sticking probability s_m is, therefore,

$$\begin{aligned} s_m &= s_s + (1 - s_s) f s_s + (1 - s_s)^2 f^2 s_s \ldots \\ &\approx s_s/[1 - (1 - s_s) f]. \end{aligned} \tag{4.31}$$

Analytical forms of f have been derived from various geometric models. One of these [10] is

$$f = 1 - 1/R, \tag{4.32}$$

where R is the roughness factor, i.e. the ratio of the actual surface area available for adsorption to the area of the geometrically plane surface. For $R = 3$, $s_m = 0{\cdot}7$ when $s_s \approx 0{\cdot}44$. The capture probability by an evaporated film is, therefore, higher than that by a polycrystalline filament. Thus, for the nitrogen–tungsten system at zero coverage and a temperature of 300 K, s_m for an evaporated film is 0·75, compared with values ranging from 0·60 to 0·30 for a polycrystalline filament.

4.4.4. Effect of the geometry of the crystal planes

Initial sticking probabilities of nitrogen on different crystal planes of tungsten have been recorded within fairly large experimental errors. In Table 4.1,

TABLE 4.1. Values of s_0 at 300 K on various W planes

Plane	s_0	Reference
(110)	$<10^{-3}$	Tamm and Schmidt [12]
(211)	$<10^{-2}$	Delcher and Erhlich [13]
(321)	$<10^{-2}$	Adams and Germer [14]
(111)	$\sim 10^{-2}$	Adams and Germer [15]
(310)	0·28	Adams and Germer [15]
(411)	0·31	Adams and Germer [14]
(210)	0·25	Adams and Germer [15]
(311)	0·29	Adams and Germer [14]
(100)	0·4	Delchar and Erhlich [13]

values of s_0 at 300 K are grouped into two categories, viz. $s_0 < 10^{-2}$ and $s_0 \sim 0{\cdot}3$. Planes for which s_0 is very low are closely packed with threefold symmetry sites [as for the (110) plane], whereas those having the fourfold symmetry sites characteristic of the more open (100) plane are associated with higher s_0 values (see Fig. 4.8). On the other hand, the s_0 values for CO approach unity and are not sensitive to crystal structure. For H_2 on W(100, 111, 211) planes, s_0 is initially independent both of coverage and temperature; it seems that passage into a precursor state depends on the energetics of bond

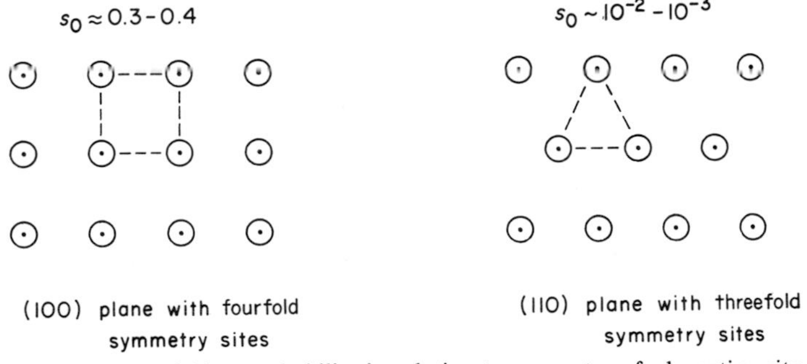

FIG. 4.8. Sticking probability in relation to symmetry of adsorption site.

formation rather than as a result of phonon-assisted thermal accommodation.

4.4.5. Information derived from sticking probabilities

A constant value of s, independent of coverage, indicates the presence of a precursor state of sufficiently long life that there is a high probability of transformation to the chemisorbed state before desorption ensues. Favourable conditions are low temperatures and small coverages. At higher temperatures and larger coverages, the precursor is, for the most part, desorbed before it can find an unoccupied chemisorption site.

In the absence of a precursor, the incoming molecules return to the gas phase after collision with an occupied site and s, therefore, decreases as the coverage increases. On collision with an unoccupied site, s is invariably less than unity, probably because a particular configuration is involved with respect to the colliding molecule with an adsorption site (i.e. a steric factor as in homogeneous gas reactions), or because the adsorption sites are not directly related to the *ideal* crystallography of the surface. For associative chemisorption and s constant at the correct adsorption site (i.e. independent of occupation of neighbouring sites by other chemisorbed molecules), the measured sticking probability should be a linear function of coverage, or

$$s_\theta = s_0(1 - \theta). \tag{4.33a}$$

For dissociative chemisorption of a homonuclear diatomic molecule involving pairs of adjacent unoccupied sites,

$$s_\theta = s_0(1 - \theta)^2. \tag{4.33b}$$

However, chemisorption of a molecule at a single site associated with a particular co-ordination number, followed by dissociation into adatoms due to strong repulsive interactions between them, would also give a linear dependence of s with θ [9].

For N_2 on W(111), Eq. (4.33a) describes the results [9], indicating the absence of a precursor state and direct molecular chemisorption at a specific site. The total number of sites N_s is obtained by extrapolation to $s = 0$ to give $N_s = 1{\cdot}25 \times 10^{14}\ \text{cm}^{-2}$, corresponding to a collision cross-section s_0/N_s of 6 Å^2, i.e. close to the cross-sectional area of a surface tungsten atom. The N_s value is not related to the crystallography of the W(111); hence, the site is assigned to a vacancy in the otherwise ideal surface. From field-ion microscopy, the number of disordered W atoms on the W(111) region of a tip thermally annealed at 2500 K is probably of order $10^{14}\ \text{cm}^{-2}$. It is, therefore, suggested [9] that chemisorption only take place at these (four-coordination) sites and that the adjoining sites are not engaged in the adsorption process.

4.5. THERMAL ACCOMMODATION AND MOLECULAR BEAM SCATTERING

The initial act of adsorption involves transfer of kinetic energy of the colliding gas molecule to the adsorbent, which thereby undergoes phonon excitation. A high capture probability into an initial physisorbed state indicates a high efficiency of energy transfer. A classical-mechanical calculation [16–18], based on a one-dimensional linear chain of atoms as a model for the solid, predicts a condensation coefficient of unity; this result has been experimentally confirmed [19] for the collisions of gaseous metal atoms with the metal surface.

A unit-capture efficiency would appear to show that gas molecules incident on the surface are scattered in thermal equilibrium with that surface, whereas in fact a fraction are specularly reflected. This incomplete accommodation of energy is expressed in terms of an accommodation coefficient of less than unity; but a more detailed investigation of the microscopic initial and final energy-states can now be accomplished by molecular beam scattering.

4.5.1. The accommodation coefficient [20, 21]

The accommodation coefficient is a measure of the efficiency of energy and momentum transfer of gas molecules on collision with a solid surface. The accommodation coefficient for energy transfer, A_c, may be defined as

$$A_c = (\bar{E}_1 - \bar{E}_0)/(\bar{E}_s - \bar{E}_0), \qquad (4.34)$$

in which $\bar{E}_0$ is the average molecular energy of the molecules colliding with the surface, and $\bar{E}_1$ is the average molecular energy of molecules re-emitted from the surface after impact. The term $\bar{E}_s$ is the value of $\bar{E}_1$ for gas molecules emitted as a Maxwellian beam characteristic of the temperature of the filament. Since a monatomic gas has zero internal energy, its total energy is its kinetic energy and A_c may be formulated alternatively in terms of temperatures, i.e.

$$A_c = (T_1 - T_0)/(T_s - T_0). \qquad (4.35)$$

In practice, the temperature of the surface is maintained at constant temperature T_s, which is different from that T_0 of the incident gas molecules. Invariably, $T_s > T_0$, and $T_s - T_0$ is small.

For gaseous monatomic molecules of molecular mass m having velocities between v and $v + dv$, the total energy E_0 incident on unit area of surface per second is

$$\int_0^\infty \tfrac{1}{4} f(v) v \, . \tfrac{1}{2} m v^2 \, dv, \tag{4.36}$$

in which, for a Maxwellian distribution,

$$f(v) = \frac{4nv^2}{\pi^{\frac{1}{2}}} \left(\frac{m}{2kT_0}\right)^{\frac{3}{2}} \exp\left(\frac{-mv^2}{2kT_0}\right). \tag{4.37}$$

Integration gives

$$E_0 = n(2k^3 T_0^3/\pi m)^{\frac{1}{2}}, \tag{4.38}$$

where n is the total number of molecules cm^{-3} in the gas phase. Since

$$n\left(\frac{kT_0}{2\pi m}\right)^{\frac{1}{2}} = \frac{P}{(2\pi mkT_0)^{\frac{1}{2}}}, \tag{4.39}$$

$$E_0 = 2PkT_0/(2\pi mkT_0)^{\frac{1}{2}}. \tag{4.40}$$

But molecules leaving the surface have a mean energy corresponding to $T_1 (\neq T_s)$. When $T_s - T_1$ is small (~ 20 K), the distribution departs only slightly from Maxwellian, so that

$$n_v = Av^3 \exp(-mv^2/2kT_1). \tag{4.41}$$

The number of molecules leaving unit area per second is

$$\int_0^\infty n_v \, dv = \frac{2Ak^2 T_1^2}{m^2} \tag{4.42}$$

and, in the steady state, this number is equal to the collision number, i.e.

$$P/(2\pi mkT_0)^{\frac{1}{2}} = 2Ak^2 T_1^2/m^2. \tag{4.43}$$

The energy E_1 of molecules re-emitted from the surface after collision is, therefore,

$$E_1 = \int_0^{\infty} Av^3 \exp\left(\frac{-mv^2}{2kT_1}\right) \tfrac{1}{2}mv^2 \, dv = \frac{4Ak^3T_1^3}{m^2} = \frac{2PkT_1}{(2\pi mkT_0)^{\frac{1}{2}}}. \quad (4.44)$$

Hence, the energy loss per unit area per second is

$$E_1 - E_0 = 2kP(T_1 - T_0)/(2\pi mkT_0)^{\frac{1}{2}} \quad (4.45)$$

or, in units of J cm^{-2} s^{-1}, using Eq. 4.35,

$$E_1 - E_0 = 7.3 \times 10^3 \frac{A_c P(T_s - T_0)}{(MT_0)^{\frac{1}{2}}}. \quad (4.46)$$

For a system comprising a metal filament of known surface area and an inert gas of molecular weight M at a pressure P (dyn cm^{-2}), the energy loss $E_1 - E_0$ is equated to the electrical energy supplied to the filament in order to maintain it at a constant temperature T_s, which is evaluated from its measured resistance and its temperature coefficient of resistance. It is essential to confirm that the value of A_c obtained using Eq. (4.46) is independent of the magnitude of $(T_s - T_0)$ and that T_s is sufficiently high to prevent adsorption of the gas on the filament. Various experimental methods are given in Knudsen [20] and Goodman [21].

At lower temperatures of the system, adsorption of the inert gas on the filament takes place. The accommodation coefficient, therefore, increases not only because the kinetic energy of the colliding gas atoms is smaller, but also because of the increase in the number of adsorbed atoms which evaporate with a Maxwellian distribution of energies characteristic of the temperature of the solid. Similarly, at a low temperature ($\sim$78 K) and for the heavier inert gases, A_c should increase as the pressure is raised because the number of adsorbed molecules is larger.

Little or no dependence of A_c on the pressure has in fact been reported. The reason is the approximate cancellation of two opposing effects: (i) A_c is increased because the relative number of collisions with occupied sites becomes larger with increasing extent of adsorption; but (ii) it is simultaneously decreased because the relative number of trapped molecules decreases with increase of coverage.

In general, A_c increases with: (i) decrease of the mass of the gas molecule; (ii) decreasing surface temperature; (iii) increasing attractive potential between gas and surface atoms; (iv) presence of contaminant, particularly when its atomic mass is less than that of the surface atoms.

Classical-mechanical calculations for head-on encounters of a gas atom with a linear chain of harmonically bound particles at rest predict that A_c

is unity for low-energy collisions of a gas atom identical with the atoms of the solid; smaller values are obtained when the incident kinetic energy of the gas atom is some 25 times greater than the energy required to dissociate an atom from the solid. The effect of including the lattice vibrations of the surface atoms is to increase the magnitude of A_c, particularly so at higher temperatures of the solid [22]. More sophisticated calculations indicate that the surface of a three-dimensional solid is less efficient [23] than the linear chain model in accommodating the energy transfer; also that A_c depends on the angle of impact of the gas molecule, such that energy transfer is about 40% less efficient at glancing impact than for a head-on encounter.

An additional feature is that the incoming molecule may not be captured in its ground state. After transference of sufficient kinetic energy by phonon excitation of the solid, the molecule can be bound in a higher vibrational state. These excited molecules make transitions to lower states in order to attain the Maxwellian distribution characteristic of the temperature of the solid.

The time interval between the initial capture and reversion of the excited molecule to the ground state is extremely rapid [24, 25]; for condensation of W atoms at a mean temperature of 3000 K on a solid tungsten lattice at 20 K, localization is observed to occur in the field-ion microscope within a period of a few oscillations.

At the initial capture, the maximum energy that can be accommodated by the lattice in a one-quantum transition is $\sim RT_D$, where T_D is the Debye temperature of the solid and is usually less than 500 K. The energy transferred is <4 J mol^{-1}, but atoms with kinetic energies corresponding to a gas-phase temperature of 500 K are, in fact, captured. Indeed, condensation with high efficiency occurs for temperatures well in excess of this value. A probable explanation is that the lattice is sufficiently perturbed by the impact of the colliding molecule to allow relaxation of the one-quantum restriction and multiple phonon transitions can take place.

4.5.1.1. *Application to chemisorption*

One application of the measurement of accommodation coefficients to chemisorption investigations was by Roberts [26]. Using a tungsten filament, the surface of which had been freed of contaminants by high-temperature flashing, the accommodation coefficient of an inert gas was found to be substantially increased after small amounts of hydrogen had been chemisorbed on the filament. The increase is large for chemisorbed molecules, particularly for hydrogen having a mass considerably smaller than that of the atoms of the filament, having a strong interaction with the solid. This work was largely responsible for the recognition that contaminants on metal surfaces in many previous chemisorption investigations render the results of little value; even so, accommodation techniques have rarely [27, 28] been used in chemisorption studies since the pioneering work of Roberts. More detailed knowledge can now be obtained by molecular beam scattering [29–31].

4.5.2. Molecular beam scattering

The energy accommodation coefficient is theoretically well defined and the conversion of the energy terms to the corresponding effective temperatures is valid for monatomic gas molecules. For polyatomic molecules, the translational, vibrational and rotational energies are probably accommodated to different extents; partial energy accommodation coefficients and partial effective temperatures can, thus, be defined, but little work on such systems has been done.

Molecular beam scattering can provide very detailed information, but the experimental techniques involve considerable practical difficulties. In this procedure, a beam of gas molecules, with a well-defined Maxwellian energy distribution, corresponding to the source temperature T_g, is impinged upon a single crystal plane maintained at a constant temperature T_s. Ideally, the flux, density and velocity of the scattered beam is measured as a function of scattering angle. The source of the incident molecules is a temperature-controlled oven having a small orifice through which the thermally equilibriated gas effuses randomly towards the crystal face. The angular distribution in the beam varies as the cosine of the angle with the normal to the orifice when the mean free path of the molecules is greater than diameter of the opening. The gas pressure in the oven must be very small and is maintained constant by use of high-speed pumps and controlled leaks. The beam intensity is extremely low; it can, however, be increased by using a multichannel array of capillaries [32].

The geometry of the crystal surface is determined by LEED. Freedom from contaminants is usually effected by ion-bombardment and annealing, and the cleanliness is monitored by Auger spectroscopy. The residual gas pressure must be less than 10^{-7} Pa during these operations.

The density of the scattered beam is measured by a detector comprising an electron beam source to effect ionization of the gas particles and an ion gauge, or quadruple mass spectrometer [33]. To obtain the angular density distribution, the detector or crystal must be rotatable. The signal-to-noise ratio is small, but can be increased by chopping the incident beam at a particular frequency and detecting and amplifying the signal by tuning to the same frequency. For the determination of the velocity distribution, a slotted-disc velocity selector or a variable-frequency time-of-flight velocity analyser [34] is usually employed.

4.5.2.1. *Inelastic scattering–hard-cube theory*
Information about the gas/surface energy exchange is derived from inelastic scattering. Surface atoms vibrate about their equilibrium positions with an average energy $3RT_s$, where T_s is the surface temperature, whereas the incident beam thermally equilibriated at the oven temperature T_0 has an average kinetic energy of $3RT_0/2$ for a monoatomic gas. Since the two energies have similar magnitudes, direct energy transfer between incident molecules and surface atoms takes place and the molecules in the scattered beam have different energy and angular distributions.

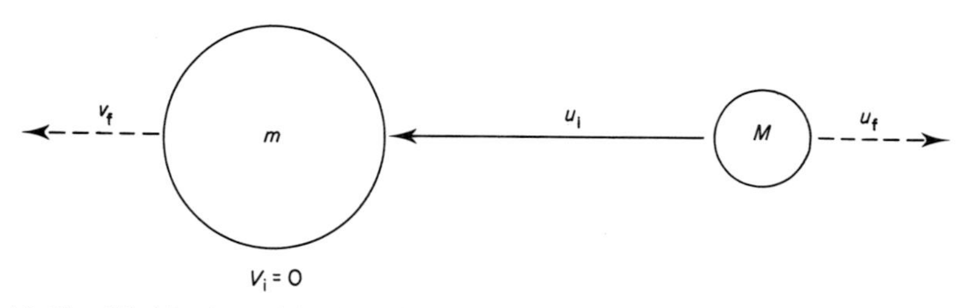

FIG. 4.9. Simplified Baule model. The surface atom of mass m is initially at rest. Gas atom of mass M impacts by head-on collision with initial velocity v_i. After collision, velocity of m is v_f and that of M is $-u_f$.

For the present purpose, the hard-cube theory (and its later extensions) is the most instructive. Some of the main features were already contained in Baule's original theory [35]; a simplified version follows. Large repulsive and negligible attractive forces are assumed, and each collision involves a gas atom (of mass M) and a single surface atom (of mass m). The surface atom is assumed to be initially at rest, i.e. its thermal motions and its temperature T_s are zero. On collision, the tangential momentum component of the gas atom is assumed to be unchanged as for a "head-on" collision. The gas atom is a point particle and the surface atom a sphere of radius equal to the sum of the two radii, i.e. the potential is taken to be a hard-sphere spherically-symmetrical one. From Fig. 4.9, conservation of energy and momentum gives

$$\tfrac{1}{2}Mu_i^2 = \tfrac{1}{2}mv_f^2 + \tfrac{1}{2}Mu_f^2, \tag{4.47}$$

$$Mu_i = -Mu_f + mv_f, \tag{4.48}$$

so that

$$v_f = \mu(u_i + u_f) \tag{4.49}$$

where $\mu = M/m$,the mass ratio, and must be less than unity.

The energy accommodation coefficient $A_c = \Delta E/\Delta E_i$, where ΔE is the energy transferred from gas to surface atoms and ΔE_i for complete energy transfer; hence,

$$A_c = 4\mu/(1 + \mu)^2. \tag{4.50}$$

By considering all collision angles θ and a density distribution function $f(\theta) = \sin 2\theta$ over the range $0 < \theta < \pi/2$, the average value is

$$\overline{A}_c = 2\mu/(1 + \mu)^2. \tag{4.51}$$

Hence, $\overline{A}_c$ is independent of E_0 and T_s, and depends only on the mass ratio for which $\mu < 1$.

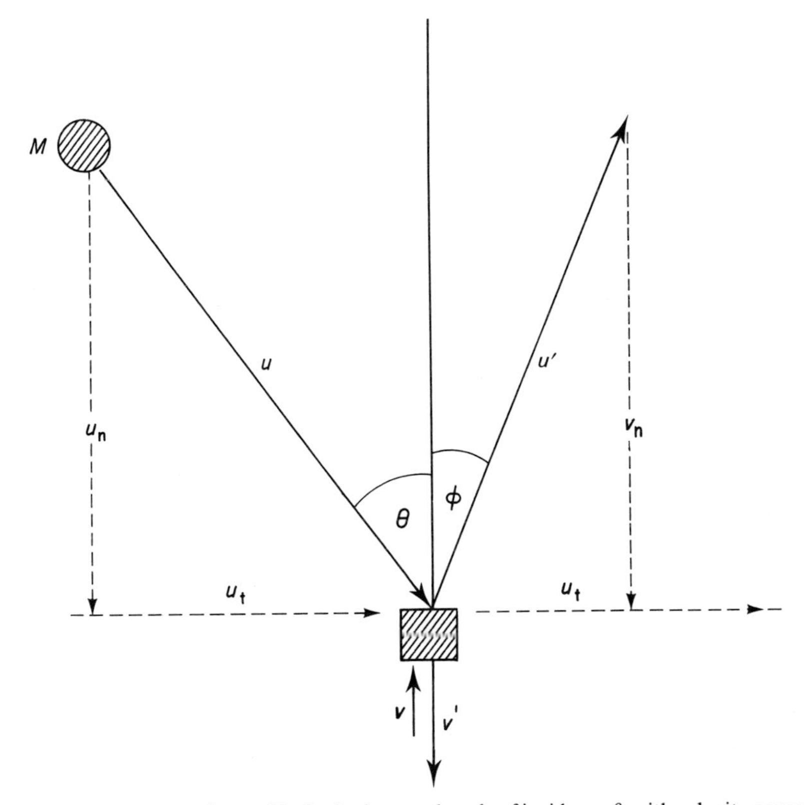

FIG. 4.10. A gas atom of mass M of velocity u and angle of incidence θ, with velocity components u_n normal to the surface and u_t tangential to the surface is scattered by surface atom (a hard cube) of mass m with a velocity v normal to the surface before collision and v' after collision (with zero tangential components). The final velocity of M is u' at a scattering angle ϕ with a normal component v_n and an unchanged tangential velocity component of u_t.

In the hard-cube model [36, 37], similar assumptions are made, viz.: (i) gas–surface interactions are weak, i.e. the residence time of the gas atom at the surface is equal to a surface atom vibration-period; (ii) the collision is an impulsive one between two rigid elastic masses, again with $\mu < 1$; and (iii) the momentum component of the gas molecule along the surface is unchanged by collision. The main difference is that the surface atom is represented by a small cube confined by a square-wave potential, with movement normal to the surface. Figure 4.10 gives a representation of the details [38]. Conservation of energy and of momentum normal to the surface requires that

$$\tfrac{1}{2}Mu_{\mathrm{n}}^2 + \tfrac{1}{2}mv^2 = \tfrac{1}{2}Mu_{\mathrm{n}}'^2 + \tfrac{1}{2}mv'^2, \tag{4.52}$$

and

$$M(u_{\mathrm{n}} + u_{\mathrm{n}}') = m(v' + v). \tag{4.53}$$

Solution of Eqs (4.52) and (4.53) gives

$$u_{\mathrm{n}}' = f(u_{\mathrm{n}}, v, \mu = M/m). \tag{4.54}$$

The scattering angle ϕ obeys the geometrical relationship

$$\cot\phi = (u_{\mathrm{n}}'/u_{\mathrm{n}})\cot\theta, \tag{4.55}$$

which, with Eq. (4.54), gives

$$v = B_1 u,$$

with

$$B_1 = \tfrac{1}{2}[(1 + \mu)\sin\theta\cot\phi - (1 - \mu)\cos\theta], \tag{4.56}$$

and

$$\mathrm{d}v/\mathrm{d}\phi = B_2 u,$$

with

$$B_2 = -\tfrac{1}{2}[(1 + \mu)\sin\theta\,\mathrm{cosec}^2\,\phi]. \tag{4.57}$$

The collision rate R of gas atoms (density $= n_\xi$) with surface atoms (density $= n_\mathrm{s}$) is

$$Z = n_\mathrm{s} n_\xi (u\cos\theta + v). \tag{4.58}$$

The probability distribution of scattered molecules $P(\phi)$ at an angle ϕ is then

$$P(\phi) = \frac{1}{n_s n_g \bar{u}_n} \frac{dR}{d\phi} = \frac{1}{\bar{u}_n} \int_{u=0}^{u=\infty} (u \cos\theta + v) f_3(u) f_1(v) \left(\frac{\partial v}{\partial \phi}\right)_u du, \quad (4.59)$$

with $\bar{u}_n$ the average normal velocity and $n_s n_g \bar{u}_n$ the number of collision of gas atoms with the surface per unit time; $f_3(u)$ is the three-dimensional Maxwellian velocity distribution of gas atoms at temperature T_g and $f_1(v)$ is the one-dimensional Maxwellian velocity distribution of surface atoms normal to the surface at temperature T_s. Hence,

$$P(\phi) = \frac{3}{4}\left(\frac{mT_g}{MT_s}\right)^{\frac{1}{2}} B_2(1 + B_1 \sec\theta)\left(\frac{(1 + mT_g)B_1^2}{MT_s}\right)^{\frac{2}{3}}. \quad (4.60)$$

Hence, $P(\phi)$ depends on: (i) the mass ratio $\mu = M/m$; (ii) the incidence angle θ; and (iii) the temperature ratio T_g/T_s. The theory is restricted to: (i) monatomic molecules, although changes of internal degrees of freedom of polyatomic molecules may not be significant for very weak interactions; (ii) an ideally smooth surface to avoid multiple collisions; and (iii) residence times $\leqslant 10^{-12}$ s; (iv) incident angles $> 10°$, otherwise the assumption that the tangential momentum component is unchanged after collision is not strictly valid. Agreement with experiment is good (see Fig. 4.11), and is

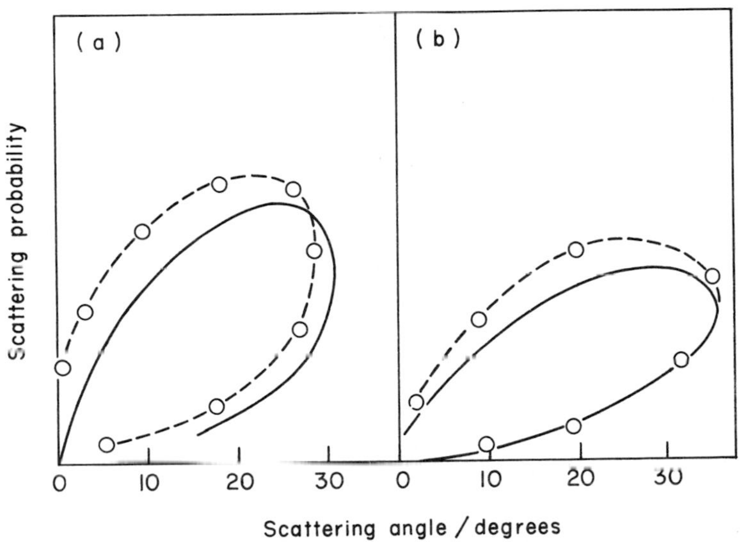

FIG. 4.11. Scattering probability as function of scattering angle. Comparison of experimental results (O---O) and calculated values (—) from the hard cube theory [Eq. (4.60)] for the scattering of argon atoms at a platinum surface at a fixed angle of incidence $\theta = 67{\cdot}5°$). Gas temperature 295 K; $M/m = 0{\cdot}2$. Surface temperature: (a) 1085 K, (b) 365 K.

improved by later extensions (soft-cube model [39], and modified rigid rotor model [40] [41].

4.5.2.2. *Elastic scattering*

For the lightest inert gas, helium, the de Broglie equation gives a particle wavelength of

$$\lambda = h/(2M\bar{E})^{\frac{1}{2}}, \tag{4.61}$$

where M is the atomic mass and $\bar{E}$ its average translational energy $= 3kT/2$; λ is, therefore, of the order of 1 Å at room temperature. Hence, for a mono-energetic beam, obtained by rotation of two slotted rotating discs with angular velocity similar in magnitude to that of the beam velocity, a narrow velocity range can be transmitted. The angular distribution of the scattered beam is obtained by rotation of the detector or of the sample, and the measured intensity is plotted as a function of the scattering angle. The surface forms a two-dimensional grating and diffraction peaks from (100)LiF were obtained as early as 1929 [42]. However, the large free-electron density in metals modifies the amplitude of the periodicity and diffraction has not been observed except for He from a W(112) plane [43].

Elastic scattering does not involve energy transfer between gas and surface atoms and, therefore, does not give information about interaction potentials.

4.5.2.3. *Efficiency of energy transfer*

Information about energy transfer by collision of gas atoms with surface atoms is derived from inelastic scattering. Only weak interactions have been considered in the various theoretical treatments, and helium is, therefore, usually employed. In the scattered beam, a specular reflection component arising from elastic scattering with a peak at $\phi = \theta$ often appears. With the heavy inert gas Xe, thermal equilibration with the surface at T_s is approached (greater interaction and longer residence time) and the main component is a broad cosine distribution. For neon, argon and krypton, there is an increasing degree of energy transfer and a broader distribution with peaks at smaller values of ϕ (which is less than that of the incident angle θ) as the mass is increased [44]. For Ar at $T_g \sim 300$ K, a peaked distribution with a large cosine distribution and a near-Maxwellian distribution of velocities corresponding to T_s is observed; but, with a high-energy incidence beam, the scattered distribution velocity is wider than that of the incident beam; and, in the specular lobe, there is low energy accommodation. Hence, the energy transfer is less efficient and the scattering more elastic as T_g is decreased.

4.5.2.4. *Reactive scattering*

Some interesting results have been obtained from scattering of H_2, HD and D_2 by silver [Ag(111)] [45]. Hydrogen gives a large elastic specular component, whereas both HD and D_2 have broad peaked distributions. It seems that the energy of a single Debye phonon ($\sim 2\ \mathrm{kJ\ mol^{-1}}$) is sufficient to effect rotational transitions in HD and D_2, but insufficient for H_2. Thus, with polyatomic molecules, changes of internal energy probably take place under certain circumstances.

Reactive scattering is becoming a rewarding technique for studying surface reactions; e.g. it has provided some information about the dissociative chemisorption of hydrogen. Scattering of hydrogen molecules by a tungsten surface at sufficiently high temperature produces hydrogen atoms; the process is favoured by increasing T_s, and thermal accommodation is complete before hydrogen atoms are evolved [46]. Similarly, H–D exchange has been found to take place on nickel [47] and on high-index (stepped) planes of Pt, but not on Pt(111) [48]. Again, fairly complete thermal accommodation takes place before emission of the product.

The importance of molecule beam techniques for investigating energy transfer is evident, but, as with accommodation coefficients, their application to chemisorption problems is not as yet clearly defined.

REFERENCES

1. G. Ehrlich, *J. Chem. Phys.*, 1961, **34**, 29.
2. J. S. Wagener, *Z. Angew. Phys.*, 1954, **6**, 433.
3. G. Ehrlich, *J. Phys. Chem.*, 1956, **60**, 1388.
4. D. O. Hayward, D. A. King and F. C. Tompkins, *Proc. Roy. Soc. A*, 1967, **297**, 305 and 321.
5. J. G. Taylor and I. Langmuir, *Phys. Rev.*, 1933, **44**, 423.
6. G. Ehrlich, *J. Phys. Chem.*, 1955, **59**, 473; *J. Phys. Chem. Solids*, 1956, **1**, 3.
7. L. R. Clavenna and L. D. Schmidt, *Surface Sci.*, 1970, **22**, 365.
8. P. Kisliuk, *J. Phys. Chem. Solids*, 1957, **3**, 95; 1958, **5**, 78.
9. D. A. King and M. G. Wells, *Surface Sci.*, 1972, **29**, 454.
10. C. S. McKee and M. W. Roberts, *Trans Faraday Soc.*, 1967, **63**, 1418.
11. D. A. King and F. C. Tompkins, *Trans. Faraday Soc.*, 1968, **64**, 496.
12. P. W. Tamm and L. D. Schmidt, *Surface Sci.*, 1971, **26**, 286.
13. T. A. Delchar and G. Ehrlich, *J. Chem. Phys.*, 1965, **42**, 2686.
14. D. L. Adams and L. H. Germer, *Surface Sci.*, 1971, **27**, 21.
15. D. L. Adams and L. H. Germer, *Surface Sci.*, 1971, **27**, 109.
16. N. Cabrera, *Disc. Faraday Soc.*, 1959, **28**, 16.
17. R. W. Zwanzig, *J. Chem. Phys.*, 1960, **32**, 1173.
18. B. McCarroll and G. Ehrlich, *J. Chem. Phys.*, 1963, **38**, 523.
19. S. Wexler, *Rev. Mod. Phys.*, 1958, **30**, 402.
20. M. Knudsen, *Ann. Physik*, 1911, **34**, 593.

21. F. O. Goodman, *Surface Sci.*, 1971, **24**, 667; *Prog. Surface Sci.*, 1975, **5**, 261.
22. F. O. Goodman, *J. Phys. Chem. Solids*, 1965, **26**, 85.
23. E. T. Kinzen and C. M. Chambers, *Surface Sci.*, 1965, **3**, 261.
24. G. Ehrlich, *in* "Metal Surfaces: Structures, Energetics and Kinetics", American Society of Metals, Metals Park, Ohio, 1963, Vol. 7, p. 221.
25. R. D. Young and D. C. Schubert, *J. Chem. Phys.*, 1965, **42**, 3943.
26. J. K. Roberts, "Some Problems in Adsorption", Cambridge University Press, Cambridge, 1939.
27. B. McCarroll, *J. Chem. Phys.*, 1963, **39**, 1317.
28. P. Feur, *J. Chem. Phys.*, 1962, **39**, 1311.
29. N. F. Ramsey, "Molecular Beams", Oxford University Press, London, 1956.
30. R. P. Merrill, *Catalysis Rev.*, 1970, **4**, 115.
31. J. N. Smith, *Surface Sci.*, 1973, **34**, 613.
32. D. R. Olander, *J. Appl. Phys.*, 1970, **41**, 2769.
33. J. N. Smith and W. L. Fite, *in* "Rarefied Gas Dynamics" (J. A. Laurman, ed.), Academic Press, New York, 1963, Vol. 1, Suppl. 2.
34. A. E. Dabiri, T. J. Lee and R. E. Stickney, *Surface Sci.*, 1971, **26**, 522.
35. B. Baule, *Ann. Phys.*, 1914, **44**, 145.
36. R. M. Logan and J. C. Keck, *J. Chem. Phys.*, 1966, **44**, 195.
37. R. E. Stickney, *in* "The Structure and Chemistry of Solid Surfaces" (G. A. Somorjai, ed.), Wiley, New York, 1969, Vol. 1, p. 41.
38. G. A. Somorjai, "Principles of Surface Chemistry", Prentice Hall, Inc., Englewood Cliffs, New Jersey, 1972, pp. 201–205.
39. R. M. Logan and J. C. Keck, *J. Chem. Phys.*, 1968, **49**, 860.
40. R. E. Forman, *J. Chem. Phys.*, 1971, **55**, 2839.
41. J. D. Doll, *J. Chem. Phys.*, 1973, **59**, 1038.
42. I. Eastermann and O. Stern, *Z. Phys.*, 1930, **61**, 95.
43. D. V. Tendulkar and R. E. Stickney, *Surface Sci.*, 1971, **27**, 516.
44. J. N. Smith, Jr, H. Saltsburg and R. L. Palmer, *J. Chem. Phys.*, 1968, **49**, 1287.
45. R. L. Palmer, H. Saltsburg and J. N. Smith, Jr, *J. Chem. Phys.*, 1969, **50**, 4661.
46. J. Smith and W. Fite, *J. Chem. Phys.*, 1962, **37**, 1962.
47. R. L. Palmer, J. N. Smith, Jr, H. Saltsburg and D. R. O'Keefe, *J. Chem. Phys.*, 1970, **53**, 1666.
48. S. L. Bernasek, W. J. Siekhaus and G. A. Somorjai, *Phys. Rev. Letters*, 1973, **30**, 1202.

5

Desorption Phenomena

5.1. METHODS OF EFFECTING DESORPTION

The desorption of chemisorbed molecules may be accomplished by various procedures. It takes place when the temperature of the gas/metal system is raised to a sufficiently high value such that an appreciable fraction of the adspecies acquires a Maxwellian distribution of thermal energies from the lattice vibrations of the adsorbent equal to or greater than the desorption energy. This process is termed *thermal desorption*.

Photodesorption is effected by excitation of the adspecies by light of sufficiently high frequency. *Field desorption* is the application of a strong external electric field of around 10^8 V cm^{-1}; the width of the surface energy barrier is reduced and electron tunnelling from the metal to the adsorbate occurs. The ground state of the adspecies is converted to an ionic state and desorption of the adsorbate ion then occurs.

Excitation by the impact of electrons, ions, or neutrals of sufficiently high energy effects transitions from the adsorbate ground state to an excited or ionic state, and both neutral and ionic forms of the adspecies are desorbed. However, quantitative information can best be obtained by *electron-impact* (or *electron-stimulated*) *desorption*. The use of inert-gas ions for excitation is normally restricted to removing surface contaminants in the preparation of clean single crystal surfaces.

From desorption studies, information about the activation energy of desorption, the kinetic order of the evolution process, the detection of different states of a single adsorbed species, and the population of these states, may be obtained.

5.2. THERMAL DESORPTION

5.2.1. General kinetic equation

An ideal system comprises: (i) a uniform surface with all adsorption sites identical; (ii) negligible interaction between adsorbed molecules on the surface; and (iii) the molecular form of the adspecies the same as that of the gaseous adsorbate. Desorption occurs when the adsorbate–adsorbent bond acquires the activation energy for desorption in the form of vibrational energy of the chemisorption bond. The vibrational frequency ν_1 is taken to be *ca* 10^{13} s^{-1} at the temperatures at which desorption proceeds at a convenient rate; the frequency of bond rupture is, therefore, $\nu_1 \exp(-e_d/kT)$, e_d being the activation energy for desorption of a single adsorbed molecule. For a surface concentration of n_a adsorbate molecules cm^{-2}, the rate of desorption $-dn_a/dt$ is

$$-\mathrm{d}n_a/\mathrm{d}t = n_a \nu_1 \exp(-E_d/RT) \text{ molecules cm}^{-2}\,\text{s}^{-1}, \tag{5.1}$$

where E_d is the activation energy of desorption per mole of the adspecies. The (first-order) desorption rate constant is

$$k_d = \frac{\mathrm{d}n_a}{n_a \mathrm{d}t} = \frac{\mathrm{d}(\ln n_a)}{\mathrm{d}t}; \tag{5.2}$$

it is a function of T but not of n_a.

For simple diatomic molecules (e.g. H_2, O_2, etc.), dissociative chemisorption takes place at the surface of most transition metals and adatoms are formed. These adatoms are mobile at the temperature at which desorption of the adsorbate molecule takes place; bimolecular collisions between the adatoms in the two-dimensional surface phase occur, followed by desorption of adsorbate molecules. The rate of loss of adatoms by this second-order process is

$$-\mathrm{d}n_a/\mathrm{d}t = n_a^2 \nu_2 \exp(-E_d/RT), \tag{5.3}$$

and the rate of evolution of molecules is $-\frac{1}{2}(\mathrm{d}n_a/\mathrm{d}t)$, where n_a is the surface concentration of adatoms. The frequency ν_2 is the number of collisions per second for unit concentration of adatoms in a two-dimensional gas, i.e. $(\pi kT/m)^{\frac{1}{2}}d$, or *ca* 10^{-2} cm^{-2} s^{-1}, where m is the mass of the adatom and d its collision diameter. Mobility of adatoms involves surmounting an energy barrier equal to the activation energy for mobility, but this term is contained in the experimental value of E_d.

The general kinetic equation for desorption is, therefore,

$$-\mathrm{d}n_a/\mathrm{d}t = n_a^x \nu_x \exp(-E_d/RT), \tag{5.4}$$

where x is the kinetic order in terms of the surface concentration of the appropriate adspecies. From desorption rates, the activation energy of desorption E_d, and the kinetic order x of the evolution process can, therefore, be determined.

5.3. DETERMINATION OF KINETIC ORDER AND DESORPTION ENERGIES

5.3.1. Limitations of measurements

For a first-order desorption from a full monolayer of adsorbate (n_m molecules cm^{-2}) on a filament of area $A \approx 1$ to $2\ cm^2$, the rate per cm^2 surface is given by

$$-dn_a/dt = n_m \nu_1 \exp(-E_d/RT), \tag{5.5}$$

in units of number of molecules $cm^{-2}\ s^{-1}$.

A typical activation energy of desorption is $125\ kJ\ mol^{-1}$; with $\nu_1 \approx 10^{13}\ s^{-1}$, and $n_m \approx 10^{15}\ cm^{-2}$, the rate at 500 K is $\approx 10^{28} \exp(-30) \approx 10^{15}$ molecules $s^{-1}\ cm^{-2}$. An apparatus of $500\ cm^3$ can easily accommodate the filament and an ionization gauge. Hence, the rate of increase of pressure in this closed system is initially $\sim 9 \times 10^{-3}\ Pa\ s^{-1}$ decreasing to $9 \times 10^{-5}\ Pa\ s^{-1}$ at 1% coverage, assuming a zero re-adsorption rate. Evolution rates can, therefore, be measured without difficulty with an accuracy of $<1\%$ over a wide range of E_d and T, and in the presence of a substantial rate of readsorption.

5.3.2. Closed system – single adsorbed state

5.3.2.1. *Isothermal method* [1]

The clean filament of surface area $A\ cm^2$ is first covered with $N_a (= n_a A)$ admolecules at a pressure P_0 of the gaseous adsorbate and at a temperature T_0. The temperature of the filament is then raised rapidly to a constant higher temperature T_f and the increase of pressure P_t developed in the adsorption system of volume V at the temperature T_0 is recorded as a function of time t. The desorption rate is

$$-\frac{dn_a}{dt} = \frac{V}{AkT_0}\frac{dP_t}{dt} = n_a^x \nu_x \exp(-E_d/RT_f) \text{ molecules cm}^{-2}\ \text{s}^{-1} \tag{5.6}$$

when the rate of re-adsorption is negligible. The amount remaining on the surface at any time t is $An_a(t) = An_0 - (P_t V/kT) = (P_m - P_t)V/kT$, where n_0 is the initial concentration of admolecules cm^{-2} and P_m is the final pressure corresponding to complete desorption. Tangents to the $n_a(t)$ versus t plot at a series of points provide $dn_a(t)/dt$ values. The plot of $\ln(dn_a(t)/dt)$ against $n_a(t)$ gives a linear plot of slope of one (first-order) or two (second-order process). Measurements are repeated for different constant temperatures T_f of the filament. The plot of $\ln(dn_a/n_a^x\,dt)_{T_f}$ against $-1/RT_f$ is a straight line of slope E_d.

The isothermal method has disadvantages, viz.: (i) the use of a temperature step-function means that the time t_0 of starting the desorption at T_f is ill-defined—rate-analysis procedures for time $t > t_0$ can, however, be employed; (ii) the increase of P_t as desorption proceeds introduces increasing errors arising from the concurrent rate of re-adsorption; and (iii) for each constant-temperature determination of the desorption rate, a clean surface has to be regenerated—high-temperature flashing in high vacuum is suitable and effective for only a few metals. The alternative method, of using a series of temperature jumps as desorption proceeds, is subject to considerable uncertainty but has value when a single-crystal plane is the adsorbent.

In general, the direct determination of P_t is the normal procedure. Measurement of the amount of adsorbate on the surface as a function of time is less accurate. Various procedures, such as the change of: (i) field-emission current from a field-emission tip [2]; (ii) the ion current from electron-impact desorption [3]; and (iii) surface potential, or work function, have been employed [4], the assumption being that these quantities vary linearly with coverage. However, the "instantaneous" pressure increase produced by high-temperature flashing after a specific interval of time gives a good measure of the amount of adsorbate on the surface [5].

5.3.2.2. *Temperature-programmed method* [6–8]

A non-isothermal method is usually preferred. After adsorption of a measured quantity of gas by the clean filament at a temperature T_0, its temperature is raised at a prescribed rate. Two temperature programmes have been commonly employed: (i) a linear rate ($\mathrm{K\,s^{-1}}$), such that $T_t = T_0 + \beta t$, where $\beta = dT/dt$; and (ii) a hyperbolic rate given by $1/T_t = 1/T_0 + bt$, with b constant. (Attention is largely confined to (i), since extension to (ii) is straightforward.)

In order to minimize errors arising from re-adsorption, a rapid heating rate (β large) is employed, so that departures from adsorption equilibrium are as large as is conveniently practical over the whole temperature range. The negligible effect of readsorption is experimentally tested by confirming that the evolution rate is independent of the heating rate β. For a linear rate, $\beta = dT/dt$, Eq. (5.4) can be written as

$$-\beta dn_a/dT_f = n_a^x v_x \exp(-E_d/RT_f). \tag{5.7}$$

The kinetic order ($x = 1$, or 2) is that for which the plot of $\ln(dn_a/n_a^x\, dT_f)$ against $-1/RT_f$ is linear and its slope gives the value of E_d. Hence, both x and E_d are evaluated from a single experiment. The frequency term ν_x can be estimated from Eq. (5.7) [or from Eq. (5.6) for the isothermal method with $\beta = 0$] with a knowledge of x, E_d, β, and the value of $\ln(dn_a/n_a^x\, dT_f)$ at a particular temperature T_f.

The procedure is to measure P_T as a function of temperature (which for a linear heating rate is directly proportional to time). From Eqs (5.6) and (5.7),

$$\frac{V}{\beta A k T_0}\frac{dP_T}{dT_f} = -\frac{dn_a}{dt} = n_a \nu_1 \exp\left(\frac{-E_d}{RT_f}\right)$$

for a first-order desorption, the rate of increase of pressure is determined by the values of n_a and the temperature T_f. The initial increase of rate reflects the dominance of the exponential term, but the rate decreases in the later stages as the surface is denuded.

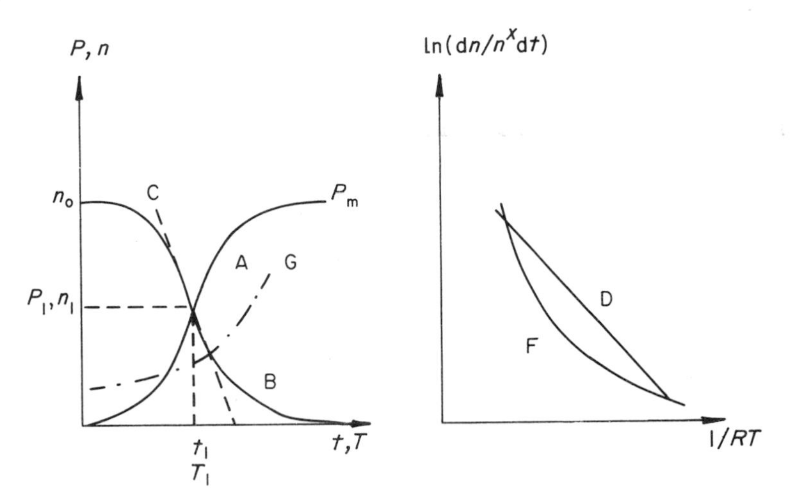

FIG. 5.1. Determination of the kinetic order and desorption energy for a temperature-programmed procedure.

Curve A, pressure developed in a constant volume (closed) system as a function of time. G (·—·—), filament temperature T as function of time t for a hyperbolic heating rate.

Curve B, amount of gas remaining on surface given by $N_0 - P_i V/kT_i$, where N_0 is the initial amount on the surface at t_0, and P_i, T_i are the pressure P_i and temperature T_i at t_i (curve G); V is the volume of the system.

Curve C, tangent to curve B at the point n_1, t_1 at a temperature T_1 giving the rate dn/dt at a surface concentration n_1.

Curve D, plot of $\ln(dn/n\, dt)$ as a function of $1/T$. The linearity shows that a first-order desorption process is being measured; the slope gives the value of $-E_d/R$, the activation energy of desorption.

Curve F, non-linear plot of $\ln(\ln(dn/n^2\, dt))$ as $f(1/T)$ for a second-order desorption process.

A typical P_t against t plot for a hyperbolic heating rate is shown in Fig. 5.1, in which details of the procedure for determining the kinetic order x and activation energy of desorption E_d are given.

5.3.2.3. *Kinetic order and variable E_d*

The isothermal rate of desorption given by Eq. (5.6) is valid for constant E_d and x, and the slope of the linear plot of $\ln(dn_a/dt)$ against $\ln n_a$ gives an unambiguous value for the order of the evolution process. However, ambiguity arises when E_d varies with surface concentration.

To illustrate this difficulty, E_d is assumed to decrease linearly with increase of $\theta(=n_a/n_m)$, i.e.

$$E_d^\theta = E_d^0 - b\theta, \tag{5.8}$$

where the constant $b = E_d^0 - E_d^m$, and E_d^0, E_d^θ and E_d^m are, respectively, the desorption energy per mole for $\theta = 0$, θ, and 1. The rate of a first-order desorption process is then given by

$$-d\theta/dt = a\theta \exp(b\theta/RT), \tag{5.9}$$

where $a = \nu_1 \exp(-E_d^0/RT)$.

The mean slope of the $\ln(d\theta/dt)$ against $\ln \theta$ plot between the limits θ_1 and θ_e is, therefore,

$$1 + [b(\theta_1 - \theta_e)/RT] \ln(\theta_1/\theta_e). \tag{5.10}$$

For a small θ-interval, $\theta_1 = 0{\cdot}3$, $\theta_2 = 0{\cdot}2$, the $\ln(d\theta/dt)$ versus $\ln \theta$ plot is reasonably linear of unit slope; but at 300 K the second term in (5.10) is unity when $b = E_d^0 - E_d^m = 10{\cdot}2\ \text{kJ mol}^{-1}$, i.e. the slope is 2 as for a second-order process. Hence, a first-order desorption with a small linear decrease of E_d can closely simulate a second-order evolution at constant E_d. The possible variation of E_d at various coverages should, therefore, be always examined. Figure 5.2 shows a P_t versus t plot for a hyperbolic heating rate for a first-order desorption with E_d constant (curve A) and with E_d linearly decreasing with coverage (curve B). Curve B is virtually indistinguishable from curve B in Fig. 5.4 for a second-order desorption with E_d constant.

In the non-isothermal method, x is selected (either 1 or 2) in an attempt to obtain a linear plot of $\ln(dn_a/n_a^x\, dT_f)$ against $-1/RT_f$. However, in general, linearity cannot be achieved when $E_d = f(n_a)$. The procedure adopted is therefore, to determine a series of desorption curves, starting with different initial coverages of the filament but retaining a constant heating rate β. The slopes $-dn_a/dT_f$ (or $-dn_a/dt$ for the isothermal procedure) at various

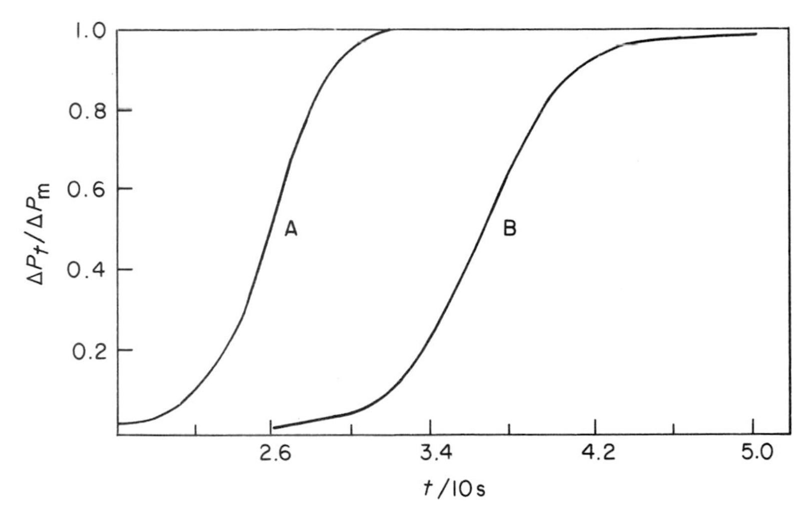

FIG. 5.2. Desorption curves for first-order desorption. Curve A, constant value for desorption energy (E_d = 340 kJ mol^{-1}); curve B, with variable desorption energy (E_d = 460 − 50θ kJ mol^{-1}). Initial surface concentration in each case = 4 × 10^{13} molecules cm^{-2}. Curve B is virtually identical with curve B (initial concentration = 4 × 10^{13}) for the second-order desorption (E_d = 340 kJ mol^{-1}) in Fig. 5.4.

temperatures T_f at a specific surface concentration n'_a are then interpolated from the set of curves and $\ln(\mathrm{d}n_a/\mathrm{d}T_f)_{n'_a}$ is plotted against $-1/RT_f$. The slope of this linear plot gives E_d at the concentration n'_a. For different fixed values of n'_a, $E_d(n'_a)$ as a function of surface coverage can then be obtained as well as the true kinetic order of the desorption process.

5.3.3. Flow system [9–11]: desorption kinetics and temperature displacement methods

The non-isothermal method is also employed using a flow system. At an initial temperature T_0, gas is introduced at a constant rate (r_i molecules s^{-1}) into the system containing the clean filament, and a constant rate of pumping through the exhaust port is applied. After a short interval of time, a steady-state pressure P_0 and an associated equilibrium surface concentration $n_m\theta_0$ on the filament are established; at this stage $r_i = P_0 r_o$, where r_o is the outflow rate at unit pressure.

For a linear heating rate β and a first-order desorption, the following two relationships are obeyed. The first condition is

$$-n_m \frac{\mathrm{d}\theta_t}{\mathrm{d}t} = -n_m\beta\frac{\mathrm{d}\theta_T}{\mathrm{d}T} = [k_d n_m \theta_T - k_a P_T(1-\theta_T)], \tag{5.11}$$

i.e. the rate of decrease of surface concentration is the difference between the rates of desorption and adsorption. The second condition is

$$P_T r'_o = A[k_d n_m \theta_T - k_a P_T(1 - \theta_T)], \tag{5.12}$$

i.e. the net loss of molecules s^{-1} at the outflow port equals the net rate of desorption of adsorbed gas from the filament, where $P_T r'_o = r_o(P_T - P_0)$.

Two limiting cases can be considered [6–8]: (i) the kinetics of desorption are rate-determining; and (ii) desorption occurs by temperature displacement of a steady-state surface concentration approximating to adsorption equilibrium at the temperature of the filament and ambient pressure of gaseous adsorbate over it. In case (i), a high heating rate (β large) and a high pumping speed r_o are required so that rate of re-adsorption $k_a P_T(1 - \theta_T)$ can be neglected. Since the rate of desorption $-\mathrm{d}n_a/\mathrm{d}t = n_a v_1 \exp(-E_d/RT)$ for a first-order process then, at constant n_a and T, the constancy of this rate should be confirmed experimentally. In case (ii), the pressure P_t at time t is increased by the rise of temperature, causing displacement of the steady-state concentration to lower coverages. Since $\mathrm{d}n_a/\mathrm{d}t = (\mathrm{d}n_a/\mathrm{d}T)(\mathrm{d}T/\mathrm{d}t) = \beta\,\mathrm{d}n_a/\mathrm{d}t$, the slope of the n_a versus t curve at a given temperature should be proportional to β. Again, experimental confirmation that the slope of the desorption curve at a given temperature increases linearly with increase of the slope of the temperature-time curve is essential.

Case (i). The effect of readsorption is negligible; this condition requires a high value of β and a high pumping speed r_0. From Eq. (5.11), $k_d = v_1 \exp(-E_d/RT)$ and, neglecting the term $k_a P_T(1 - \theta_T)$,

$$\frac{\mathrm{d}\theta_T}{\mathrm{d}T} = -\frac{k_d \theta_T}{\beta} = -\frac{\theta_T v_1}{\beta} \exp\left(\frac{-E_d}{RT}\right); \tag{5.13}$$

from Eqs (5.12) and (5.13),

$$P_T = -\frac{A n_m \beta}{r'_o} \frac{\mathrm{d}\theta_T}{\mathrm{d}T}. \tag{5.14}$$

The typical shape of the P_T versus T plot for a first-order process is shown as curve A in Fig. 5.3. From Eq. (5.14), P_T is directly proportional to $\mathrm{d}\theta_T/\mathrm{d}t$, i.e. the rate of desorption at any temperature T is recorded in terms of the measured increase of pressure. The shape of the plot may be deduced from Eq. (5.13); the rate in the early stages increases with rise of temperature, since the exponential term is dominant, but in the later stages, as the surface is being denuded, the decrease in surface concentration begins to have the

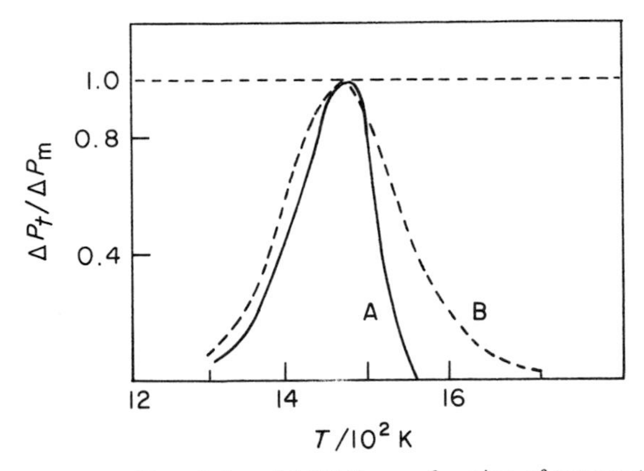

FIG. 5.3. Desorption profiles of plot of $\Delta P_t/\Delta P_m$ as a function of temperature for a hyperbolic heating rate and high pumping speed. Curve A, first-order desorption with $E_d \simeq 385$ kJ mol^{-1} with profile asymmetric about T_m. Curve B, second-order desorption with $E_d \simeq 465$ kJ mol^{-1} with symmetric profile.

greater effect and a maximum rate is attained at a particular (peak) temperature T_m. From a knowledge of T_m and the linear heating rate β for a first-order desorption with a "normal" frequency factor ($\nu_1 \approx 10^{13}$ s^{-1}), E_d may be estimated by the following procedure.

At the peak temperature T_m, $\mathrm{d}P_T/\mathrm{d}T_m = -\mathrm{d}^2\theta_T/\mathrm{d}T_m^2 = 0$. Differentiation of Eq. (5.13) gives

$$\exp\left(-\frac{E_d}{RT_m}\right)\frac{\mathrm{d}\theta_T}{\mathrm{d}T_m} - \frac{\theta_T E_d}{RT_m^e}\exp\left(-\frac{E_d}{RT_m}\right) = 0, \qquad (5.15)$$

which may be transformed to

$$\frac{\nu_1}{\beta}\exp\left(-\frac{E_d}{RT_m}\right) = \frac{E_d}{RT_m^2}. \qquad (5.16)$$

The term E_d/RT_m^2 is small and its logarithm may be approximated to $[\ln(\nu_1/\beta)]/T_m$. Equation (5.16) can, therefore, be written in approximate form as

$$E_d/RT_m \approx \ln(\nu_1/\beta) + \ln[T_m/\ln(\nu_1/\beta)]. \qquad (5.17)$$

A typical heating rate is $\beta = 10$ K s^{-1}, and $\nu_1 \approx 10^{13}$ s^{-1}; hence,

$$E_d/\text{kJ mol}^{-1} \approx 230\ T_m/\text{K}. \qquad (5.18)$$

A peak temperature of 500 K, therefore, corresponds to a desorption energy of about 116 kJ mol^{-1}.

Alternatively, E_d can be obtained by repeating the experiment with a different heating rate β', when a new peak temperature T'_m will be observed. From Eq. (5.16),

$$\frac{E_d}{R}\left(\frac{1}{T'_m} - \frac{1}{T_m}\right) = \ln\left(\frac{\beta}{\beta'}\right) + 2\ln\left(\frac{T'_m}{T_m}\right), \tag{5.19}$$

from which E_d may be evaluated. For reasonable accuracy, β/β' should be large (*ca* 10^2), but practical difficulties ensue; for small β/β', extraneous evolution from other surfaces of the system can be comparatively large; and at high β/β', the desorption rate from the filament may be too rapid for accurate measurement (and multiple peaks are not resolved) [12].

Case (*ii*). The re-adsorption rate is rapid, a condition favoured by a sufficiently small value of β and a low pumping speed. In this case, $P_T r'_o \ll Ak_a(1 - \theta_T)P_T$ and Eq. (5.12) can be approximated to

$$k_d n_m \theta_T = k_a P_T(1 - \theta_T), \tag{5.20}$$

in which P_T is the pressure at any given temperature T in equilibrium with the surface coverage θ_T. Transformation of Eq. (5.20) gives

$$P_T = K_T n_m \theta_T/(1 - \theta_T), \tag{5.21}$$

in which

$$K_T = (k_d/k_a)_T = \exp(-\Delta G^{\ddagger}/RT) = \exp(\Delta S^{\ddagger}/R)\exp(-\Delta H^{\ddagger}/RT),$$

where $\Delta G^{\ddagger}$, $\Delta S^{\ddagger}$, $\Delta H^{\ddagger}$ are the standard free energy, entropy and enthalpy changes of formation of the transition state which is in thermodynamic equilibrium with the adsorbate, and K_T is the dimensionless equilibrium constant (see Section 5.3.5) at the temperature T.

The desorption energy $E'_d(\simeq -\Delta H^{\ddagger})$ can then be evaluated from Eq. (5.22),

$$P_T = n_m \frac{\theta_T}{1 - \theta_T}\exp\left(\frac{\Delta S^{\ddagger}}{RT}\right)\exp\left(\frac{-\Delta H^{\ddagger}}{RT}\right). \tag{5.22}$$

A set of evolution curves for different initial coverages θ_i, at the same heating rate, is determined. Values of P_T and T corresponding to a specific coverage are interpolated from these curves. The plot of $\ln[P_t(1 - \theta)/\theta_T]$ against $-1/RT_f$ is linear of slope E_d for a first-order process. For a second-order evolution, $\ln[P_T(1 - \theta_T)^2/\theta_T^2]$ against $-1/RT_f$ gives the corresponding linear plot.

The value of $\theta_T = (n_a/n_m)_T$ can be derived from the continuity condition describing the conservation of mass within the desorption chamber, i.e.

$$\frac{V\mathrm{d}(P_t - P_0)}{kT_0\,\mathrm{d}t} = \frac{A\,\mathrm{d}n_a}{\mathrm{d}t} - r_o(P_t - P_0), \tag{5.23}$$

which, on integration, gives

$$n_a(t) = (1/A)\int_t^{\infty}\left[(P_t - P_0)r_o + \frac{V\,\mathrm{d}(P_t - P_0)}{kT_0\,\mathrm{d}t}\right]\mathrm{d}t. \tag{5.24}$$

This equation is also applicable to case (i).

For the more general case [7], Eq. (5.14) may be written as

$$-\frac{\mathrm{d}\theta_T}{\mathrm{d}t} = \frac{P_T r'_o}{An_m\beta} = \frac{K\theta_T r'_o}{(1-\theta_T)A\beta}. \tag{5.25}$$

At the peak temperature T_m, $\partial P/\mathrm{d}T_m = -\mathrm{d}\theta_T^2/\mathrm{d}T^2 = 0$ and the differentiation of Eq. (5.21) gives

$$\frac{\mathrm{d}K}{\mathrm{d}T}\frac{\theta_T}{1-\theta_T} = \frac{K}{(1-\theta_T)^2} = \frac{\mathrm{d}\theta_T}{\mathrm{d}t} = 0. \tag{5.26}$$

Since $K = \exp(\Delta S^{\ddagger}/R)\exp(-\Delta H^{\ddagger}/RT)$, and $-\Delta H^{\ddagger} = E_d$ for zero activation of adsorption, then

$$-\frac{\theta}{T_m^2}\left(\frac{A\Delta H(1-\theta_T)^2}{Rr'_o\exp(\Delta S^{\ddagger}/R)}\right) = \exp\left(\frac{-\Delta H^{\ddagger}}{RT_m}\right), \tag{5.27}$$

or

$$2\ln T_m - \ln\beta - \ln\left\{\frac{(1-\theta_T)^2 A\Delta H}{Rr'_o\exp(\Delta S^{\ddagger}/R)}\right\} = \frac{\Delta H^{\ddagger}}{RT_m} = \frac{-E_d}{RT_m}. \tag{5.28}$$

The coverage θ_m at T_m is a function of the initial coverage θ_i and E_d/RT_m, and is usually only slightly dependent on the value of E_d. Hence, for constant θ_i and r'_o, the slope of the plot of $(2\ln T_m - \ln\beta)$ against $-1/RT_m$ for different values of the heating rates β gives the value of E_d.

5.3.4. Characteristic features of desorption profiles

The desorption rate [Eq. (5.4)] is

$$-\frac{dn_a}{dT} = \frac{n_a^x v_x}{\beta}\exp\left(\frac{-E_d}{RT}\right)$$

and at the temperature T_m,

$$\frac{d}{dT}\left(\frac{dn_a}{dT}\right) = 0.$$

For case (i), using Eq. (5.16),

$$\ln(T_m^2/\beta) = (E_d/RT_m) + \ln(E_d/v_1 R) \tag{5.29}$$

for a first-order desorption, and

$$\frac{d}{dT}\left(\frac{dn_a}{dT}\right) = 2(n_a)_m \frac{v_2}{\beta}\exp\left(\frac{-E_d}{RT_m}\right) - \left(\frac{E_d}{RT_m^2}\right) = 0, \tag{5.30}$$

or

$$\ln(T_m^2(n_a)_m/\beta) = E_d/RT_m + \ln(E_d/2v_2 R), \tag{5.31}$$

for a second-order process, where $(n_a)_m$ is the surface concentration at T_m and is approximately equal to half the initial concentration $(n_a)_0$. From Eq. (5.29), T_m is independent of $(n_a)_0$, and the profile is asymmetric about T_m for a first-order desorption. From Eq. (5.31) for $x = 2$, T_m decreases with increase of $(n_a)_0$ and the profile is symmetric about T_m (see Fig. 5.3). Alternatively, the plot of $\ln(T_m^2(n_a)_0/\beta)$ against $1/RT_m$ gives a line parallel to the ordinate for $x = 1$, but one of slope E_d for $x = 2$. One other difference is that the time interval to effect complete desorption is independent of $(n_a)_0$ for $x = 1$, but varies inversely as the initial concentration for a second-order desorption (Fig. 5.4).

5.3.5. Application of transition-state theory [13, 14]

Desorption occurs when the adsorbed molecule acquires sufficient energy E_d to escape from the surface. The activation energy for desorption is $E_d = E_A + q_c$, where E_A is the activation energy of chemisorption, assumed to be non-dissociative, and q_c is the heat of chemisorption, all energies being with respect to one mole of adsorbate. The desorption is described in terms of a transition-state complex in equilibrium with the adsorbed molecule.

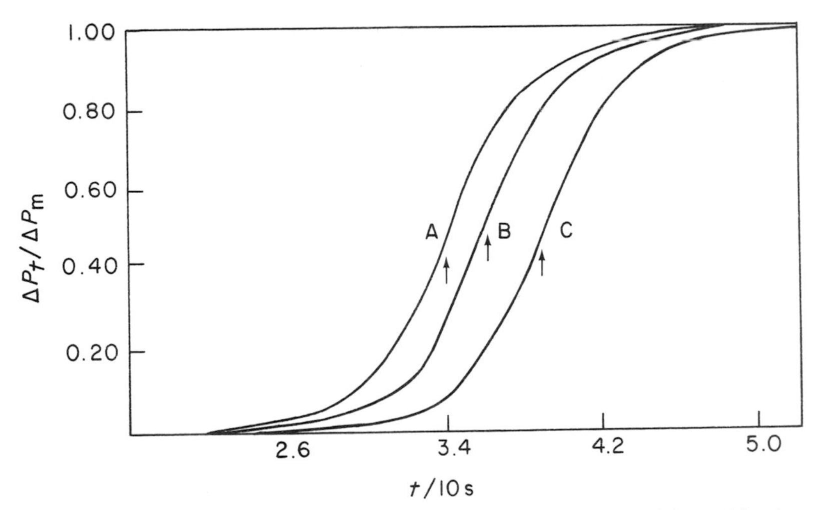

FIG. 5.4. Desorption curves for second-order desorption as a function of time with a hyperbolic heating rate for different initial surface concentrations. Curve A, 10×10^{13}; curve B, 4×10^{13}, curve C, 1×10^{13} molecules cm^{-1} with increase of initial surface concentration; the time for complete desorption ($\Delta P/\Delta P_m \rightarrow 1$) increases with decrease of initial surface concentration.

The equilibrium constant $K^{\ddagger}$ is given by

$$K^{\ddagger} = \frac{n^{\ddagger}}{n_a} = \frac{f^{\ddagger}}{f_a} \exp\left(-\frac{E_d}{RT}\right), \tag{5.32}$$

where $n^{\ddagger}$ is the number of transition-state complexes per cm^2 surface, and $f^{\ddagger}$, f_a are, respectively, the complete partition functions of the complex and adsorbed molecule. In the transition state, the complex vibrates along the reaction co-ordinate perpendicular to the surface with a frequency ν_d but, since, contrary to a normal vibration, there is no restoring force, each vibration results in desorption. The frequency of desorption is, therefore, ν_d and the rate of desorption per cm^2 surface is $r_d = \nu_d n^{\ddagger}$ molecules $cm^{-2}\ s^{-1}$. Since there is no restoring force, ν_d is small and the vibrational partition function $[1 - \exp(-h\nu_d/kT)]^{-1}$ approximates to $kT/h\nu_d$. This term is separated from $f^{\ddagger}$ to give $f_{\ddagger}\ kT/h\nu_d$, where $f_{\ddagger}$ is the partition function of the complex without this vibrational mode. Hence,

$$\frac{n^{\ddagger}}{n_a} = \frac{kT}{h\nu_d}\frac{f_{\ddagger}}{f_a} \exp\left(-\frac{E_d}{RT}\right), \tag{5.33}$$

and

$$r_d = \nu_d n^{\ddagger} = n_a\left(\frac{kT}{h}\right)\frac{f_{\ddagger}}{f_a} \exp\left(-\frac{E_d}{RT}\right). \tag{5.34}$$

With the assumption that the complex and adsorbed molecule have identical internal degrees of freedom, and that the vibration frequency of the adsorbed molecule perpendicular to the surface is high so that its associated partition function can be approximated to unity, then $f_{\ddagger} = f_a$. The frequency term kT/h is of order 10^{13} s^{-1} and $n_a = n_m\theta$, where $n_m \sim 10^{15}$ cm^{-2}; hence,

$$r_d^\theta \approx 10^{28}\theta \exp(-E_d/RT) \tag{5.35}$$

and E_d, the activation energy of desorption, equals the heat of chemisorption for a non-activated adsorption process.

However, symmetrical diatomic molecules, such as hydrogen, undergo dissociative chemisorption and the adsorbed layer comprises adatoms that are freely mobile at the desorption temperature. Bimolecular collision of adatoms with the necessary activation energy is followed by desorption of molecular hydrogen. Equilibrium is, therefore, established between two adatoms at adjacent single sites and the diatomic molecular complex attached to a pair of neighbouring sites, i.e.

$$2\mathrm{Hs} + \mathrm{s}_2 \rightleftharpoons \mathrm{H_2s_2} + 2\mathrm{s}$$

in which s and s_2 represent a single and a pair of neighbouring sites. Hence

$$K^{\ddagger} = \frac{n^{\ddagger}n_s^2}{n_a^2 n_{s_2}} = \frac{f^{\ddagger}f_s^2}{f_a^e f_{s_2}} \exp\left(-\frac{E_d}{RT}\right), \tag{5.36}$$

and

$$r_d = \frac{n_a^2 n_{s_2}}{n_s^2}\frac{kTf_{\ddagger}f_s^2}{f_a^2 f_{s_2}} \exp\left(-\frac{E_d}{RT}\right). \tag{5.37}$$

For a surface co-ordination number z,

$$n_{s_2} = zn_s^2/2n_m. \tag{5.38}$$

Both f_s and f_{s_2} can be assumed to be unity. For $z = 4$,

$$r_d = \frac{2kT}{h} n_m\theta^2 \frac{f_{\ddagger}}{f_a^2} \exp\left(-\frac{E_d}{RT}\right). \tag{5.39}$$

For mobile adatoms and mobility of the complex, two translational degrees of freedom are transformed to vibrational and rotational modes of the complex, so that $f_{\ddagger}/f_a^2$ is rather less than unity. The approximate value of

r_d is, therefore,

$$r_d \approx 10^{28}\theta^2 \exp(-E_d/RT) \text{ molecules cm}^{-2} \text{ s}^{-1}. \tag{5.40}$$

5.3.6. The frequency factor

The desorption profile of a single-state adspecies can sometimes be equally well fitted by a kinetic order of either 1 or 2 by simultaneously adjusting the numerical values of E_d and ν in Eq. (5.4). This flexibility is, however, not possible when, as is usually so, independent calorimetric values of E_d are available. Nevertheless, figures as low [15] as 10^5 s^{-1} and as high [16] as 10^{15} s^{-1} have been reported for experimental rates of desorption of physisorbed, or weakly chemisorbed, molecules, despite the fact that the rates apparently conform, as expected, to a well-defined first-order process (see Table 5.1).

TABLE 5.1. Frequency factors and desorption energies for first-order desorption of some adsorbates from single planes of tungsten

Plane	Adsorbate	State	ν_1/s^{-1}	E_d/kJ mol^{-1}	References
Group A: Systems with normal values of ν_1 (10^{12} to 10^{13} s^{-1})					
W(100)	N_2	γ	10^{13}	40	Clavena and Schmidt [17]
	CO	$\beta_1, \beta_2, \beta_3$	10^{13}	240, 260, 310	Clavena and Schmidt [18]
	H_2	β_1	$10^{13\pm1}$	110	Tamm and Schmidt [5]
W(110)	CO	β_1, β_2	10^{12}	235, 285	Kuhrt and Gomer [15]
W(poly)	CO	$\beta_1, \beta_2, \beta_3$	10^{12-13}	210–320	Winterbottom [19]
Group B: Systems with abnormal values of ν_1					
W(110)	CO	γ	10^5	41	Kuhrt and Gomer [15]
W(111)	Xe	—	10^{15}	40	Dresser *et al.* [16]

Three possible explanations of this disparity can be proposed in terms of transition-state theory [13, 14]. In the application to desorption rates, the transmission factor κ has been assumed to be unity, i.e. transformation of the complex across the saddle point to form the final state takes place during each vibration of the frequency mode ν_d. There is, however, a certain probability of reflection at the saddle point with a consequent return of the complex back to the reactant state. A transmission factor of small magnitude (10^{-3}) compared with unity can be justified to account for a frequency factor 10^{10}

s^{-1}, but it is difficult to accept the high values ($\sim 10^{-8}$) for the probability of reflection that are sometimes required.

Other explanations are concerned with departures of the ratio $f_{\ddagger}/f_a$ from unity. The frequency factor is the product of $\kappa kT/h(\approx 10^{13}\ s^{-1})$ and this ratio; the latter closely approximates to unity for simple adsorbates, provided that the complex and adspecies are both immobile, or both have the same mobility. Immobility of the adspecies (and probably of the complex) at the desorption temperature is extremely unlikely, but it is reasonable to assume that the mobilities of these two species are not the same. A higher mobility of the complex would cause $f_{\ddagger}/f_a$ to exceed unity and, in the theoretical limit, the frequency factor could be increased to $10^{17}\ s^{-1}$. Thus, the experimental value of $v_1 = 10^{15}\ s^{-1}$ for the desorption of Xe from W(111) can be obtained by assuming a highly-mobile complex and an immobile reactant [16]. A frequency factor of less than $10^{13}\ s^{-1}$ would, likewise, be obtained for an immobile complex in equilibrium with a mobile reactant.

Finally, the equilibrium constant $K^{\ddagger}$ can be written as

$$K^{\ddagger} = \exp\left(-\frac{\Delta G^{\ddagger}}{RT}\right) = \exp\left(-\frac{\Delta H^{\ddagger}}{RT}\right)\exp\left(\frac{\Delta S^{\ddagger}}{R}\right)$$

$$\approx \exp\left(-\frac{E_d}{RT}\right)\exp\left(\frac{\Delta S^{\ddagger}}{R}\right); \qquad (5.41)$$

hence, $\Delta S^{\ddagger} = 0$ when $f_{\ddagger}/f_a = 1$. For more complex adsorbate molecules, the transition state may possess more (or less) structure than that of the reactant, with the result that $\Delta S^{\ddagger}$ can be greater (or less) than zero [20, 21].

In these ways, small departures from the normal frequency factor can be rationalized by appeal to transition-state theory; but a more reasonable approach is to deny its basic postulate and assume that equilibrium between the transition state and the reactant is not established. A rate constant describing the energy transfer from the adsorbent to the adsorbate can then be introduced [22]. Should it be associated with a long time-constant, energy transfer becomes rate-determining and the frequency factor is decreased from its normal value. For rapid transfer of energy, motion along the reaction co-ordinate is impeded and, again, low values are predicted. Only at some intermediate values of this time constant does the frequency factor approach $10^{13}\ s^{-1}$.

This theory has been further developed [23, 24] by the inclusion of an additional rate constant to account for the relaxation of the surface population from a non-equilibrium to an equilibrium state. It is then found that the pre-exponential term becomes slightly temperature dependent and a compensation effect is operative. In this quantum-mechanical approach, multiphonon processes were included and (when bound vibrational states of the adsorbate were excluded) a frequency factor of $10^5\ s^{-1}$ was calculated for the low-temperature desorption of neon from the surface of solid xenon in agreement with the experimental value [25].

5.4. RATES OF DESORPTION FOR VARIABLE ACTIVATION ENERGY OF DESORPTION [26]

The heats of chemisorption of simple gases on transition metals normally decrease with increasing coverage owing to the non-uniformity of the adsorption sites and net repulsive interactions between the adsorbed molecules. Since the chemisorption is usually non-activated, the activation energy of desorption increases with decrease of coverage and the rate of evolution falls rapidly as the surface is denuded. The rate equation can be formulated when the functional dependence of E_d on θ is specified. The simplest relationship is a linear one, i.e.

$$E_d^\theta = E_d^0 - \alpha\theta, \tag{5.42}$$

where E_d^0 refers to zero coverage and E_b^θ to a coverage θ, both expressed in terms of one mole of adsorbate. The term $\alpha = dE/d\theta = E_d^0 - E_d^m$, and E_d^m is the desorption energy as $\theta \to 1$.

The rate of desorption (molecules cm^{-2} s^{-1}) at a coverage θ_t and time t is

$$r_d^\theta = -\frac{d\theta_t n_m}{dt} = \nu_1 n_m \theta_t \exp - \left[\frac{(E_d^0 - \alpha\theta_t)}{RT}\right]. \tag{5.43}$$

Since the variation of θ_t is negligible with respect to that of $\exp(-\alpha\theta_t/RT)$ (except when θ approaches zero),

$$-\frac{d\theta_t}{dt} = -\nu_1 \exp\left(-\frac{E_d^0}{RT}\right)\exp\left(\frac{\alpha\theta_t}{RT}\right), \tag{5.44}$$

or

$$-d\theta_t/dt = a\exp(\alpha\theta_t/RT),$$

where

$$a = \nu_1 \exp(-E_d/RT). \tag{5.45}$$

Integration between the limits $\theta = 1$ at $t = 0$ and $\theta = \theta_t$ at $t = t$ and rearrangement gives

$$\exp[\alpha(1 - \theta_t)/RT] = 1 + at\alpha/RT \tag{5.46}$$

or

$$(1 - \theta_t) = \frac{RT}{\alpha} \ln\left(\frac{t + t_0}{t_0}\right) \tag{5.47}$$

where $t_0 = RT/\alpha a = RT/\alpha v_1 \exp(-E_d/RT)$.

The linear relationship $E_d^\theta = E_d^0 - \alpha\theta$ could apply to a mobile adlayer on a uniform surface with pairwise lateral repulsive interactions between the adspecies at lower θ values.

An equation of the same form can also be derived for a non-uniform surface comprising a large number of small uniform patches of equal area ds for which [27]

$$E_s = E_d^0 - \beta s,$$

where s is the sequential reference number of the patch and β is a constant $= \mathrm{d}E_s/\mathrm{d}s$.

The total desorption rate $-\mathrm{d}\Theta_t/\mathrm{d}t$ at time t may be expressed in the form

$$\frac{-\mathrm{d}\Theta_t}{\mathrm{d}t} = v_1 \int_0^1 \theta_s \exp\frac{(-E_d^0 - \beta s)}{RT}\,\mathrm{d}s, \tag{5.48}$$

where Θ_t is the fraction of the total surface still covered at time t: θ_s is the fractional coverage of the s patch at this time and, for small ds, integration replaces the summation. The limits of integration imply that all patches with $E_s < E_t$ are bare and those with $E_s > E_t$ are fully covered. The integration can, therefore, be carried out over the covered surface (for which $\theta_s = 1$). between the limits $s = 0$ to $s = \Theta_t$. Equation (5.48) can then be written as

$$-\frac{\mathrm{d}\Theta_t}{\mathrm{d}t} = v_1 \int_0^{\Theta_t} \exp\frac{-(E_d^0 - \beta s)}{RT}\,\mathrm{d}s \tag{5.49}$$

$$= \left[\frac{v_1 RT}{\beta}\exp\left(-\frac{E_d^0}{RT}\right)\right]\left[\exp\left(\frac{\beta\Theta_t}{RT}\right) - 1\right]$$

and, since $\exp(\beta\Theta_t/RT) \gg 1$, except when Θ_t approaches zero,

$$-\mathrm{d}\Theta_t/\mathrm{d}t = b\exp(\beta\Theta_t/RT), \tag{5.50}$$

where $b = (\nu_1 RT/\beta)\exp(-E_d^0/RT)$. On integration, Eq. (5.50) gives

$$(1 - \Theta_t) = \frac{RT}{\beta}\ln\left(\frac{t + t_0}{t_0}\right) \tag{5.51}$$

with $t_0 = RT/\beta b$.

Values for E_d^0 and for α (or β) can be evaluated from the slope of the linear plots of $\ln(d\theta_t/dt)$ [or $(\ln d\Theta_t/dt)$] against $1/RT$ to give $E_d^0 - \alpha\theta_t$ [or $E_d^0 - \beta\Theta_t$], respectively, together with the slope of the linear plot of $(1 - \theta_t)$ [or $(1 - \Theta_t)$] against $\ln(t + t_0)$ at constant temperature T to provide values of α/RT [or β/RT], respectively.

5.4.1. Continuous distribution [28]

No account has yet been taken of the fact that there is a Maxwellian distribution of probabilities for desorption, the maximum probability being centred at E_t at time t. However, this omission does not invalidate Eqs (5.47) and (5.51) except at small t. Thus, for a constant distribution of desorption activation energies on a set of uniform patches of equal area, the desorption rate from the x patch is

$$-d\theta_x/dt = k_x\theta_x, \tag{5.52}$$

where $k_x = \nu_1 \exp(-E_x/RT)$, and θ_x is the fractional coverage on this patch at time t. On integration between the limits $t = 0, \theta_x = 1$, and $t = t, \theta_{xt}$,

$$\theta_{xt} = \exp[-\nu_1 t \exp(-E_x/RT)].$$

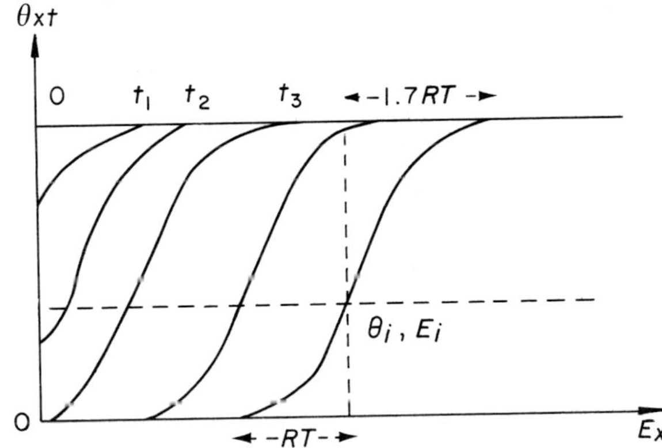

FIG. 5.5. Establishment of a desorption profile. The relationship between θ_{xt} as $f(E_x)$ for increasing time t_1, t_2, t_3 is indicated. Areas to the left of the desorption front have been denuded. A constant profile is established at time t_3, after which the Elovich equation is applicable. The vertical dashed line shows the profile for $\theta_x = 0$ for $E_x < E_t$, and $\theta_x = 1$ for $E_x > E_t$ (step function).

The relationship between θ_{xt} and E_x for increasing times, depicted in Fig. 5.5, shows that a constant desorption front is established after a time t_3. On writing θ as a continuous function of E and omitting the subscript x, logarithmic differentiations at constant t give

$$\frac{1}{\theta}\frac{\partial\theta}{\partial E} = \frac{v_1 t}{RT}\exp\left(-\frac{E}{RT}\right), \tag{5.54}$$

and

$$-\frac{1}{\theta^2}\left(\frac{\partial\theta}{\partial E}\right)^2 + \frac{1}{\theta}\frac{\partial^2\theta}{\partial E^2} = -\frac{v_1 t}{(RT)^2}\exp\left(-\frac{E}{RT}\right). \tag{5.55}$$

At the inflection point (E_i, θ_i), $\partial^2\theta/\partial E^2 = 0$ and

$$E_i = RT \ln v_1 t. \tag{5.56}$$

Combination of Eqs (5.56), (5.53) and (5.54) give $\theta_i = 1/\mathrm{e}$, and

$$\partial\theta/\partial E_i = 1/RT\mathrm{e}. \tag{5.57}$$

It follows that

$$\int_0^1 \mathrm{d}\theta_i = \frac{1}{RT\mathrm{e}}\int_{E_{min}}^{E_{max}} \mathrm{d}E, \tag{5.58}$$

i.e.

$$\Delta E = E_{max} - E_{min} = 2{\cdot}7\,RT.$$

The ordinate θ_i and the slope of the tangent at (θ_i, E_i) are, therefore, constant and independent of time. The desorption profile is unchanged as the process continues and may, therefore, be replaced by a vertical line through the inflection point.

The total bare surface at time t is then given by

$$(1 - \Theta) = \int_{E_{min}}^{E_{max}} \frac{(1 - \Theta)}{\alpha}\mathrm{d}E, \tag{5.59}$$

where α is a constant. For the vertical profile, $\theta = 0$, $E_{min} = E_d^0$, and $E_{max} = RT \ln v_1 t$; hence,

$$(1 - \Theta) = (RT \ln v_1 t - E_d^0)/\alpha$$

or

$$\frac{-\mathrm{d}\Theta}{\mathrm{d}t} = \frac{v_1 RT}{\alpha} \exp\left\{\frac{-E_{\mathrm{d}}^0 + \alpha(1 - \Theta)}{RT}\right\}. \tag{5.60}$$

The equation has the same form as that derived for a discontinuous distribution; it is valid after the constant desorption profile has been established, i.e. it represents the effect of the distribution of site energies after a short initial interval corresponding to t_3.

5.4.2. The Elovich equation

The rate equation of the general form

$$-\mathrm{d}q_t/\mathrm{d}t = b \exp(\beta q_t/RT), \tag{5.61}$$

for a linear decrease of desorption energy with increase of coverage is commonly referred to as the Elovich equation. Normally, the effect of repulsive interactions is too small to account for the magnitude of the energy decrease; and the change is largely attributed to a non-uniformity of site energies which conforms to a constant distribution function. This type of distribution is improbable, yet the Elovich equation is obeyed over a wide range of conditions for a variety of different gas/solid systems [26].

A possible explanation [28] is that a linear plot of q against $\ln(t + t_0)$ can be obtained for a surface made up from as few as three uniform patches with small differences in their respective activation energies. Thus, for a particular patch j,

$$-\mathrm{d}\theta_{jt}/\mathrm{d}t = k_j\theta_{jt}, \qquad \text{or} \quad \theta_j = \exp(-k_j t), \tag{5.62}$$

in which k_j is a first-order rate constant of the form

$$k_j = \text{const.} \exp(-E_j/RT). \tag{5.63}$$

For the whole surface, comprising n patches with different k values,

$$\frac{-\mathrm{d}\Theta}{\mathrm{d}t} = \frac{1}{n}\sum_1^n \frac{\mathrm{d}\theta_{jt}}{\mathrm{d}t} = \frac{-1}{n}\sum_1^n k_j \exp(-k_j t) \tag{5.64}$$

and

$$\Theta = \frac{1}{n}\sum_1^n \theta_j = \frac{1}{n}\sum_1^n k_j \exp(-k_j t). \tag{5.65}$$

For three patches, described by desorption energies E_1, E_2, E_3 which increase in increments of *ca* 6 kJ mol^{-1}, the plot of Θ against $\ln(t + t_0)$, where t_0 is a disposable constant, is virtually linear with a very slight wavy character.

For an evaporated film as adsorbent, the surface predominantly comprises the three low-index planes, usually assumed to be of equal area. The planes are to some extent non-ideal and some higher-index planes are undoubtedly present in small concentrations, so that the wavy character would be eliminated. Such a model would, therefore, be a physically acceptable one as an explanation of the applicability of the Elovich equation.

5.4.3. Other distribution functions [29]

The distribution function describes the product of the number of sites and their corresponding associated activation energies. Other distribution functions can be postulated and different forms of the rate equation can then be derived. Thus, the rate of desorption at time t is

$$-\mathrm{d}\theta_x/\mathrm{d}t = k\theta_x \exp(-E_x/RT),$$

or

$$\theta_{E,t} = \exp[-kt\exp(-E/RT)], \tag{5.66}$$

where $\theta_{E,t}$ is the fractional coverage of the surface having an activation energy of E at time t; hence, for the complete surface,

$$\Theta_{E,t} = \int_{E_{\min}}^{E_{\max}} \exp[-kt\exp(-E/RT)]\, f(E)\,\mathrm{d}E, \tag{5.67}$$

where $f(E)$ is the distribution function over the energy range $E_{\min}$ to $E_{\max}$. As examples, for:

$$\begin{aligned} f(E) &= \text{constant}, & \Theta_t &= A_1 \ln t + B_1; \\ f(E) &= C\exp(\delta E), & \Theta_t &= A_e t(1/n) + B_e; \\ f(E) &= D/E, & \Theta_t &= A_3 \ln(kt) + B_3; \end{aligned} \tag{5.68}$$

where the undefined terms are constants. In principle, the distribution function can be derived from the corresponding rate equation; but this derivation is more easily carried out from the corresponding adsorption isotherms.

5.5. MULTIPLET ADSORBED STATES

The activation energy for desorption and the kinetic order of the evolution process has been derived from the profile of the thermal desorption spectrum of an adspecies in a single adsorbed state. In most systems, however, the chemisorption heat of a single adsorbate varies from one crystallographic plane to another; consequently, the activation energy of the overall desorption process from a polycrystalline filament may comprise more than one energy term. Moreover, different adsorbed states of a single adsorbate can frequently co-exist even on the surface of a single crystal plane, and each state is associated with a specific activation energy and kinetic order.

For a single state,

$$E_d/RT_m = \ln(v_1/\beta), \tag{5.69}$$

and for a typical linear heating rate β of 10 K s^{-1} and a normal frequency factor $v_1 = 10^{13}\ s^{-1}$,

$$E_d/\mathrm{kJ\ mol^{-1}} = 230\ T_m/\mathrm{K} \tag{5.70}$$

For two different states of an adspecies having chemisorption heats differing by ~20 kJ mol^{-1}, there are two T_m values separated by about 100 K on the temperature axis, or, at the heating rate of 10 K s^{-1}, by about 10 s on the time axis. Since the width of a desorption profile at half-peak height

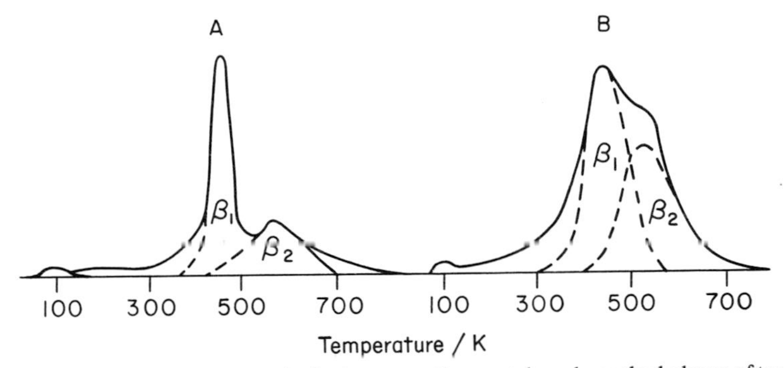

FIG. 5.6. Thermal desorption spectra for hydrogen on the most densely packed planes of tungsten. A, W(100) and B, W(110) after exposure to H_2(g) at a pressure of ~10^{-6} Pa and 78 K. Two states, β_1 and β_2 indicated by dashed lines are present on both planes, and are evident as separate peaks (A), or as one peak with shoulder (B). Desorption of β_1 from W(100) is first order with E_d ~ 110 kJ mol^{-1}, and v_1 ~ $10^{13}\ s^{-1}$; all other desorptions are second-order with v_2 ~ 10^{-2} $cm^2\ s^{-1}$ (at low coverages) with E_d ~ 135 for β_2 on W(100); 113 kJ mol^{-1} for β_1 on W(110); and ~140 kJ mol^{-1} for β_2 on W(110).

is sometimes 2 or 3 times greater than this separation, the two individual profiles often overlap to a large extent. However, the presence of the two states can be recognized by the appearance of two maxima or of a shoulder on the temperature profile, and both desorption energies can be estimated from the T_m values (see Fig. 5.6). Moreover, from the change of positions, if any, of the peak temperatures with initial coverage, the order of the process may be derived.

The experimental procedure follows that given in Section 5.3.5. Conservation of mass within the desorption system is given [cf. Eq. (5.23)] by

$$\frac{V}{kT_0}\frac{\mathrm{d}(P_t - P_0)}{\mathrm{d}t} = A\frac{\mathrm{d}}{\mathrm{d}t}(\Sigma n_{ai}) - r_o(P_t - P_0), \tag{5.71}$$

where r_o is the outflow conductance (molecules s^{-1}) and P_0 is the steady-state pressure at T_0; n_{ai} is the concentration (molecules cm^{-2}) of the ith state on the surface at time t. The total concentration of adsorbed gas is therefore

$$A\Sigma n_{ai}(t) = \frac{V}{kT_0}\int_t^{\infty}\frac{\mathrm{d}(P_t - P_0)}{\mathrm{d}t}\,\mathrm{d}t + \int_t^{\infty} r_o(P_t - P_0)\,\mathrm{d}t \tag{5.72}$$

and the net desorption flux of the ith state is

$$\mathrm{d}n_i/\mathrm{d}t = -v_1 n_i^x \exp(-\Delta H_i/RT) + k_a P_t(1 - \Sigma n_{ia}/n_m), \tag{5.73}$$

the last term being negligible for high pumping speed and fast heating rate.

In an analysis of the results, the lowest peak temperature T_{mi} is first located as a shoulder or maximum, and the change of position, if any, of T_{mi} with change of the total initial concentration on the filament is then determined; the kinetic order x and an initial estimate of $-\Delta H_i$ are thereby obtained. From the theoretical shape of the evolution profile for a single surface state, an approximate characterization of the first state is then made by fitting the initial rise of the experimental P_t versus t plot, and an approximate theoretical shape of the profile is constructed; this is then restructured to obtain the best fit of the initial rise to a large number of experimental P_t versus t plots obtained by variation of the heating rate, initial total coverages, pumping speed, etc. The contributions of each other state in order of increasing ΔH_i values are then included in turn. The complication arising from contributions of states of lower binding energy to the later sections of the spectrum can be minimized by using low total coverages and higher temperatures during the initial adsorption. In the later stages of synthesis of the complete plot, slower heating rates and faster pumping speeds can be used to obtain better resolution of states differing by about 20 kJ mol^{-1} in their desorption energies. The general iteration procedure of restructuring is continued until there is satisfactory agreement between the numerical synthesis to provide the total desorption profiles of a complete series of experimental P_t versus t plots (Fig. 5.7).

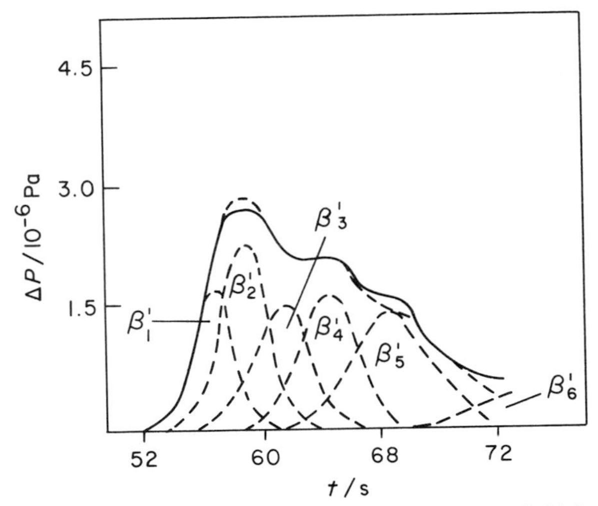

FIG 5.7. Desorption spectrum for β-CO on polycrystalline tungsten. Initial adsorption at 400 K; heating rate 35 K s^{-1}. Profiles calculated for individual states and the complete profile for the sum of single state curves are indicated by dashed lines (– – –). [19].

5.5.1. Interpretation of thermal desorption spectra

The interpretation of the profiles observed in the thermal desorption spectrum of multiplet states is difficult and often ambiguous. As has been noted, a first-order process with a small linear decrease of E_d can closely simulate a second-order desorption with a constant E_d. Even on a single crystal surface, at the high coverages that are usually employed, repulsive interactions between the adspecies can be responsible for the small decrease of E_d that is envisaged. In addition, it is assumed: (i) that the individual contributions to the measured total rate of evolution from states of different binding energies are independent of one another; and (ii) that the states of the adspecies evolved at the higher temperatures are those present in the chemisorbed layer formed at the initial lower temperature of adsorption. Thermal reorganization of, and conversion between, multiplet states may, in fact, occur. When either of these assumptions is invalid, then the total rate is not, as is accepted, a simple summation of the rates of desorption from each adsorbed state.

In flash desorption, reorganisation is minimized by completing the heating programme within about 0·1 s; but a slower rate of around 30 K s^{-1} is necessary to provide greater resolution of the peaks and to obtain the finer details for separation of states of similar binding energies.

It is, therefore, essential to examine with great care the desorption profiles of multiplet states and to accept with caution the detailed conclusions that are often drawn. Nevertheless, results of considerable value and interest for various gas/metal systems have been reported, as is briefly exemplified for the CO/W system. Other systems are discussed in Chapter 27.

5.5.1.1. *The carbon monoxide polycrystalline tungsten system*

There are two main states: (i) α-state, T_m = 200 to 500 K; E_d = 80 to 120 kJ mol^{-1}; and (ii) β-state, T_m = 1100 to 1800 K; E_d = 200 to 320 kJ mol^{-1}. Within the β-state there are three substates β_1, β_e, β_3, the rate parameters of which have been reported by various workers (see Table 5.2).

TABLE 5.2

State	E_d/kJ mol^{-1}					$\nu_1/10^3\ s^{-1}$
β_1	210–250[a]	248[b]	265[c]	217[d]	(206–252)[e]	–, 1, 2·1,–(0·1)
β_2	315	292	323	234	(280)	3, 1, 0·7, 3·3 (0·7)
β_3	~420	317	378	324	(315–328)	–, 1, 1, 1·1, 1 (1)

[a] Ehrlich [10]. [b] Redhead [30]. [c] Rigby [31]. [d] Degras [32].
[e] Winterbottom [19].

The agreement between the E_d values is within 25 to 60 kJ mol^{-1} (β_1, 210 to 260; β_2, 290 to 315; β_3 320 to 380) with T_m values, respectively, of around 1000, 1300 and 1500 K.

In recent work by Winterbottom [19], the β_1 state has been resolved into three states β'_1, β'_2, β'_3, and β_3 into β'_5 and, possibly, β'_6. The rate parameters resolved using the analysis outlined previously are collected together in Table 5.3 and included in parentheses in Table 5.2. Results and theoretical analysis are shown in Fig. 5.7.

TABLE 5.3

	β'_1	β'_2	β'_3	β'_4	β'_5	β'_6(?)
E_d/kJ mol^{-1}	206	222	252	280	3·5	328 (?)
$\nu_1/10^{13}\ s^{-1}$	0·1	0·1	0·2	0·7	1·0	(2nd order ?)

5.6. ELECTRON-STIMULATED DESORPTION [33, 34]

When a chemisorbed layer on a metal is bombarded with a beam of electrons, part of their kinetic energy is expended in excitation of the adspecies. Since the impact time is much less than the period of vibration of the chemisorption bond, excitation of the adspecies is brought about by a Frank–Condon process and a vertical transition results. The probability of such excitation is greatest at the maximum displacement of the vibrating adspecies at all vibrational levels except the ground state; and the amplitude increases with increase of the vibrational quantum number. In the absence of excitation of electronic states of the adspecies, sufficient energy can rarely be transferred to the vertical vibration of the chemisorption bond to effect desorption of the neutral adspecies. However, neutrals (and adsorbate ions) are freely desorbed following electronic excitation.

5.6.1. Desorption following electronic excitation

Desorption of ions results from electronic excitation to an antibonding state by incident electrons of sufficient energy (>10 eV). A transition of an electron from the electronic ground-state of the adsorbate–adsorbent interaction potential to the repulsive part of the ionized adsorbate–adsorbent potential curve (or of some other excited-state potential curve) takes place followed by desorption of the adsorbate with a range of kinetic energies between 0 and 10 eV [35, 36].

A set of potential energy curves are depicted in Fig. 5.8, in which $U(x)$ denotes the potential energy of the system and x is the distance co-ordinate of the adsorbate from the metal surface. Curve N shows the potential energy of the system $U(x)$ as a neutral molecule M approaches the surface of the metal adsorbent A. M is chemisorbed in the ground state at the equilibrium distance x_0 with a chemisorption heat q_c. Curve D represents the variation of $U(x)$ as the ion M^+ is moved towards the surface, the energy difference between curve A and D at $x = \infty$ being eI ,where I is the ionization potential of the molecule M. At some value of $x = x_c$, the potential barrier separating the unoccupied energy level of M^+ and the occupied electron levels of the metal becomes sufficiently narrow for electron tunnelling through the barrier, and the energy difference between curve C and A at $x = \infty$ is reduced to $e(I - \Phi)$, where Φ is the effective work function of the metal. Curve B is an excited antibonding state.

On impact of the primary electron with the chemisorbed molecule in its ground state, promotion to an excited state (antibonding, ionic, or excited neutral state) is effected. The excitation can take place at any point x between

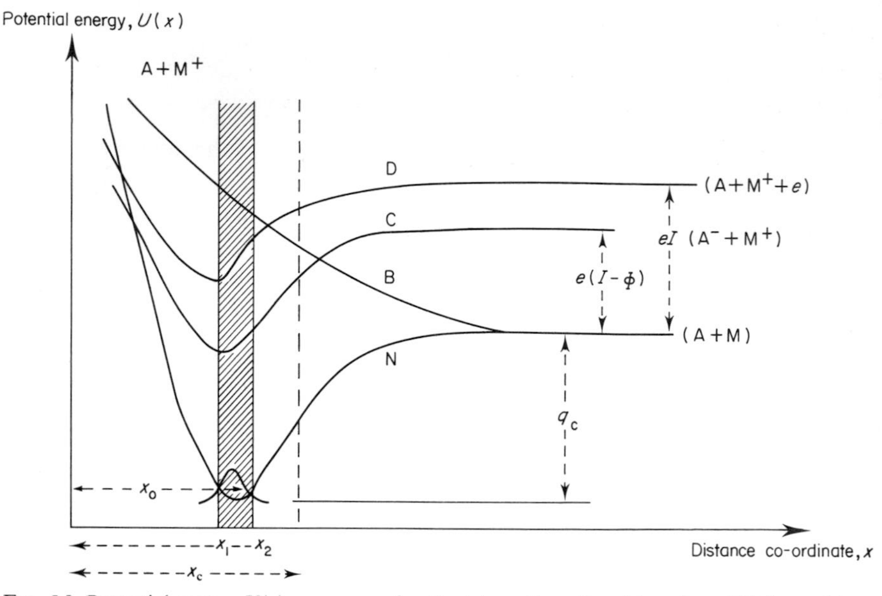

FIG. 5.8. Potential energy $U(x)$ vs. x curve for the interaction of an M, and an M^+ ion, with a metal adsorbent. Some excited states of the metal adsorbate system are: curve N, neutral state (A + M): curve B, antibonding state A + M^+ (excited adsorbate molecule); curve C, ionic state (A^- + M^+). Curve D represents the approach of an M^+ ion to the metal surface (A + M^+ + e) should no electron tunnelling from the metal occur. The heat of chemisorption of M is q_c, and its ionization potential is I; the work function of the metal A is Φ. The equilibrium distance for M chemisorbed in the ground state is x_0. Electron tunnelling to metal occurs at $x = x_c$. Excitation of chemisorbed molecules takes place within the Frank–Condon vertical band (x_1, x_2).

x_1 and x_2 as M oscillates in one of the vibrational levels, i.e. within the shaded Frank–Condon vertical band. The energy distribution of the desorbing species therefore reflects, in part, its vibrational energy distribution. A neutral molecule may be desorbed from the antibonding state with kinetic energy in the range $U(x_1)$ to $U(x_2)$. Alternatively, excitation may create the ionic state A^- + M^+, and M^+ moves away from the surface. However, for distances $x < x_c$, electron tunnelling from the metal to the electron vacancy in the M^+ ion can occur. Should this takes place after the moving ion has acquired kinetic energy greater than the neutral ground-state binding energy, desorption of the neutral molecule occurs. But when the kinetic energy is smaller, neutralization leads to the recapture of the M molecule into the original ground-state (or other such states when multiplet states exist). The recapture rate depends (approximately exponentially) on the height and width of the potential barrier and is extremely rapid because of the high density of occupied electron states in the metal and the high transparency of the barrier.

5.6.2. Ionization cross-sections

The ionization of an adspecies is analogous to the dissociation ionization of a gas-phase molecule by electron impact. The probability of effecting ionization, Q, is defined as the ionization cross-section in cm^2. Dissipation of the excitation energy of an adspecies is brought about by phonon excitation of the metal lattice and is much more efficient than the energy-loss experienced by a gas molecule by molecular collisions. Thus, Q is of the order 10^{-17} to 10^{-22} cm^2 for an adspecies, compared with the much larger value of 10^{-15} to 10^{-16} cm^2 (for 100 eV electrons) for the free molecule. The probability for ionization of the adspecies decreases with the increase in its heat of adsorption because the perturbation of the electronic and vibrational energy levels is larger and the lifetime of the metastable state of the adspecies is, thereby, reduced.

5.6.3. Theoretical treatment of desorption [35, 36]

The charge on an adion within a distance x_c from the surface can be rapidly neutralized by electrons tunnelling from the metal to the adion through the surface potential barrier. In an one-dimensional approximation, the total probability of desorption $P(x)$ is given by

$$P(x) = \exp\left\{-\int_{x_0}^{x_c} R(x)\,\mathrm{d}x/v(x_c)\right\}, \tag{5.74}$$

in which $R(x)$ is the tunnelling probability as a function of barrier width and is approximately proportional to $\exp(-x)$. The term $v(x_c)$ is the velocity of the ion at the distance x_c away from the surface. The threshold value x_c is the distance at which the kinetic energy of the ion equals the energy of desorption, i.e.

$$\tfrac{1}{2}m[v(x_c)]^2 = U(x_0) - U(x_c) = q. \tag{5.75}$$

In this equation, q denotes the desorption energy, x_0 is the equilibrium distance of the adspecies in its ground state from the surface, and m is the mass of the ion. The function $U(x)$ represents a repulsive potential of the Born–Mayer type and is also approximately proportional to $\exp(-x)$.

Equation (5.74) may, therefore, be transformed to

$$P(x) = \exp\left\{-m^{\frac{1}{2}}\int_{x_0}^{x_c} \text{const.}\,\mathrm{e}^{-x}\,\mathrm{d}x/(q/2)^{\frac{1}{2}}\right\}. \tag{5.76}$$

The lower limit of integration x_0 is largely dictated by the molecular size of the adspecies, and the effect of isotopic substitution is determined by the term $m^{\frac{1}{2}}$. At higher temperatures, the population of the higher vibrational states of the chemisorbed bond increases and the mean value of x_0 (the distance of the adspecies from the surface) is lengthened. As a consequence, the rate of neutralization of adions is smaller, their cross-section for ionization Q^+ is somewhat larger and the adion concentration increases at the expense of that of neutral molecules.

5.6.4. Multiplet states

The magnitudes of the cross-sections for ionization of different binding states of a single adsorbate may differ considerably, although the cross-section is approximately constant within one state for various temperatures of the substrate and different incident energies of the primary beam. Thus, for CO chemisorbed on tungsten, values range from 10^{-17} to 10^{-21} cm^2. The larger cross-section is usually associated with the more weakly adsorbed state; and for incident electrons of constant energy, the state of lower binding energy is preferentially desorbed at lower temperatures. Figure 5.9 gives a plot of the ion current as a function of temperature for the α-state of CO on tungsten. Two substates, from one of which CO^+ ions are evolved and from the other O^+ ions at the higher temperatures, can be distinguished.

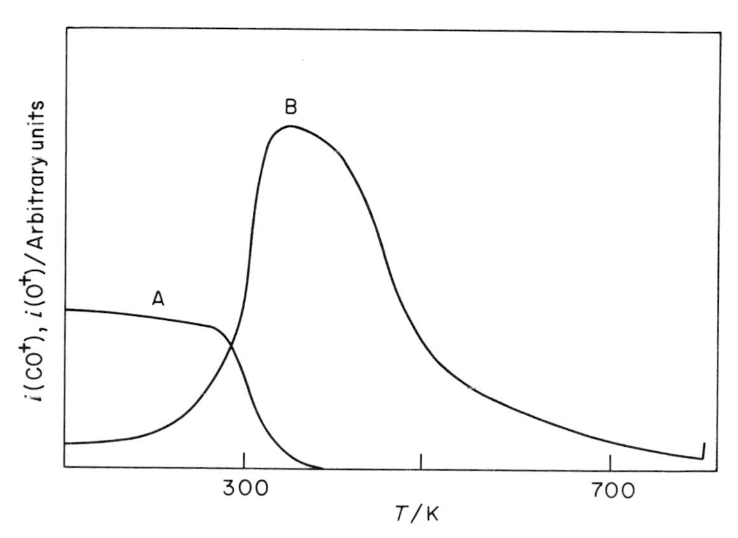

FIG. 5.9. Electron-stimulated desorption of CO^+ and O^+ ions from α-CO on tungsten as temperature is raised. Ionic currents $i(CO^+)$, $i(O^+)$ as a function of temperature; curve A, CO^+; curve B, O^+.

Another procedure is to plot the derivative of the ion current with respect to the retarding potential and so obtain the energy distribution of the desorbed ions as shown in Fig. 5.10, which displays the same two α-substates. The lower-energy peak corresponds to emission of CO^+ ions and the higher one to that of O^+ ions [37, 38]. The population of the two states varies with temperature in accord with their different binding energies. Furthermore, by bombardment for different periods of time with electrons of energy close to that required for desorption of the neutral molecule from its chemisorbed ground state, interconversion between the two sub-states can be effected [37, 38]. For incident energies above 10 eV, dissociation of chemisorbed CO into C and O adatoms becomes the dominant process.

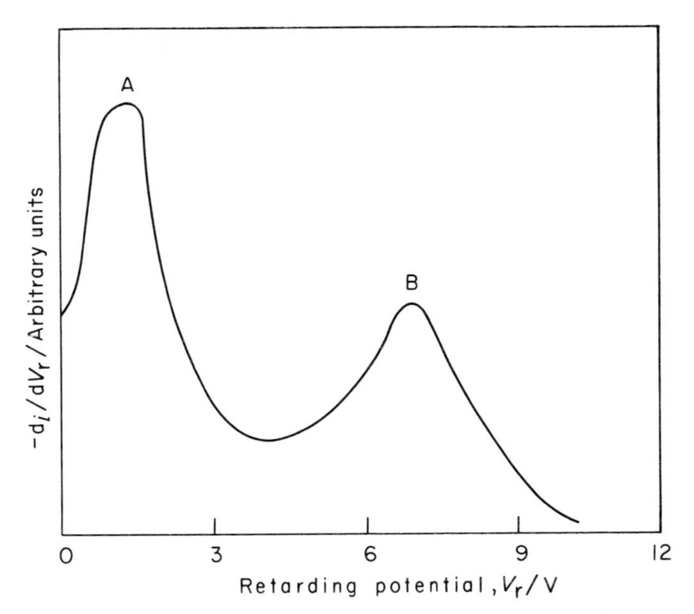

FIG. 5.10. Plot of first derivative of the ionic current with respect to applied retarding voltage V_r. Energy distribution for ions evolved from α-CO adsorbed on tungsten. Substrate temperature, 195 K. The two substrates correspond to those shown in Fig. 5.9: curve A, CO^+; curve B, O^+.

5.6.5. Rate measurements

The rate of desorption of neutral adspecies is obtained by measuring the increase of pressure in a constant volume by an ionization gauge or a mass spectrometer. This rate is given by

$$\frac{dN_t}{dt} = Qi_e N_t,$$

or

$$N_t = N_0 \exp(Qi_e t), \tag{5.77}$$

where i_e is the incident electron flux, N_0 is the initial concentration of adspecies (number of admolecules cm^{-2}) and N_t is the concentration at time t. The surface is assumed to be uniform and Q to be independent of N_t. From the number of molecules desorbed during the time period t and the measured electron flux per cm^2 surface, Q can be evaluated.

The rate of desorption of ions is obtained by measurement of the ionic current that is developed: the current i_t at time t is given by

$$i_t = i_0 \exp(Q^+ i_e t) \tag{5.78}$$

in which i_0, the initial current at $t = 0$, is equal to $Q^+ i_e N_0$.

Desorption rates have also been obtained by monitoring the surface concentration of the adspecies as a function of time e.g. from the variation of the work function, or the field emission current [39], or the Auger peak intensity [40], etc.

5.6.6. Desorption energies [41]

The impact energy of primary electrons that have been accelerated through a potential difference V is $eV + e^2/4x_0$, the last term being the contribution of the image forces; and x_0 is the distance of the adspecies from the surface. This energy is equal to the sum of: (i) the desorption energy E_d of the adspecies (ii) the kinetic energy E_k of the desorbed species and its excitation energy E_e; and (iii) the residual kinetic energy E'_k of the incident electron after impact with the adspecies. The minimum incident energy eV_m corresponds to the adspecies in its ground state $E_e = 0$ and kinetic energy approaching zero ($E_k \to 0$), hence E_d can be evaluated from the relationship ($E'_k \to 0$),

$$eV_m = E_d - e^2/4x_0. \tag{5.79}$$

In ionic desorption, $E_e = e(I - \Phi)$, where I is the ionization potential of the chemisorbed molecule and Φ the effective work function of the metal. Consequently,

$$eV_m = E_d^i - e^e/4x_0 + e(I - \Phi). \tag{5.80}$$

The minimum energy (eV_m or eV_m^i) is derived from a plot of the total desorption flux of desorbed particles against the accelerating voltage applied to the incident electrons by extrapolation to zero flux. Unfortunately, the plot approaches the voltage axis asymptotically so that the values of E_d and E^i_d are ill-defined and highly approximate (Fig. 5.11).

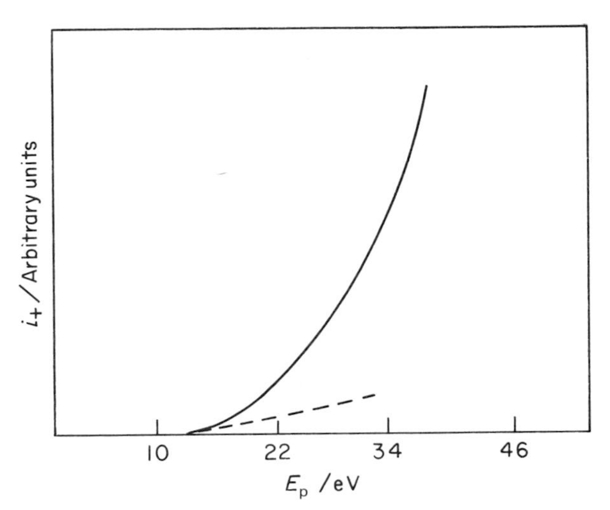

FIG. 5.11. Plot of ion current as a function of the energy E_p of the incident primary electrons for hydrogen chemisorbed on Ni(111). The extrapolated value of E_p obtained from the dashed curve is eV_m from which a crude estimate of E_d can be calculated [Eq. (5.79)].

5.6.7. Determination of the vibrational energy of adspecies [42, 43]

The kinetic energies of desorbed ions have a range of 0 to 10 eV and the spectral peak is roughly Gaussian of half-width of 2 to 4 eV. The energy distribution of desorbed ions is determined using a hemispherical retarding potential analyser, and is represented as a plot of the differential di^+/dV_r against V_r, the retarding potential. The shape of the curve is determined by the shape of the repulsive part of the excited-state plot, the squares of the vibrational wave functions of the surface oscillations, the population in the ground and excited states, and the rate of neutralization of the adions. At room temperature, the adspecies are predominantly in the vibrational ground state, but with increase of temperature, excited states (predominantly the first) begin to be populated. The fractional contribution of these states is directly proportional to the difference in the half-width of the distribution curves at various temperatures. From the slope of the plot of half-width against temperature, assuming higher states than the first excited state do not contribute, the ground-state frequency of the adspecies can be evaluated. For oxygen on tungsten, its magnitude is close to the stretching frequency of an adsorbed O_2 molecule, and might indicate that the β_1 state is a molecular one.

5.6.8. The value of electron-stimulated desorption studies [30, 44]

In principle, information about: (i) excitation and de-excitation of the adsorbate–surface complex; (ii) electron tunnelling through the surface barrier; and (iii) energy transfer in the formation of the chemisorption bond might be obtained. As yet, only qualitative conclusions are possible.

However, desorption studies have demonstrated that surface disturbances may frequently be induced by impact of low-energy electrons and highlights the difficulty of avoiding surface damage in electron spectroscopic investigations. At present, its main value has been in the investigation of multiplet states of adsorbates since within each state the ionization cross-section is fairly constant but is often substantially different for the different states.

5.7. FIELD DESORPTION

By application of a high electrostatic field, desorption of an adspecies in an ionic state can be effected. The (positive) applied field substantially decreases the potential energy of the ionic state, whereas the lowering of that of the neutral state due to polarization of the adatom is small [45–49]. The forms of the $U(x)$ versus x curves for the neutral $(M + A)$ and ionic states $(M^- + A^+)$ in the absence and in the presence of the applied field are diagrammatically sketched in Fig. 5.12 (dashed and full lines, respectively). At x_c, the potential energies of the two states are the same in the presence of the field $F(x_c)$ and an adiabatic transition from the neutral ground state to the ionic state takes place when the neutral state acquires the activation energy Q_2 with subsequent desorption of the ion.

In its simplest form (the image-force model), the process is the escape of the ion over the potential barrier formed by the superposition of the potential $-eFx_c$ (for a singly-charged ion) created by the applied field $F(x_c)$ at a distance x_c from the image plane, and the image potential energy $-e^e/4x_c$ which attracts the ion to the surface. The net effect is the reduction of the potential energy of the ion by an amount $e^{\frac{3}{2}}F(x_c)^{\frac{1}{2}}$ at the maximum of the adatom curve at x_c. The activation energy is given by

$$Q_1(F(x_c)) = Q_0 - e^{\frac{3}{2}}F(x_c)^{\frac{1}{2}}, \tag{5.81}$$

where Q_0 is the energy to remove the ionized adatom to infinity in the absence of the applied field and is equal to

$$Q_0 = q_c + (I - \Phi) \tag{5.82}$$

for desorption of singly-charged ions, where q_c is the binding energy of the

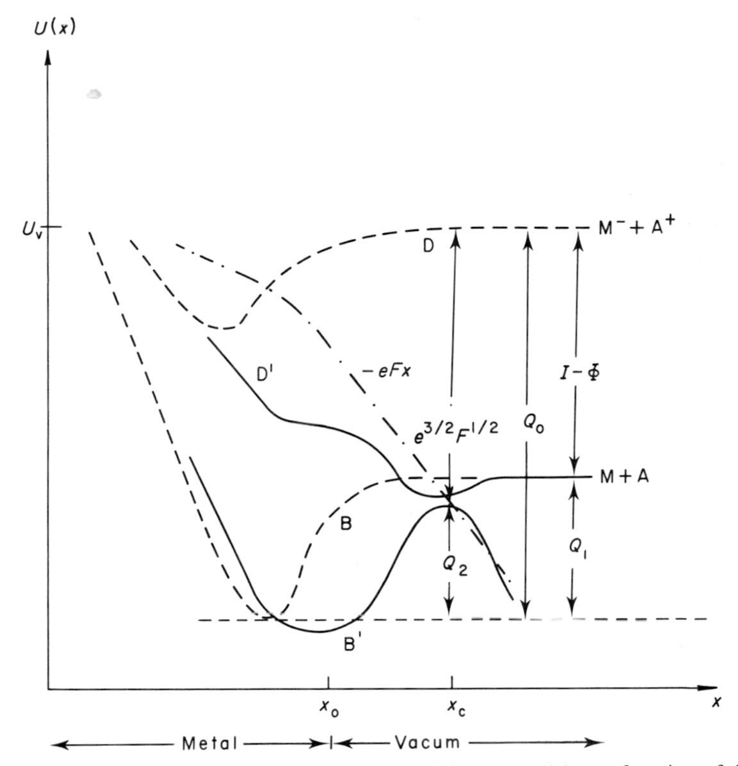

FIG. 5.12. Diagrammatic representation of the potential energy $U(x)$ as a function of distance co-ordinate x: plots B and D (dashed lines) are for neutral $(M + A)$ in the absence of a field, respectively. Corresponding plots in the presence of a field F are shown as B′, D′ (full lines), respectively. Application of the field F to give potential variation of $-eFx$ is denoted by (·—·). U_v is the vacuum level, Q_1 is the desorption energy of A in the absence of the field, Q_0 is the vacuum level, Q_1 is the desorption energy of A in the absence of the field, Q_0 is the energy to remove A as the ion A^+ to inifinity in the absence of the field, and Q_2 is the activation energy to surmount the Schottky saddle at x_c.

adatom to the surface. At high temperatures (at which tunnelling through the barrier is small) the desorption rate can be approximated to

$$r_d = \nu \exp\{-Q_1(F)/kT\}, \tag{8.83}$$

where ν is the vibrational frequency of the surface atoms; $Q_1(F)$ can, therefore, be evaluated from the relative rates at different temperatures. From this value and use of Eqs (5.81) and (5.82), the adatoms binding energy q_c can then be obtained.

For atoms of the third transition series on various tungsten planes, an almost direct proportionality was found experimentally between the binding energy and the number of bonding *d*-electrons of the adatoms [50, 51]. However, even with refinements of the simple model to include field penetration into the surface, ion-core repulsions, polarization and hyper-polarization [the polarization correction alone may amount to several eV for adatoms on W(110)], the binding energies should be viewed with caution because the values of the parameters used in the calculations are subject to considerable uncertainty.

5.8. PHOTODESORPTION [52–55]

Photodesorption of adsorbates from metal surfaces is a controversial subject. The main problem is to distinguish it from the thermal desorption brought about by heating of the metal on absorption of the incident photon energy. Despite precise automatic temperature control of the bulk metal, the surface temperature may still be increased because the depth of penetration of photons in a metal is very small (~ 2 to 10 Å).

Most investigations have been concerned with carbon monoxide as adsorbate, since it appears to be most easily desorbed adspecies under photon irradiation. One explanation [55] to account for this result is that the photons are absorbed by the adsorbate complex; it is thereby excited to a repulsive state by an electron transition to a low-lying, previously unoccupied, orbital, Should the bonding of α-CO on W be similar to that of the CO ligands in the carbonyl $W(CO)_6$, then, since the energy range in which photodesorption occurs is almost coincident with that effective in the photodissociation of the carbonyl, photodissociation of the adsorbate complex could occur. Alternatively, absorption of photons by the metal could lead to excitation of its conduction electrons to unoccupied states just above the Fermi level, with subsequent tunnelling into an unoccupied electronic level in the adsorbate complex. The threshold energy has about the right magnitude for such an electron transition and the result might again be to produce a repulsive state of the adsorbate. The yield is, as expected, small ($<10^{-7}$ molecules per photon).

REFERENCES

1. J. A. Becker and C. D. Hartman, *J. Phys. Chem.*, 1953, **57**, 157.
2. C. Kohrt and R. Gomer, *J. Chem. Phys.*, 1970, **52**, 3283.
3. W. Jelland and D. Menzel, *Surface Sci.*, 1974, **42**, 1974.
4. J. C. Tracy, *J. Chem. Phys.*, 1972, **56**, 2736 and 2748.
5. P. W. Tamm and L. D. Schmidt, *J. Chem. Phys.*, 1969, **51**, 5352.

6. G. Ehrlich, *J. Appl. Phys.*, 1961, **32**, 4.
7. P. A. Redhead, *Vacuum*, 1962, **12**, 203.
8. G. Carter, *Vacuum*, 1962, **12**, 245.
9. J. S. Wagener, *Proc. Inst. Elect. Eng.* (*London*), *Part III*, 1952, **99**, 135; *Z. Angew. Phys.* 1954, **6**, 433.
10. G. Ehrlich, *J. Chem. Phys.*, 1961, **34**, 29 and 39.
11. P. T. Dawson and Y. K. Peng, *Surface Sci.*, 1972, **33**, 565.
12. C. Pisani, G. Rabino and F. Ricca, *Surface Sci.*, 1974, **41**, 277.
13. S. Glasstone, K. J. Laidler and H. Eyring, "The Theory of Rate Processes", McGraw Hill, New York, 1941.
14. K. J. Laidler, *in* "Catalysis" (P. H. Emmett, ed.), Reinhold Publ. Corp., New York, 1954, Vol. 1, Chap. 3.
15. C. Kurht and R. Gomer, *Surface Sci.*, 1971, **24**, 77.
16. M. J. Dresser, T. E. Madey and J. T. Yates, Jr, *Surface Sci.*, 1972, **42**, 533.
17. L. R. Clavena and L. D. Schmidt, *J. Chem. Phys.*, 1969, **51**, 5352.
18. L. R. Clavena and L. D. Schmidt, *Surface Sci.*, 1970, **22**, 365.
19. W. L. Winterbottom, *J. Vac. Sci. Tech.*, 1972, **9**, 936.
20. C. G. Goymour and D. A. King, *J. Chem. Soc., Faraday Trans. I*, 1972, **68**, 289; 1973, **69**, 736.
21. A. K. Muzumdar and H. W. Wassmuch, *Surface Sci.*, 1973, **34**, 249.
22. H. A. Kramers, *Physica*, 1940, **7**, 284.
23. P. J. Pagni and J. C. Keck, *J. Chem. Phys.*, 1973, **58**, 1162.
24. P. J. Pagni, *J. Chem. Phys.*, 1973, **58**, 2940.
25. B. Bendow and S. C. Ying, *J. Vac. Sci. Tech.*, 1972, **9**, 804; *Phys. Rev. B*, 1973, **7**, 622.
26. C. Aharoni and F. C. Tompkins, *Adv. Catalysis*, 1970, **21**, 1.
27. S. Brunauer, K. S. Love and R. G. Keenan, *J. Amer. Chem. Soc.*, 1942, **64**, 751.
28. E. R. S. Winter, *J. Catalysis*, 1965, **4**, 134.
29. S. Z. Roginski, "Adsorption und Katalyse an Inhomogenen Oberflachen", Akad. Verlag, Berlin, 1958.
30. P. A. Redhead, *Trans. Faraday Soc.*, 1961, **57**, 641.
31. L. J. Rigby, *Can. J. Phys.*, 1964, **42**, 1256.
32. D. A. Degras, *Nuovo Cimento* (*Suppl.*), 1967, **5**, 408.
33. T. E. Madey and J. T. Yates, Jr., *J. Vac. Sci. Tech.*, 1971, **8**, 525.
34. J. H. Leck and B. P. Stimpson, *J. Vac. Sci. Tech.*, 1972, **9**. 293.
35. D. Menzel and R. Gomer, *J. Chem. Phys.*, 1964, **41**, 3311, 3329.
36. P. A. Redhead, *Can. J. Phys.*, 1964, **42**, 886.
37. D. Menzel, *Ber. Bunsenges Phys. Chem.*, 1968, **72**, 591.
38. J. T. Yates, Jr and D. A. King, *Surface Sci.*, 1972, **32**, 479.
39. D. Menzel, *Surface Sci.*, 1975, **47**, 384.
40. R. G. Musket, *Surface Sci.*, 1970, **21**, 440.
41. P. A. Redhead, *Nuovo Cimento* (*Suppl.*), 1967, **5**, 586.
42. M. Nishijima and F. M. Propst, *Phys. Rev.*, 1970, **82**, 2368.
43. D. Lightman, F. N. Simon and T. R. Kurst, *Surface Sci.*, 1968, **9**, 325.
44. J. T. Yates, Jr, T. E. Madey and J. K. Payne, *Nuovo Cimento* (*Suppl.*), 1967, **5**, 558.
45. M. G. Ingham and R. Gomer, *Z. Naturforsch.*, 1955, **10a**, 836.
46. E. W. Müller, *Phys. Rev.*, 1956, **102**, 618.
47. R. Gomer, *J. Chem. Phys.*, 1959, **31**, 341.
48. R. Gomer and L. W. Swanson, *J. Chem. Phys.*, 1963, **38**, 1613.
49. E. W. Muller and T. Tsong, *Prog. Surface Sci.*, 1974, **4**, 1.
50. G. Ehrlich and C. F. Kirk, *J. Chem. Phys.*, 1968, **48**, 1465.

51. E. W. Plummer and T. N. Rhodin, *J. Chem. Phys.*, 1968, **49**, 3479.
52. R. O. Adams and E. E. Donaldson, *J. Chem. Phys.*, 1965, **42**, 770.
53. H. Moesta and H. D. Brewer, *Surface Sci.*, 1969, **17**, 439.
54. P. Génequand, *Surface Sci.*, 1971, **25**, 643.
55. P. Kronauer and D. Menzel, *in* "Adsorption–Desorption Phenomena" (F. Ricca, ed.), Academic Press, London, 1972, p. 313.

6

Adsorption Isotherms

6.1. REPRESENTATION OF EQUILIBRIUM DATA

Since the amount adsorbed per unit area of surface is uniquely determined by the equilibrium pressure of the gaseous adsorbate and the temperature of the adsorbent–adsorbate system, equilibrium adsorption data can be expressed in the form of isotherms, isosteres or isobars. An isotherm depicts the amount adsorbed at constant temperature as a function of pressure; an isobar expresses the functional relationship between the amount adsorbed and the temperature at constant pressure; an isostere relates the equilibrium pressure of the gaseous adsorbate to the temperature of the system for a constant amount of the adsorbed phase. It is usually convenient, from an experimental point of view, to determine isotherms at a series of constant temperatures. The co-ordinates of pressure at the different temperatures for fixed amount adsorbed can then be interpolated to construct a set of isoteres; a similar procedure is followed to obtain an isobaric series.

6.2. THE LANGMUIR MODEL

The simplest model for the derivation of a theoretical isotherm is that of Langmuir [1]; with modifications, it also forms the basis of the Freundlich [2] and the Temkin [3] isotherms. Postulates of the model common to all three isotherms are as follows [4].†

† See also Chapter 9.

(i) The adsorbent surface is planar, and the total number of adsorption sites is constant under all experimental conditions.
(ii) The adsorption is restricted to the monolayer.
(iii) On impact of the gaseous adsorbate molecule with an unoccupied or free site, it is adsorbed with zero activation energy.
(iv) An occupied site comprises one adsorbate molecule adsorbed at a single site. Collision of a gaseous adsorbate molecule with an occupied site is elastic; the adsorbed molecule remains unperturbed at the site and the gaseous molecule is returned to the gas phase with no loss of kinetic energy.
(v) Desorption of an adsorbed molecule occurs as soon as it has acquired sufficient thermal energy from the lattice vibrations of the adsorbent to equal the heat of adsorption.
(vi) At equilibrium, the rate of adsorption on the unoccupied sites equals that of desorption from the occupied sites.

6.2.1. The Langmuir equation

Two additional postulates are included: (i) the adsorptive properties of all sites are identical, i.e. the surface is ideally uniform; and (ii) there are no lateral interactions between neighbouring adsorbed molecules.

An adsorbate–adsorbent system at equilibrium at a temperature T K and at a pressure P Pa of gaseous adsorbate is considered; N sites cm^{-2} of the total number of sites N_s cm^{-2} are occupied by adsorbed molecules. The rate of adsorption per cm^2 surface s^{-1} is then

$$dn/dt = k_a P(N_s - N), \tag{6.1}$$

where $(N_s - N)$ is the number of unoccupied sites cm^{-2}, and k_a is the rate constant per site for unit pressure; $k_a N_s$ = the corresponding collision rate Z, given by the Herz–Knudsen equation, i.e. $Z = 2{\cdot}63 \times 10^{20}\ (MT)^{\frac{1}{2}}$ molecules $cm^{-2}\,s^{-1}$ at a pressure of 1 Pa for a gas of molecular weight M. However, the capture efficiency on collision is rarely unity; k_a is, in fact, the product $s'Z$, where s' may be termed the (constant) sticking probability per unoccupied site.

The rate of desorption per cm^2 surface per second is $k_d N$, where k_d (in s^{-1}) is the rate constant per adsorbed molecule for desorption. It includes a term $\nu \exp(-q/kT)$, in which q is the heat of adsorption per admolecule and is independent of N, and ν is the vibrational frequency of the chemisorbed bond.

At equilibrium,

$$k_a P(N_s - N) = k_d N, \tag{6.2}$$

or

$$k_a P(v_m - v) = k_d v, \tag{6.3}$$

where v is the volume of gas adsorbed and v_m is the volume to give a complete monolayer, both in units of cm^3 s.t.p. Hence,

$$v = k_a P v_m/(k_a P + k_d). \tag{6.4}$$

Since the fractional coverage is $\theta = N/N_s = v/v_m$, then

$$\theta = k_a P/(k_a P + k_d) \tag{6.5}$$

or

$$P = \frac{k_d}{k_a}\frac{\theta}{(1-\theta)} = \frac{K\theta}{1-\theta}, \tag{6.6}$$

where $K = k_d/k_s$ is the equilibrium constant for adsorption.

6.3. DEPARTURES FROM THE LANGMUIR MODEL

Departures from the Langmuir model may be grouped into two main categories: (i) the surface is uniform, and the departures are formulated in geometric terms; (ii) the effective adsorption potential of sites varies with the coverage as a consequence of heterogeneity of the surface or/and the existence of lateral interactions between the adsorbed molecules.

6.3.1. Geometric factors

6.3.1.1. *Dual-site occupancy*

The admolecule is bound to more than one adsorbent site. Thus, in the associative chemisorption of ethylene on nickel at low temperatures, the two C atoms are bonded to two adjacent sites. The number of adjacent doubly-unoccupied sites as a function of fractional coverage must, therefore, be evaluated. Each unoccupied site has z neighbours; for a random arrangement, each site has a probability $(N_s - N)/N_s$ of being unoccupied. The average number of unoccupied neighbours of any given unoccupied site is, therefore, $z(N_s - N)/N_s$. Since there is a total of $(N_s - N)$ unoccupied sites, each with $z(N_s - N)/N_s$ unoccupied neighbours, then the average number of pairs of neighbouring unoccupied sites is $\frac{1}{2}z(N_s - N)^2/N_s$, where the numerical factor is included to prevent the counting of each pair twice.

The rate of adsorption is, therefore,

$$k_a P(z/2)(N_s - N)^2/N_s, \tag{6.7}$$

in which k_a, the rate constant, includes a sticking probability for molecular adsorption on unoccupied pair-sites.

Desorption of an admolecule frees two occupied sites; the number of admolecules is, therefore, given by $N/2$. The rate of desorption is $k_d N/2$, and, at equilibrium,

$$k_a P(z/2)(N_s - N)^2/N_s = k_d N/2 \tag{6.8}$$

or

$$P = K\theta/z(1 - \theta)^2. \tag{6.9}$$

6.3.1.2. *Dissociative adsorption*

In this process, the initial act is a two-site adsorption of the adsorbate molecule; the admolecule then undergoes dissociation into two fragments, each of which occupies one adsorption site. The two products may be identical, as in the chemisorption of a homonuclear diatomic molecule such as hydrogen, or, as in the chemisorption of ammonia, they may comprise two different adspecies, hydrogen adatoms and $—NH_2$ adradicals. For a completely immobile layer, with the two products occupying adjacent sites, the situation corresponds to dual-site occupancy. However, since the adspecies are normally mobile, desorption of the admolecule involves a bimolecular recombination of two adatoms and the rate is proportional to $\frac{1}{2}N(N/N_s)$, where N/N_s is the probability of finding an adatom at an occupied site, and N the number of sites occupied by adatoms at equilibrium. Hence

$$k_a P(z/2)(N_s - N)^2/N_s = k_d N^2/2N_s, \tag{6.10}$$

or

$$P^{\frac{1}{2}} = (K/z)^{\frac{1}{2}}\,\theta/(1 - \theta). \tag{6.11}$$

The analytical form is similar to that for the Langmuir equation, but the exponent of P is one-half instead of unity.

6.3.1.3. *Single-site adsorption of large admolecules with exclusion of adjacent sites* [5]

An admolecule of large molecular dimensions attached to a single site can render the adjacent sites around it inaccessible to other adsorbate molecules. For example, on a square lattice, four nearest-neighbour sites may become inactive in adsorption, i.e. the adsorbed molecule effectively occupies a group of five sites at low coverage. However, as the coverage increases, some of the inactive sites are shared by more than one admolecule and the average number $\bar{n}$ of excluded sites per admolecule decreases. If the number of blocked sites per admolecule on a surface packed to maximum capacity in a regular array is denoted by n', then the fractional coverage $\theta = nn'/N_s$: where n is the number of admolecules and N_s the number of adsorption sites, each per unit area. The isotherm, therefore, has the form

$$P = \text{const.}\frac{n}{N_s - n\bar{n}} = \text{const.}\frac{\theta/n'}{(1 - \theta\bar{n}/n')}, \tag{6.12}$$

which reduces, as it should do, to the Langmuir equation for $\bar{n} = n' = 1$. The magnitude of n' is obtained from simple geometric considerations; for a square lattice with exclusion of the four nearest-neighbour sites, $n' = 2$. The average exclusion number $\bar{n}$ can be evaluated by a Monte Carlo procedure; it is found that $\bar{n}$ is an almost linearly decreasing function of θ, such that $\bar{n}_\theta \simeq (4 - \theta)$ up to $\theta \sim 0{\cdot}6$. The isotherm then approximates at lower coverages to

$$P = \frac{\text{const.}\,\theta}{2 - 4\theta + \theta^2}. \tag{6.13}$$

Thus, with increase of p, θ increases more slowly than for a Langmuir equation and, particularly as the monolayer capacity is approached, the amount of free surface is significantly greater.

6.3.2. Energetic factors

6.3.2.1. *Excess occupancy of adsorption sites*

At a metal surface, the conduction electrons tend to smooth out the fluctuations of potential energy by partially screening the field of the positive ion-cores; consequently, the energy difference between the peak and trough of the potential energy function is comparatively small. For small adsorbate molecules, the distance between the adsorption sites (or the lattice constant) of the metal is large compared with the diameter of the adspecies. At high coverages, particularly on a heterogenous surface, the decrease of free energy brought about by the adsorption of an adsorbate molecule at one of the few available unoccupied sites (which are necessarily of lower adsorption potential) can be less than that brought about by its adsorption into an energy trough already occupied by a single admolecule. Adsorption can then take place by lateral displacement of the original admolecule from the bottom of the trough with the two admolecules symmetrically located at each side of the initially occupied adsorption site. The binding energy of each adspecies is less than that of the single admolecule at the bottom of the trough, but the difference between their combined energy and that of single occupancy may be greater than the binding energy of the adsorbate molecule at one of the unoccupied low-energy sites still remaining on the surface. Multi-occupancy is, therefore, favoured; and, since the free space between adspecies on adjacent sites is invariably less than the diameter of the adsorbate molecule, the adsorbed layer undergoes compression [6, 7]. The number of adspecies then exceeds the number of sites derived from surface crystallography; the Langmuir concept of a monolayer associated with a constant number of adsorption sites is no longer valid. An example is the chemisorption of CO at higher pressures on some transition metals.

Multi-occupancy can also occur when dimerization of a gaseous monomer is energetically possible in the adsorbed state. The process is favoured at lower coverages, since steric factors can inhibit dimerization in a concentrated layer.

6.3.2.2. *Exclusion of sites*

Nearest-neighbour sites can be rendered inactive by the presence of a strongly adsorbed molecule on the central site. Thus, a layer of adatoms, with occupation of alternate sites, is formed when nitrogen is adsorbed on the (111) plane of tungsten; the complete monolayer then corresponds to the occu-

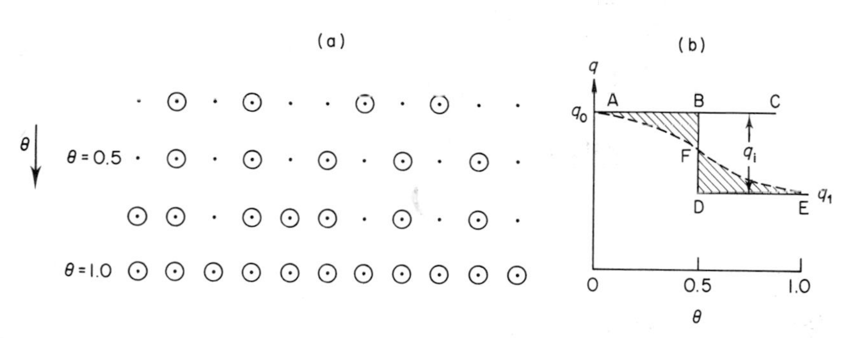

FIG. 6.1. (a) Adsorbate configuration; (b) variation of adsorption heat with coverage.

Schematic representation of the configuration of adsorbed CO on the (100) plane of nickel as the coverage is increased, assuming a mobile layer, a pairwise repulsive interaction V and a co-ordination number z so that $q_1 = zV$. Up to $\theta = 0{\cdot}5$, only alternative sites are occupied with $k_d = \nu \exp(-q_0/kT)$; above $\theta = 0{\cdot}5$, repulsive interactions become effective and $k_d = \nu \exp(-q_0 + q_i)/kT$, where q_0 = differential adsorption heat at $\theta = 0$ and $q_1 = q_0 - q_i$ at $\theta = 1$. The isotherm can be constructed assuming that ABDE is the shape of the q versus θ plot; in practice, thermal migration of adatoms opposes the attainment of the minimum energy configuration and the q versus θ plot is given by AFE with the shaded sections ABF and FDE increasing in area with increase of temperature.

pation of only one-half of the total number of sites expected from surface crystallographic considerations. A probable explanation is that some of the bonding orbitals of trivalent nitrogen atoms interact with those of nearest-neighbour tungsten atoms, the residual adsorption potential of which is then not sufficiently high to effect dissociative chemisorption involving the rupture of the strong N≡N bond of the gaseous molecule.

In the chemisorption of CO on nickel, alternative site occupation is also predominant up to half-coverage, because the lateral repulsion between the admolecules reduces the adsorption heat on the alternate unoccupied sites. However, with increasing pressure of gaseous CO the monolayer is completed, but the heat of adsorption at coverages for $\theta > \frac{1}{2}$ is smaller (see Fig. 6.1). On a uniform surface, two Langmuir equations, with different values of k_d, one below and the other above half-coverage, may be combined to give a fair representation of the total isotherm.

6.3.2.3. *Existence of multiplet states*

A single adsorbate can commonly exist in more than one adsorbed state with different binding energies and in different molecular forms. As has already been noted, nitrogen adatoms occupy only alternative sites on the (111) plane of tungsten, but nitrogen molecules can be chemisorbed on the single unoccupied sites. Two different adsorbed states, an atomic β and a molecular α state, are then simultaneously present in the adsorbed layer. A similar circumstance arises when dissociative chemisorption, involving simultaneous occupation of two adjacent sites, forms an immobile layer. The complete monolayer in this random array comprises about 12% of unoccupied single sites, and at higher pressures, an α-molecular adspecies can be chemisorbed at these positions.

Finally, a weak molecular chemisorption can take place on top of a primary adatom layer, particularly when hydrogen bonding between the molecular and atomic species is possible.

6.3.2.4. *Non-uniformity of the adsorption potential of sites*

A common cause for departures of the experimental isotherm from the Langmuir equation is the heterogeneity of the surface, i.e. the variation of adsorption potential of the sites. At equilibrium, sites are preferentially occupied, in order of their decreasing binding energies, according to a Boltzmann distribution of energies. The shape of the isotherm is, therefore, determined by the distribution of the number and the binding energy of the sites. Site heterogeneity may be characterized by an appropriate continuous distribution function. Its analytical form is usually unknown and, in practice, various functions that are amenable to mathematical analysis are selected in order to obtain the best fit with the experimental isotherm.

6.3.2.5. *Interaction between adsorbed molecules*
This interaction almost invariably has a net repulsive character. Consequently, the binding energy of an adsorbate molecule to an unoccupied site adjacent to an occupied one is reduced, and the differential heat of adsorption decreases with increase of coverage. Interaction between nearest-neighbour admolecules is significantly larger than that between next-nearest neighbours and, as has already been noted, the incoming molecule tends to avoid nearest-neighbour unoccupied sites. The interaction energy arises not only from the Coulombic forces between the surface dipoles of the admolecules, which may themselves be polar or have polarity induced in them by the surface field of the metal, but also because the wave functions of the bonding orbitals extend much further within the metal surface region than in free space [8]. This interaction decreases according to an inverse square law and may oscillate between repulsion and attraction as the distance between admolecules increases. Such lateral interactions are present on both uniform and non-uniform surfaces and make an increasing contribution as the coverage increases. Moreover, the heterogeneous surface may have a patchwise character, each patch having an array of uniform sites which differ in their adsorption potential from patch to patch. On the patch with the highest-energy sites, virtually all sites may be occupied at low pressures of the adsorbate, whereas, on those of lowest energy, the fractional coverage may still be essentially zero [9]. Interaction between admolecules in such circumstances may, therefore, provide a significant energetic contribution, even when the fractional coverage of the total surface is very small.

6.4. ISOTHERMS FOR VARIABLE BINDING ENERGIES

In the Langmuir model, the binding energy of the adsorbate molecule at all sites is constant and independent of the fractional coverage; consequently, the differential isosteric heat of adsorption is also invariant in magnitude at all coverages. Rigorous thermodynamic definitions of the different heats of adsorption that are obtained by different experimental techniques are discussed later.† In the present context, the energy liberated by the adsorption of a small amount of adsorbate at a particular coverage, expressed as kJ mol^{-1} adsorbate, is a sufficiently precise definition of the differential heat of adsorption. Evidently, there must be some correlation between the variation of the differential heat with coverage and the shape of the isotherm.

A considerable amount of experimental data, obtained under isothermal conditions, can be described, with reasonable adequacy, by either of two

† See Chapter 7.

equations, the Freundlich or the Temkin, and each of these can be rationalized in terms of a relationship between the heat and the coverage (cf. Fig. 6.2 for comparative plots of Langmuir, Freundlich and Temkin isotherms).

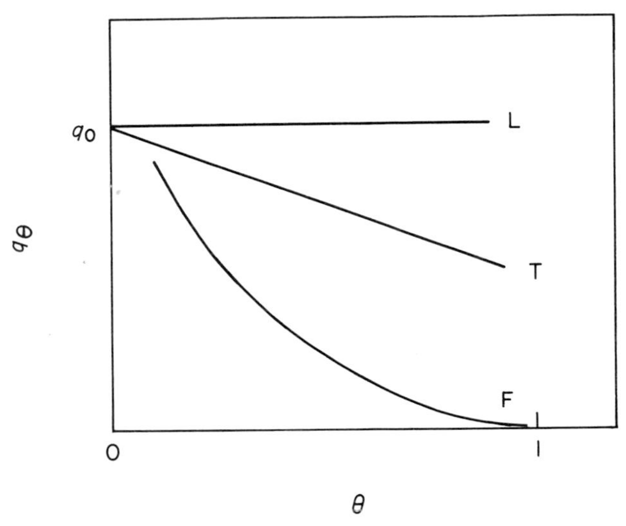

FIG. 6.2. Variation of heat of adsorption q_θ with increasing coverage for Langmuir (L), Temkin (T) and Freundlich (F) isotherms.

6.4.1. The Freundlich equation

This equation is one of the earliest relationships used to describe adsorption isotherms; it relates the amount adsorbed to a fractional power of the pressure at constant temperature, i.e.

$$v = \text{const.}\ P^{1/n}, \tag{6.14}$$

in which n is usually greater than unity.

The equation can be derived by assuming† that $q = q_m (\ln \theta - \ln \theta_m)$, i.e. that the adsorption heat decreases logarithmically with increase of θ. The term q_m is a constant of magnitude considerably larger than RT, where T is the temperature of the adsorption system.

Adsorption equilibrium is given by

$$k_3 P(1 - \theta) = v\theta \left[\exp \frac{(q_m \ln \Theta)}{RT} \right], \tag{6.15}$$

† See Chapter 9 for more detailed discussion and significance of $\Theta \theta/\theta_m$.

or

$$\ln\left(\frac{\theta}{1-\theta}\right) = \ln\left(\frac{k_3P}{v}\right) - \frac{q_m \ln\Theta}{RT}, \tag{6.16}$$

v being the frequency of the chemisorbed bond vibration, and $\Theta = \theta/\theta_m$.
Since $q_m \gg RT$, then, over a very wide range of θ values,

$$\ln\left(\frac{\theta}{1-\theta}\right) \ll \frac{q_m \ln\Theta}{RT}. \tag{6.17}$$

Hence,

$$\Theta \approx (k_5P)^{k_4}, \tag{6.18}$$

with $k_4 = (RT/q_m)$, and $k_5 = k_3/v$.

6.4.2. The Temkin equation

This equation relates the extent of coverage with the logarithm of the pressure at constant temperature, i.e.

$$v = \text{const.} \ln P + \text{const.}$$

It can be derived by assuming that the adsorption heat decreases linearly with increase of coverage, i.e.

$$q = q_0(1 - \alpha\theta), \tag{6.19}$$

in which α is a constant and q_0 is the value of q as $\theta \to 0$. Equilibrium is established when

$$k_7P(1-\theta) = v\theta\left[\exp\frac{-q_0(1-\alpha\theta)}{RT}\right]. \tag{6.20}$$

or

$$\ln P = -\ln\left[\frac{k_7}{v}\exp\left(\frac{q_0}{RT}\right)\right] + \frac{q_0\alpha\theta}{RT} + \ln\frac{\theta}{1-\theta}. \tag{6.21}$$

Again, $q_0\alpha \gg RT$ and the variation of $q_0\alpha\theta/RT$ is large compared with that of $\ln[\theta/(1-\theta)]$, provided that θ does not approach zero or unity. Hence

$$\theta = \frac{RT}{\alpha q_0}\left\{\ln P + \ln\left[\frac{k_7}{v}\exp\left(\frac{q_0}{RT}\right)\right]\right\}. \tag{6.22}$$

or

$$\theta = k_8 \ln k_6 P, \tag{6.23}$$

where $k_8 = RT/q_0\alpha$ and $k_6 = (k_7/v)\exp(q_0/RT)$.

6.5. CRITERION OF OBEDIENCE TO A PARTICULAR EQUATION

Each of the above isotherms contains two constants k_i, k_j and their validity may be tested by assessing the linearity of a derived form of the corresponding equation (see Figs 6.3 to 6.7).

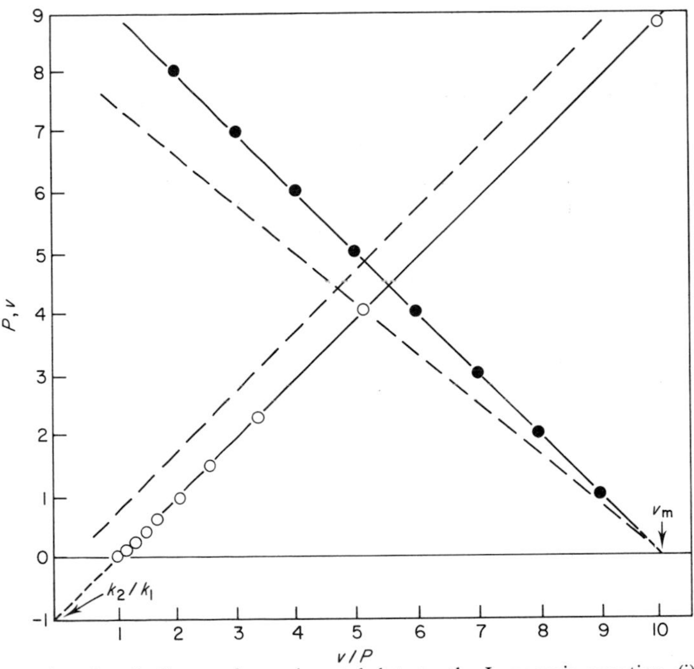

FIG. 6.3. Testing the obedience of experimental data to the Langmuir equation. (i) Plot of P against v/P (○—○); slope gives v_m independent of T; the effect of decrease of T is indicated by the dashed line giving smaller value of $-K$. (ii) Plot of v against v/P (●—●); intercept v_m is independent of T, but the slope $(-K)$ decreases at lower temperature (dashed line). In both cases, units are arbitrary.

6.5.1. The Langmuir equation

The equation in the form [$k_1 = k_a$, $k_2 = k_d$ in Eq. (6.4)]

$$v = k_1 v_m P/(k_1 P + k_2)$$

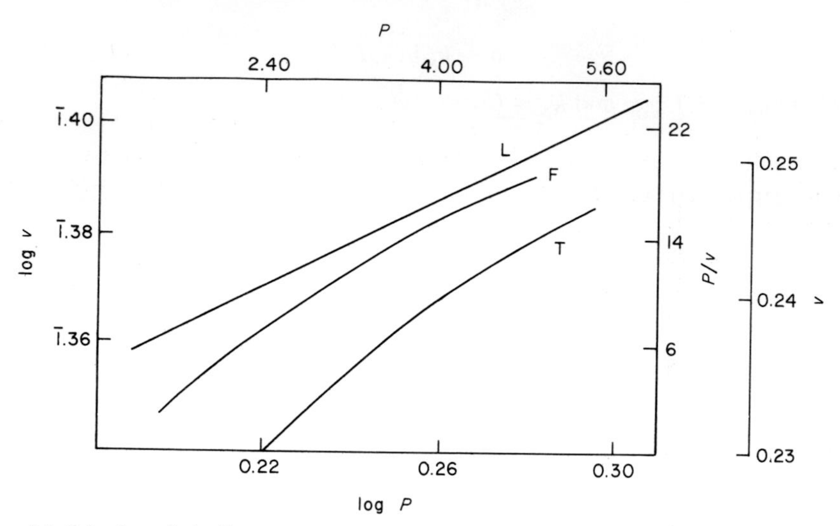

FIG. 6.4. Criterion of obedience to Langmuir, Freundlich and Temkin isotherms for the same data. L, Langmuir, P/v against P units, linear plot. F, Freundlich, log v units against log P units non-linear plot. T, Temkin, v units against log P units, non-linear plot (units arbitrary).

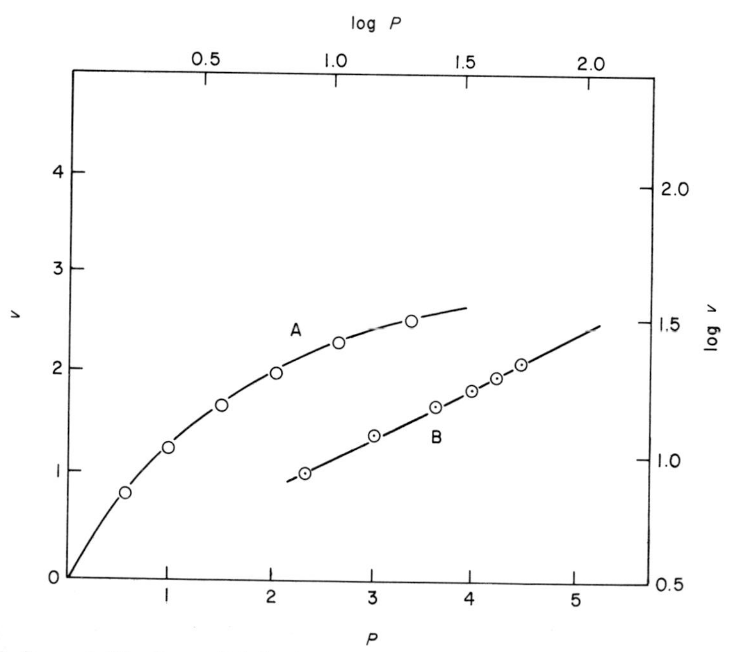

FIG. 6.5. Curve A, The Freundlich isotherm, amount adsorbed v as a function of pressure P. Curve B, plot of log v against log P.

is transformed to

$$k_1Pv + k_2v = k_1v_mP.$$

In case (i), $P = v_m(P/v) - K$. The linear plot has a slope of v_m (the monolayer volume) independent of the temperature T, and an intercept of $(-k\ /k_1)$ that varies exponentially with T (Fig. 6.3). The (desorption) heat is obtained from the slope of the $\ln K$ versus $1/T$ plot. In case (ii), $v = v_m - K(v/P)$.

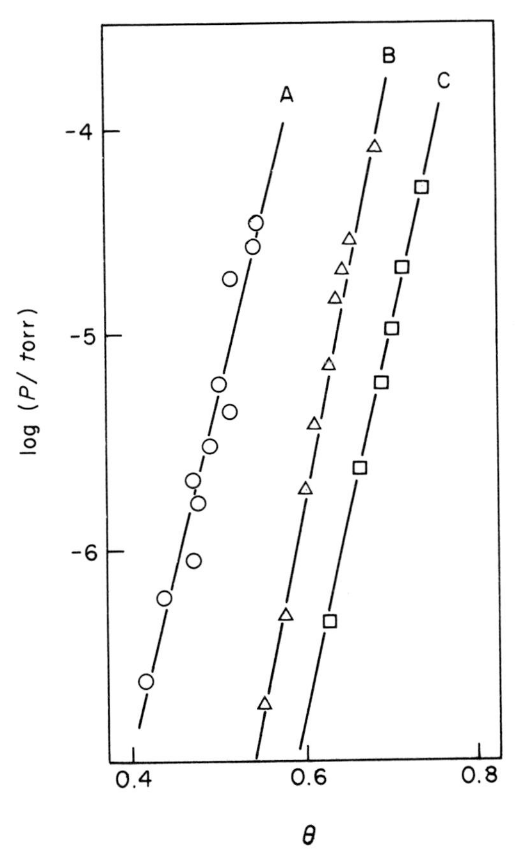

FIG. 6.6. Temkin plot of log P as a function of θ for the adsorption of hydrogen on a tungsten filament at different temperatures. Curve A, 373 K; curve B, 298 K, curve C, 273 K.

The slope $(-K)$ varies exponentially with T, but the intercept v_m is independent of temperature. The points are evenly spaced over the v-range and not, as for case (ii), crowded together at the higher range of v-values (Fig. 6.3). Case (iii), in the form $1/P = (v_m/K)/v - 1/K$, is less frequently used to evaluate

v_m as this method involves division of two temperature-dependent terms, viz. the slope (v_m/K) by the intercept $(-1/K)$.

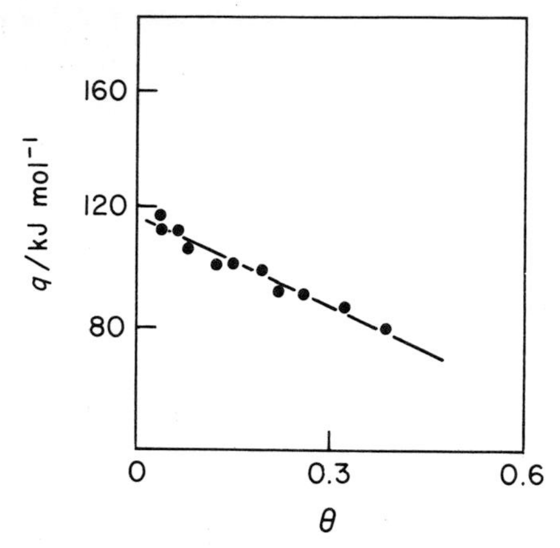

FIG. 6.7. Heat of adsorption of hydrogen on a tungsten filament as a function of coverage. The linear fall of q indicates a Temkin isotherm.

6.5.2. The Freundlich equation

The equation

$$v = (k_5 P)^{k_4}$$

is written in logarithmic form and a plot of $\ln v$ against $\ln P$ is tested for linearity.

6.5.3. The Temkin equation

The equation

$$v = k_8 \ln k_6 P$$

can be plotted directly in order to obtain a criterion of its obedience to the experimental data.

6.6. ADDITIONAL CRITERIA

The three isotherms have been derived for localized adsorption on a constant number of adsorption sites. This number is a property of the surface and, therefore, the monolayer capacity v_m corresponding to the occupation of all sites must be independent of temperature. Consequently, for the Langmuir equation,

$$Pv_m/v = (k_2/k_1) + P,$$

the slope of the linear plot of P/v against P is $1/v_m$ and should be independent of the temperature at which the isotherm is obtained.

Similarly, as is shown in the statistical–mechanical derivations of these isotherms:† (i) obedience to the Freundlich equation requires the slope and intercept of the linear plot of $\ln v$ against $\ln P$ to be proportional to the absolute temperature; (ii) for obedience to the Temkin equation, the slope of the linear plot of $\ln P$ against v (only strictly linear in the middle range of θ values) must also be proportional to the absolute temperature.

6.7. INTERPRETATION OF ISOTHERMS

The interpretation of experimental isotherms is a task of considerable difficulty and complexity. Supplementary information, such as a detailed knowledge of the interactions between adsorbed molecules, the distribution function describing the number of sites and their different adsorption potentials, etc., is usually not available. This lack is particularly apparent when the adsorbent is an evaporated film or a polycrystalline filament .Nevertheless, with increasing availability of single crystal planes, ultra-high vacuum equipment and highly precise methods of measurement, the accurate experimental determination of adsorption isotherms does provide data of value and interest.

REFERENCES

1. I. Langmuir, *J. Amer. Chem. Soc.*, 1918, **40**, 1361.
2. H. Freundlich, "Colloid and Capillary Chemistry", Methuen, London, 1926.
3. A. Frumkin and A. Slygin, *Acta Physicochim.*, 1935, **3**, 791.
4. R. H. Fowler, *Proc. Camb. Phil. Soc.*, 1935, **31**, 260.
5. B. G. Baker, *J. Chem. Phys.*, 1966, **45**, 2694.
6. J. C. Tracy and P. W. Palmberg, *J. Chem. Phys.*, 1969, **51**, 4852.
7. J. C. Tracy, *J. Chem. Phys.*, 1972, **56**, 2736, 2748.
8. T. B. Grimley, *Proc. Phys. Soc.*, 1967, **90**, 751; 1967, **92**, 776.
9. C. Aharoni and F. C. Tompkins, *Adv. Catalysis*, 1970, **21**, 1.

† See Chapter 9.

7

Heats of Chemisorption

7.1. CALORIMETRY

The heat of chemisorption, determined under precisely defined† conditions, is a quantity of fundamental thermodynamic importance. When the adsorbent is a filament or an evaporated film, heats have been invariably determined calorimetrically. In principle, they are of considerable accuracy but, in practice, the published values for any particular system are not in particularly good agreement. This situation arises because of two main difficulties, viz.: (i) the varying degrees of surface contamination of the metal specimens, particularly in the earlier work. In more recent investigations, surfaces with a high degree of cleanliness have been prepared under ultra-high vacuum conditions and contamination no longer presents a serious problem. (ii) The non-uniformity of the surface due to the presence of different crystal planes of the metal in varying proportions and because their surfaces contain micro- and macro-defects to different extents. Most of the heats of chemisorption recorded in the literature have been obtained using polycrystalline filaments and annealed evaporated films, and, hence, provide average values pertaining to surfaces of ill-defined physical characteristics. Nevertheless, the relative magnitudes for the same adsorbate on different metals, and for different adsorbates on a single metal, do have considerable value.

† See Chapter 8.

7.2. DIATHERMIC CALORIMETERS

7.2.1. Evaporated film calorimeters [1–6]

A film is deposited, under ultra-high vacuum conditions, on the internal surface of a glass cylinder (see Fig. 7.1). A wide diameter (> 3 cm) and

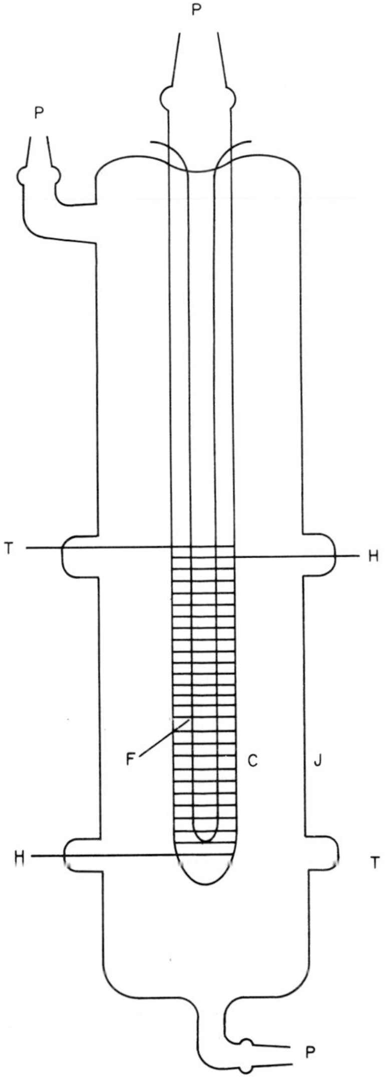

FIG. 7.1. Schematic representation of a calorimeter for measurement of heats of adsorption on a metal film. C, thin (0·3-mm thick) glass cylinder; J, glass envelope, T, resistance thermometer; ($d \sim 0{\cdot}1$ mm, $l \sim 1{\cdot}5$ m) coiled round C; H, heating coil; P, connections to pumps; F, evaporation filament to provide metal film.

short length (< 15 cm) favour: (i) uniform and rapid transport of gas to the entire external surface of the film; and (ii) the production of a film of uniform thickness and uniform degree of annealing, provided a well-designed evaporator is used. The minimum surface area of a film could be around 300 cm^2 (in practice, it is considerably larger) with a monolayer capacity of 6×10^{17} molecules, or 1×10^{-6} mol. For a chemisorption heat of 100 kJ mol^{-1} and 10% coverage, 1×10^{-5} kJ would be liberated. To maximize the accompanying rise of temperature of the calorimeter, its heat capacity is made as small as possible. The thickness of the wall of the glass cylinder is usually about 0·3 mm, which is sufficiently robust to withstand a static differential pressure of 1 atm. The thermometer normally comprises a platinum or tungsten wire of small diameter (~0·1 mm) and length 1·5 m, with a total resistance of *ca* 100 Ω. It is mechanically wound around the cylinder with constant pitch and tension; if the wire is dipped into a solution of sodium silicate, it can be cemented to the walls by careful heating. A typical calorimeter has a heat capacity of 44 J K^{-1}; its temperature rise would, therefore, be *ca* $2{\cdot}5 \times 10^{-3}$ K. The temperature coefficient of resistance of platinum is $4 \times 10^{-3}\,\Omega\,K^{-1}$. A resistance change of $1 \times 10^{-3}\,\Omega$ per 100 Ω is, therefore, obtained. A d.c. Wheatstone bridge, designed for sensitivity, linked to a high precision galvanometer, provides sufficiently accurate measurements, but a.c. bridges, particularly with automatic feedback techniques to maintain continuous balance of the bridge, are now preferred.

The calorimeter is surrounded by a glass jacket, evacuated to 10^{-3} Pa or less, and this, in turn, is enclosed in a thermostat of conventional design. Evacuation of the jacket is essential to reduce loss of heat by gas conduction to the thermostat to around 10^{-8} J s^{-1}; and radiation losses are minimized by silvering the inside wall of the jacket. Other conduction losses are: (i) along the leads of the thermometric (and calibration) windings; (ii) along the glass walls of the calorimeter; and (iii) through the residual gas phase in the calorimeter to the thermostat. Such losses are typically $< 10^{-6}$ J s^{-1} and, since radiation losses are of order 10^{-4} J s^{-1}, conduction errors can be neglected for a well-designed calorimeter.

7.2.1.1. *Calibration of the calorimeter* [7–10]

In the calibration of the calorimeter, a measured quantity of heat is generated by passing a current either: (a) through a separate heating coil wound around the cylinder between the thermometric windings; or (b) through the film itself by using electrodes such as platinum wire rings at the lower and upper boundary of the film, or movable wire brushes in contact with the film. Internal electrodes allow the generation of heat within the film, thereby producing conditions closely corresponding to the heat developed by the chemisorption process. Both steady-state and pulse methods are used;

the latter method is preferred since it more closely simulates the short time-period over which most chemisorption heats are evolved.

7.2.1.2. *Steady-state method*
A steady state is set up by application of an external e.m.f. E to the heater coil of resistance R to give a constant rate of input of electrical energy, dQ/dt. The calorimeter attains a constant temperature difference ΔT above that of the thermostat when the heat input equals the loss due to radiation. Hence, assuming Newton's law of cooling,

$$dQ/dt = E^2/R = -k\,\Delta T, \tag{7.1}$$

k, the cooling constant, being in J s^{-1} K^{-1}. Since the galvanometer deflexion Δx is proportional to ΔT, i.e. $\Delta x = a\,\Delta T$,

$$|\Delta x| = \frac{a}{k}\frac{dQ}{dt} = \frac{a}{k}\frac{E^2}{R}. \tag{7.2}$$

The quotient a/k can be evaluated since E and R are known. On termination of the heat input, the calorimeter of heat capacity C cools according to the relationship,

$$\frac{d(\Delta T)}{dt} = \frac{dQ}{C\,dt} = \frac{1}{C}(-k\,\Delta T). \tag{7.3}$$

On rearrangement and integration,

$$\int_{\Delta T_m}^{\Delta T} \frac{d(\Delta T)}{\Delta T} = \frac{-k}{C}\int_0^t dt, \tag{7.4}$$

or, since $\Delta T = \Delta x/a$,

$$\ln(\Delta x/\Delta x_m) = -(k/C)t, \tag{7.5}$$

in which ΔT_m is the temperature difference on termination of the input energy and Δx_m is the corresponding galvanometer deflexion. The term $-k/C$ is the slope of the linear $\ln(\Delta x)$ against t plot; since a/k is known, C/a can be evaluated (see Fig. 7.2).

7.2.1.3. *Pulse method*
A measured pulse of energy, insufficient for attainment of a steady-state, is fed into the heater wire. The deflexion is recorded during both heating and cooling periods; the above procedure for evaluation of C/a, appropriately modified, is then employed.

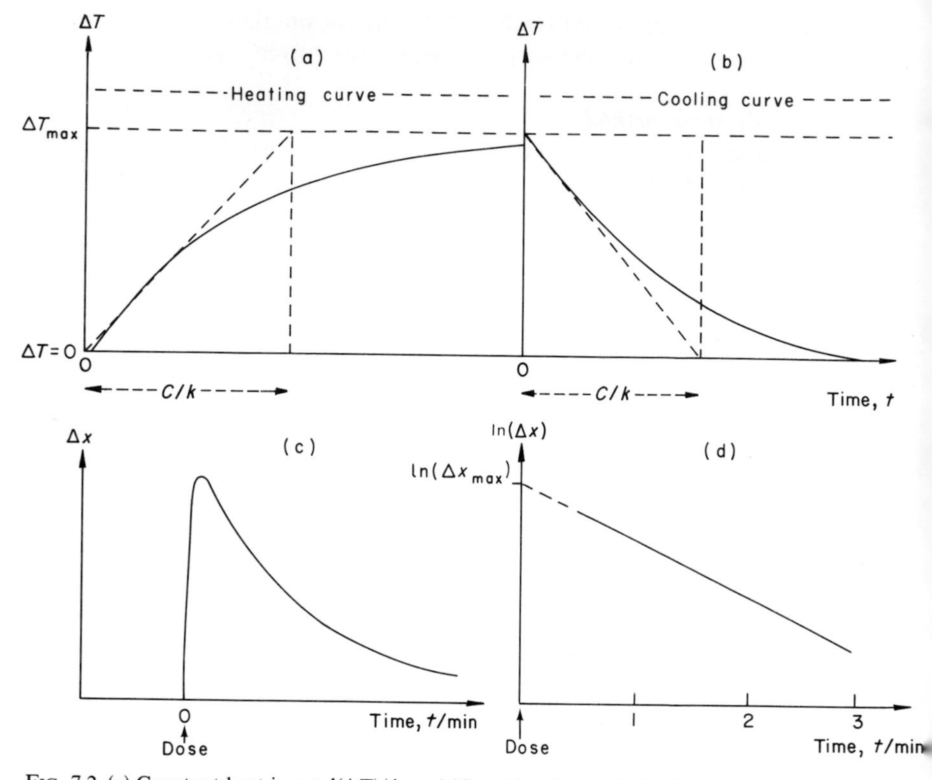

FIG. 7.2. (a) Constant heat input $d(\Delta T)/dt = k/C = 1/\tau$, where τ is the time constant or time to reach the maximum temperature rise. (b) Termination of input energy $d(\Delta T)/dt = 0$ and cooling of calorimeter; initial slope is $-k/C$, since $d(\Delta T/\Delta T_{max}) = -k\,dt/C$, or $\Delta T = \Delta T_{max}\exp(-kt/C)$. (c) Galvanometer deflection Δx as a function of time showing the form of the cooling curve. (d) Cooling curve plotted as $\ln(\Delta x)$ $[= \ln(a\Delta T)]$ against t after admission of gas at $t = 0$; a short extrapolation of the linear plot of slope $-k/C$ gives $\ln(\Delta x_{max})$.

7.2.2. Evaluation of adsorption heats

The heat liberated when a measured dose of gas is chemisorbed causes an initial rise of temperature of the calorimeter $\Delta T_i = \Delta x_i/a$, in which Δx_i is the initial deflection of the galvanometer on admission of gas to the film at $t = 0$. The deflection is less than the maximum value Δx_m which would have been attained under strictly adiabatic conditions, i.e. for zero heat losses. In practice, there is loss by radiation to the thermostat so that

$$\frac{dQ}{dt} = k\,\Delta T + \frac{C\,d(\Delta T)}{dt}. \qquad (7.6)$$

and the total amount of heat evolved is given by

$$Q = k\int_0^t (\Delta T)\,dt + C\,\Delta T_1 - \frac{k}{a}\int_0^{t_0} (\Delta x)\,dt + \frac{C}{a}(\Delta x_i), \tag{7.7}$$

where t_0 is the time taken for the calorimeter to return to its original temperature before admission of adsorbate. Hence,

$$Q = (C/a)(\Delta x_i) + (k/a)\int_0^{t_0} (\Delta x)\,dt. \tag{7.8}$$

The leakage term, given by $(k/a)\int_0^{t_0}(\Delta x)\,dt$, is evaluated by graphical integration. It is necessary to confirm experimentally that there is negligible drift in the calorimeter temperature before admission of gas, or that the drift is sufficiently small and regular to allow an accurate correction to be made for this variation.

Alternatively, $\ln(\Delta x)$ can be plotted against t over the initial part of the cooling period; a linear plot of slope k/C should result. The value of Δx_m is then obtained by a short extrapolation to $t = 0$, the time of admission of gas. Since a/k is the known calibration factor of the calorimeter,

$$Q = (k/a)\,\Delta x_m/(k/C) = (C/a)\Delta x_m \tag{7.9}$$

The method is most accurate for negligible heat losses by conduction and small losses by radiation.

For slow rates of chemisorption, e.g. by admitting gas at a very low and continuous rate, an isothermal procedure can be employed [11]. A constant rate of energy input is imposed in order to maintain the calorimeter at a steady-state temperature slightly higher than that of the surrounding thermostat. On admission of gas, evolution of heat accompanying the chemisorption takes place, but an automatic timing and relay system is incorporated which switches off the energy input during a series of time intervals; thus, the steady-state temperature difference is accurately maintained. The product of the integrated time intervals during which the heater has been switched off and the constant rate of input of energy then gives the heat liberated in the chemisorption process.

7.3. DIFFERENTIAL OR TWIN CALORIMETERS [12–15]

Two calorimeter vessels are constructed to be as nearly identical as possible; one contains the adsorbent and the other acts as a reference calorimeter.

Temperatures are measured by means of several hundred chromel–constantin thermocouples regularly arranged in series on the outside of each of the calorimetric vessels. The reference junctions of each set of thermocouples are connected to a massive metal block which is efficiently thermostatted. The e.m.f. generated by each set is arranged to be in opposition, and the differential current is fed into a continuously-recording sensitive galvanometer. Each calorimeter comprises a thin-walled glass vessel, surrounded by a thin-walled stainless-steel jacket, the space between being filled with a highly heat-conducting material. Around the jacket is a thin layer of mica in which the thermocouples are embedded. The heat liberated in the adsorption vessel is conducted from the calorimeter to the metal block by the thermocouples and the temperature of the calorimeter as it returns finally to its original value is continuously recorded. Graphical integration of the plot of galvanometer deflexion against time then permits the calculation of the heat of chemisorption. The great advantage of twin calorimeters is that the equalization of the temperature of the two cells can be achieved with considerable precision. Consequently, very high stability as well as high sensitivity can be attained. Measurements under isothermal conditions can also be made by employing some of the thermocouples to generate heat by the Peltier effect in the reference cell and, thereby, exactly to compensate that evolved in the adsorption calorimeter. Alternatively, the differential current can be nullified by generating heat within the reference cell by resistance heating.

7.4. FILAMENT CALORIMETERS

The filament is employed both as adsorbent and as a resistance thermometer [16]. The filament, typically of diameter 10^{-2} cm and 20 cm in length, is cleaned by high-temperature flashing and, after cooling, forms one arm of a Wheatstone bridge. When the zero drift of the bridge galvanometer is sufficiently small and uniform, a measured dose of gas is admitted. The heat generated rapidly by the chemisorption is accompanied by a temperature rise of 10^{-2} to 10^{-3} °C, and the galvanometer deflexion quickly attains its maximum value and then decreases as the filament cools. The gas pressure after adsorption is recorded; the vessel is then immediately evacuated to avoid thermal disturbances arising from conduction of heat by the gas from the filament to the walls of the adsorption cell. Calibration is effected by passing a small quantity of electrical energy (mA s) through the filament to generate a known amount of heat and recording the initial galvanometer deflexion and its decrease as the filament cools to its original temperature. Alternatively, the filament resistance is adjusted by shunting it, until the heating current exactly reproduces the temperature rise developed in the adsorption process. Since the surface area of a filament is only a few cm^2, the amount adsorbed is very small and the accuracy of measurement is comparatively low.

Considerable improvement in the design of the apparatus has been effected by the introduction of modern electronic and other devices. For example, an a.c. Wheatstone bridge capable of measuring relative resistance changes of $10^{-6}\,\Omega$ or a temperature rise of $2 \times 10^{-4}\,°C$ was used by Kisliuk [17],† and the heat evolved in the chemisorption of an 0·2% monolayer of nitrogen on tungsten was determined with reasonable accuracy. Similarly, the accuracy of measurement of the amount of gas adsorbed was improved by the use of an identical reference cell containing a second filament which could be isolated from, or connected to, the adsorption cell. A dose of gas was admitted to both cells to give a pressure of $\sim 10^{-7}$ Pa after both ribbons had been flashed. On cooling, adsorption took place. The reference filament was then flashed in order to generate a higher pressure in both cells, with consequent additional adsorption by the adsorbing filament. The adsorption cell was then isolated, its filament flashed, and the pressure again recorded. From these pressure changes and the volumes of the cells, the amount adsorbed was evaluated. The adsorption isotherm was traversed by varying the magnitude of the initial dose of gas.

The method is restricted to those metal filaments that can be effectively cleaned by high-temperature flashing. Furthermore, adsorption must be rapid and, preferably, associated with a high heat of chemisorption, so that the equilibrium pressures in the adsorption cell are very low; otherwise, the filament effectively operates as a Pirani gauge and large errors are introduced. Finally, strict uniformity of the filament, and stability of the surface structure at the high temperatures of flashing, are essential to obtain reproducible results of acceptable accuracy.

7.5. NON-CALORIMETRIC METHODS

7.5.1. Equilibrium measurements

A series of adsorption isotherms are determined at different temperatures. Measurement of the decrease in pressure of the gaseous adsorbate, brought about by its chemisorption in a constant volume system under isothermal conditions after introduction of a measured quantity of the gas, is the most accurate procedure when the adsorbent has a sufficiently large surface area (several 100 cm^2). For filaments with areas of 1 to 10 cm^2, complete desorption into a closed system is effected by high temperatures and the increase

† See also Eley and Norton [18].

of pressure recorded by a sensitive ionization gauge. In other methods, the amount adsorbed is directly measured, e.g. using Auger spectroscopy, surface potential changes [19–21], etc., after suitable calibration of the measuring device. These methods have been applied in investigations involving one single crystal plane of a metal adsorbent.

The basis of the volumetric method is the application of the equation,

$$q_{st} = RT^2 \left(\frac{\partial \ln P}{\partial T} \right)_{n_a}$$

which may be transformed to

$$[\partial \ln P / \partial (1/T)]_{n_a} = -q_{st}/R. \tag{7.10}$$

The isostere is usually obtained by interpolating P and T values at a constant amount adsorbed n_a from a set of isotherms determined at a series of different temperatures.

The reversibility of the chemisorption should always be experimentally confirmed by approaching the isotherm temperatures from a series of higher temperatures until reproducible results for the amounts adsorbed are obtained. The slope of the isostere, plotted as $\ln P$ as a function of $1/T$ for an amount adsorbed n_a, gives $-q_{st}/R$, where q_{st} is the isosteric, differential, heat of adsorption at the coverage n_a. From a set of isosteres, obtained by P, T interpolation at different values of n_a, the variation of heat with coverage is then obtained.

7.5.2. Kinetic measurements†

The activation energy of desorption E_d is equated to the sum of the heat of chemisorption and the activation energy E_a of the adsorption process. For many gas/metal systems, E_a does not differ greatly from zero, but, as for the chemisorption of nitrogen on iron, it is sometimes quite substantial. A separate experimental determination of E_a, i.e. of the temperature coefficient of the chemisorption rate constant, is then necessary.

The experimental methods employed to obtain rates of desorption include isothermal and non-isothermal programmed procedures in both closed and flow systems; these have been fully discussed in Chapter 5. Since E_d is frequently a function of coverage, the procedure already outlined for such situations should be followed, if only to confirm the constancy of E_d over the investigated range of coverage. The accuracy of the results is less than that attained in calorimetric and equilibrium methods and, since, of necessity,

† See Chapter 5.

the measurements are concerned with non-equilibrium states of the adsorbed layer, the chemisorption heats are ill-defined.

Other methods have been used, but the results are of poor accuracy. These include the change of emission pattern emanating from the tip of a field-emission microscope [22], but both the temperature of the tip and the amount adsorbed are uncertain. Results of an indefinite and ambiguous character have also been obtained, using field desorption in the field-ion microscope, but the theoretical basis of this procedure has been seriously criticized. The difficulties of obtaining acceptable values of desorption heats using electron-stimulated desorption have already been discussed in Chapter 5.

7.6. GENERAL CONSIDERATIONS

In the calorimetric determination of the heats of chemisorption, it is assumed that the final configuration of the adsorbed layer is the equilibrium state. However, the sticking probability of gaseous adsorbates on metal surfaces is high, particularly at low coverages, and the binding energy of the adspecies is substantial. The initial configuration is, therefore, random and only conforms to an equilibrium state for a uniform surface and in absence of interactions between the adspecies. In practice, chemisorption heats decrease with increase of coverage and often indicate that the surface is heterogeneous. The approach to equilibrium is almost entirely brought about because of the mobility of the adspecies. For a mobility activation energy of 126 $kJ\,mol^{-1}$, an approximate magnitude that would apply to adatoms of nitrogen, oxygen and carbon monoxide on a tungsten surface, there would be virtually no departure from the initial random non-equilibrium within the time-interval for calorimetric measurements at room temperature. Consequently, the measured heats would refer to a random selection of non-uniform sites and a constant heat independent of coverage would be obtained. For films, this constancy has been recorded since the sites on the outer surface are filled first and the inner layers in turn later.

7.7. THEORETICAL CALCULATIONS OF CHEMISORPTION HEATS AND BOND ENERGIES

In systems in which the molecular nature of the adspecies is the same as that in the gas phase, as for the chemisorption of carbon monoxide at room temperature on many transition metals, the binding energy of the adspecies to the surface of the metal adsorbent is derived from the plot of the experimental differential isosteric heat of chemisorption, q_{st}, as a function of the surface concentration of adspecies, by extrapolating q_{st} to zero coverages, thereby eliminating the contribution of lateral interactions between the adspecies (assuming a uniform surface). For some diatomic molecules X_2, such as hydrogen, oxygen, nitrogen, the chemisorption is dissociative and the binding energy of the adatom $E(M—X)$ is then given by

$$E(\mathrm{M—X}) = \tfrac{1}{2}[q_{st} + D(\mathrm{X—X})], \tag{7.11}$$

where $D(X—X)$ is the dissociation energy of the gaseous molecule X_2.

7.7.1. The Eley method

Eley [23] has adapted the Pauling equation [24] for calculating bond energies in free molecules and obtained theoretical estimates of adsorption heats. His basic equation is

$$E(\mathrm{M—X}) = \tfrac{1}{2}[E(\mathrm{M—M}) + D(\mathrm{X—X})] + 100(X_\mathrm{M} - X_\mathrm{X})^2. \tag{7.12}$$

The final term is the ionic contribution to the bond energy in terms of the electronegativities $X_M - X_X$ of the two free atoms M and X, the numerical factor 100 being used to convert to units of $\mathrm{kJ\,mol^{-1}}$. An approximation, discussed by Pauling [24], is to equate $X_M - X_X$ to the dipole moment μ of the M—X bond, where μ is derived from surface potential measurements. The change in work function $\Delta\Phi$ caused by the chemisorption of n adspecies per cm^2 is $\Delta\Phi = 4\pi n\mu$; $\Delta\Phi$ is measured for increasing values of n, and μ is derived from the slope of the plot as $n \to 0$. In this way, errors due to the mutual depolarization of parallelly aligned dipoles, evident at higher coverages, are eliminated.

An alternative procedure is to express the electronegativity of the surface metal atom M in terms of its electron affinity and first ionization potential [25], both of which are equal to the work function of the metal Φ so that

$$X_\mathrm{M} = 0{\cdot}355\Phi, \tag{7.13}$$

in which the numerical scaling factor converts Φ_M in eV to Pauling's scale of electronegativities. Better agreement with experimental values of the heat of adsorption of hydrogen on metals is usually given by this procedure than by the Eley method.

The single bond energy $E(M—M)$ was calculated from the latent heat of sublimation E_s. For f.c.c. metals, each atom has 12 nearest neighbours. Since two atoms are involved in each bond, $E(M—M)$ was equated to $E_s/6$. For b.c.c. metals with eight nearest neighbours and six next-nearest neighbours, the same numerical factor was proposed by Eley as a reasonable approximation, although Higuchi *et al.* [26] prefer a value of $E_s/4$. There is little theoretical justification for either assumption; moreover, the postulates are basically inconsistent with Pauling's valence bond theory.

From Eqs (7.11), (7.12) and (7.13) it follows that

$$q_{st} = E_s/6 + 200(0{\cdot}355\Phi - X_X)^2 \text{ kJ mol}^{-1}. \tag{7.14}$$

The last term represents 10 to 30% of the total contribution, so that some parallelism between the experimental heats of adsorption and the latent heats of sublimation of the metal adsorbent would be expected; but for the adsorption of hydrogen, for which there are the most extensive experimental determinations, no correlation is found.

For most transition metals, $\Phi \approx 4{\cdot}5$ eV, i.e. $X_M \approx 1{\cdot}2$ to $1{\cdot}6$ eV. Since X_H, X_O and X_N are 2·28, 3·17 and 2·33 eV, respectively, the dipole layer of these adsorbates should be negative outermost from the surface; this conclusion is confirmed experimentally in that the change of surface potential brought about by chemisorption is almost always negative. Pauling has expressed the fractional ionicity of bonds of free diatomic molecules by $[1 - \exp(0{\cdot}355\ \Phi - X_X)]$. Application of this relationship to chemisorbed bonds indicates that the ionic contribution is not high, e.g. for the H_2/W system the percentage ionicity is about 6%, i.e., the bonding is largely covalent.

7.7.2. The Higuchi, Ree and Eyring method

A different procedure for inclusion of the ionic contribution to the energy of the surface bond was devised by Higuchi *et al.* [26]. From the Schrödinger equation, they deduced that

$$\frac{E(\text{M—X}) - E(\text{M}^+\text{X}^-)_i}{E(\text{M—X}) - E(\text{M—X})_c} = \frac{1 - C}{C} \tag{7.15}$$

where C is the fractional ionic character. It was assumed equal to μ/er_{MX}, in which μ is the magnitude of the surface dipole and r_{MX} is the sum of the radius of the M atom and the covalent bond radius of the X atom. $E(\text{M—X})_c$ is the pure covalent bond energy derived from Eqn (7.12) with the final term equated to zero. $E(\text{M}^+\text{X}^-)_i$ is the pure ionic bond energy, the bond being formed by donation of an electron from the metal adsorbent to the adatom with liberation of energy equal to $A - I$ (where A is the electron affinity of the atom X, and I is the ionization potential of the M atom), to which must be added the Coulombic term $e^2/4r_{MH}$, e being the electronic charge. However, the calculated magnitudes of E(M—X) are not in better agreement with experimental values than those obtained using the Stevenson modification of the Eley procedure (see Table 7.1).

TABLE 7.1. Calometric heats[a] of chemisorption ($\theta \rightarrow 0$) for films and filaments at room temperature

Metal	H_2 Exptl	H_2 Calc.[b]	CO Exptl	CO Calc.[b]	N_2 Exptl	N_2 Calc.[b]
Cr	45	(16, 24, 21)				
Ta	45	(33, 50, 55)	[128][c]		104	(194, —, 153)
W	45–46	(37,46,64)	82	(70, —, —)	95	(191, —, 178)
Mo	40	(28, 43, —)	[60]			
Fe	32–36	(19, 32, 24)	46	(60, —, 55)	40, 70	(149, —, 79)
Ni	31–35	(19, 29, 16)	42	(53, —, 51)		
Rh	26–28	(25, 32, 24)	[44]			
Pd	27	(17, 23, —)	[40]			
Pt	—		[44]			

[a] Units are kcal mol^{-1}.
[b] Calculated values are given (in order) for: (i) Eley's method; (ii) Stevenson's modification of Eley's method; and (iii) the method of Higuchi *et al.*
[c] Values in square brackets [ ] were obtained by a non-calorimetric method.

The surface bond is invariably covalent with some ionic contribution; an exception is the alkali metal adsorbate, caesium. For a pure ionic bond, the value of q_{st} is approximately given by

$$q_{st} = \Phi - I + e^2/4r_{X^+}. \tag{7.16}$$

Energy is liberated by transference of an electron from the adatom X to the metal adsorbent in an amount $\Phi - I$, where Φ is the work function of the metal and I the ionization potential of the free X atom. The final term is the electrostatic image energy arising from the presence of the ion X^+ at a distance r_{X^+} (the radius of the ion X^+ in a crystal of the adsorbate) from the plane of the metal adsorbent. Agreement of the calculations for Cs (and to a less extent, Na) on W with experimental values is fairly good.

Despite the fact that the values of the metal–hydrogen bond energies calculated by Pauling for alkali–metal hydride molecules differ considerably from the experimental values, the calculated values of the chemisorption H—M bond show reasonable agreement with those derived from the experimental heats of chemisorption of hydrogen on various metals. However, little faith should be placed on this fact, since the theoretical deductions based on such agreement have little foundation. Similarly, attempts to extend such calculations to other adsorbates have been quite unsuccessful. For example, for N_2/metal systems, Eqn (7.2) has been modified by replacement of the term E(M—M) by E(M≡M) and given the value of 3E(M—M). Such radical additional assumptions cannot be justified without further theoretical underpinning (see Table 7.1).

7.7.3. The bond-energy bond-order (BEBO) model [27]

Since the heats of chemisorption of a single adsorbate on the different crystal planes of a metal adsorbent are usually different, the effect of surface orbital geometry, e.g. as predicted from crystal field theory, has recently been explored. In the BEBO model, the variation in bond energy with bond order of chemical bonds involved in the chemisorption process has been obtained from the spectroscopic correlation of gaseous species. For the adsorbate–adsorbent bond, a linear relationship between bond energy and bond order is assumed. The geometry of the surface orbitals is predicted from ligand field theory, and their electron occupancy is derived from the electronic configuration according to the Engel–Brewer rules, e.g. for f.c.c. metals it is $d^{n-3}sp^2$, where n is the number of electrons in the outer shell. Finally, the bonding electrons are assumed to be equally distributed over all the relevant s, p and d orbitals, and conservation of electrons (or chemical bonds) must be ensured.

An example given [27] is the chemisorption of hydrogen on the (111) plane of platinum, a f.c.c. metal having the electronic configuration $5d^7 6sp^2$. For this plane, there are two non-equivalent bonding sites, e_g and t_{2g} each comprising three orbitals, the former making an angle of 35°16′ with the surface plane and the latter an angle of 54°44′. Interaction of the hydrogen molecule is favoured at two adjacent t_{2g} sites; the H—H bond is stretched and Pt—H bonds are formed by using the available bonding electrons. This complex can dissociate and the H adatoms then diffuse to the e_g sites at which atomic chemisorption is favoured.

The binding energy in the adsorbed state is

$$V = D_{H_2} - 2E_{Pt—H, s} - E_{H_2, s} + \Delta E_{Pt—Pt}, \tag{7.17}$$

in which D_{H_2} is the dissociation energy of H_2(g) = 103·2 kcal mol^{-1}, $\Delta E_{Pt—Pt}$ is the energy to break Pt—Pt bonds, $E_{Pt—H, s}$ is the binding energy of the H adatom to the metal, and $E_{H_2, s}$ is the energy of the H_2 molecule adsorbed to the surface. From the bond-energy bond-length correlation,

$$E_{H_2} = 103{\cdot}2n_{H_2}, \tag{7.18}$$

$$E_{H_2} = E_{H_2,s}n_{H_2}^{x}, \quad \text{(where } x \simeq 1) \tag{7.19}$$

and

$$E_{Pt-H} = E_{Pt-H,s}n_{Pt-H}, \tag{7.20}$$

where n is the respective bond order, and $E_{Pt-H,s}$ is known to be 67 kcal mol^{-1}. Finally, conservation of chemical bonds requires that

$$n_{Pt-H} = 1 - n_{H_2}, \tag{7.21}$$

so that V can be calculated as a function of bond order or of the distance co-ordinate away from the surface.

Two minima, 31 kcal mol^{-1} for the binding energy of a H adatom, and 17 kcal mol^{-1} for the molecular state, are obtained. The Pt—H bond length is 1·81 Å and is close to the hard-sphere value of 1·68 Å.

Comparison of heats for the dissociative chemisorption obtained experimentally (in parentheses) and by BEBO calculations are:

H_2/Pt(111),	31 (31);
H_2/W(100),	32·2 (32·3);
O_2/Pt(111),	68 (64);
O_2/Ni(111),	114 (107);
CO/Pt(111),	28 (28).

The agreement is remarkably good, but the calculations are based on the insertion of numerical magnitudes that are closely related to the quantity that is being evaluated. The major uncertainty in the method is the value of the single bond energy to be used in calculating the gas–surface bond energy, and it is implicit in the treatment that a localized "surface molecule" is formed in chemisorption. The main advance is that the model gives the potential energy of interaction as a function of bond order or bond length, and takes into account the orbital geometry at the surface.

REFERENCES

1. O. Beeck, W. A. Cole and A. Wheeler, *Disc. Faraday Soc.*, 1950, **8**, 314.
2. M. Wahba and C. Kemball, *Trans. Faraday Soc.*, 1953, **49**, 1351.

3. J. Bagg and F. C. Tompkins, *Trans. Faraday Soc.*, 1955, **51**, 1071.
4. G. Wedler, *Z. Phys. Chem.*, (*Frankfurt*), 1960, **24**, 73; 1961, **27**, 388.
5. F. J. Bröcker and G. Wedler, *Disc. Faraday Soc.*, 1966, **41**, 87.
6. D. Brennan and J. J. Graham, *Disc. Faraday Soc.*, 1966, **41**, 95.
7. D. F. Klemperer and F. S. Stone, *Proc. Roy. Soc. A*, 1957, **243**, 375.
8. G. Ehrlich and T. W. Hickmott, *J. Chem. Phys.*, 1957, **26**, 219.
9. D. Brennan, D. O. Hayward and B. M. W. Trapnell, *Proc. Roy. Soc. A*, 1960, **256**, 81.
10. S. Cerný, V. Ponec and L. Hládek, *J. Catalysis*, 1966, **27**, 5.
11. A. V. Kiselev and G. G. Muttick, *Zhur. Fiz. Khim.*, 1961, **35**, 2153.
12. A. Tian, *J. Chim. Phys.*, 1933, **30**, 665.
13. E. Calvet and H. Prat, "Recent Progress in Microcalorimetry", Pergamon Press, Oxford, 1963.
14. P. C. Gravelle, *J. Chim. Phys.*, 1964, **61**, 455.
15. V. E. Ostrovskii, N. N. Dobrovolskii, I. P. Karpovich and F. J. Frolov, *Zhur. Fiz. Khim.*, 1968, **42**, 550.
16. J. K. Roberts, *Proc. Roy. Soc. A*, 1935, **152**, 445.
17. P. Kisliuk, *J. Chem. Phys.*, 1959, **31**, 1605.
18. D. D. Eley and P. R. Norton, *Proc. Roy. Soc. A*, 1970, **314**, 301 and 319.
19. J. C. Tracy and P. W. Palmberg, *Surface Sci.*, 1969, **14**, 274.
20. J. C. Tracy, *J. Chem. Phys.*, 1972, **56**, 2736, 2748.
21. C. S. Alexander and J. Pritchard, *J. Chem. Soc., Faraday Trans. I*, 1972, **68**, 202.
22. D. Menzel and R. Gomer, *J. Chem. Phys.*, 1964, **41**, 3111.
23. D. D. Eley, *Disc. Faraday Soc.*, 1950, **8**, 34.
24. L. Pauling, "The Nature of the Chemical Bond", Cornell University Press, Ithaca, 1960.
25. D. P. Stevenson, *J. Chem. Phys.*, 1935, **23**, 203.
26. I. Higuchi, T. Ree and H. Eyring, *J. Amer. Chem. Soc.*, 1955, **77**, 4969; 1957, **79**, 1130.
27. W. H. Weinberg and R. P. Merrill, *Surface Sci.*, 1972, **33**, 493.

8

Thermodynamics of Chemisorption

8.1. ISOSTERIC HEAT OF ADSORPTION

In physisorption, interest is almost entirely confined to the thermodynamic functions of the adsorbed layer as a separate independent phase, assuming that the adsorbent is completely unperturbed by the presence of adsorbate. For many such systems, this approximation is acceptable. In chemisorption, however, the outermost layer(s) of the solid is substantially perturbed and the system should be treated as a two-component condensed phase in equilibrium with a gaseous component at a given temperature and pressure, with the necessary assumption that the adsorbent is non-volatile. A treatment in terms of solution thermodynamics which closely follows that given by Hill [1, 2] has, therefore, been adopted in which subscripts A and 1 denote the adsorbent and adsorbate, respectively. The condensed phase comprises both adsorbent and adsorbate, and has a uniform composition comprising n_A moles of adsorbent and n_1 moles of adsorbate. The differential dn_A indicates an increase in n_A for the same state of subdivision and surface properties as the original adsorbent. Consequently, the surface area is proportional to n_A and is not an additional independent parameter.

For the condensed phase,

$$G = -S\,dT + V\,dP + \mu_1\,dn_1 + \mu_A\,dn_A, \tag{8.1}$$

which may be written in the form

$$d\mu_1 = -\bar{S}_1\,dT + \bar{V}_1\,dP + (\partial\mu_1/\partial\Gamma)_{T,P}\,d\Gamma, \tag{8.2}$$

in which μ_1, μ_A are the chemical potentials of the two components in the condensed phase at a temperature T, and a hydrostatic pressure P which is exerted by a hypothetical inert piston and equal to the equilibrium pressure of the gaseous adsorbate. S and V are the entropy and volume of the condensed phase; G is the Gibbs free energy.

$$\bar{S}_1 = (\partial S/\partial n_1)_{T,P,n_A} \quad \text{and} \quad \bar{V}_1 = (\partial V/\partial n_1)_{T,P,n_A} \tag{8.3}$$

represent partial molal derivatives; Γ is defined as n_1/n_A. At equilibrium,

$$\mu_1 = \mu_g, \tag{8.4}$$

where μ_g is the chemical potential of the gas phase. Since

$$d\mu_1 = d\mu_g,$$

then

$$-\bar{S}_1\,dT + \bar{V}_1\,dP + \left(\frac{\partial \mu_1}{\partial \Gamma}\right)_{T,P} d\Gamma = -S_g^m\,dT + V_g^m\,dP, \tag{8.5}$$

where $S_g^m = S_g/n_g$ and $V_g^m = V_g/n_g$. At constant Γ, $d\Gamma = 0$, and

$$\left(\frac{dP}{dT}\right)_\Gamma = (S_g^m - \bar{S}_1)/(V_g^m - \bar{V}_1). \tag{8.6}$$

For a perfect gas, $V_g^m \gg \bar{V}_1 = RT/P$; hence,

$$\left(\frac{\partial \ln P}{\partial T}\right)_\Gamma = (S_g^m - \bar{S}_1)/RT - (H_g^m - \bar{H}_1)/RT^2, \tag{8.7}$$

since $T(S_g^m - \bar{S}_1) = (H_g^m - \bar{H}_1)$ at equilibrium. The isosteric heat of adsorption is defined as $q_{st} = (H_g^m - \bar{H}_1)$ and is, therefore, given by

$$q_{st} = \left(\frac{\partial \ln P}{\partial T}\right)_\Gamma RT^2. \tag{8.8}$$

Since this equation has been derived in terms of n_A, it is valid although perturbation of the solid (which could include a change of the number of adsorption sites) may have been substantial during chemisorption.

The internal energy of the adsorbent in the absence of the gaseous adsorbate ($n_1 = 0$, $dn_1 = 0$) may be written as

$$dE_{0A} = T\,dS_{0A} - P\,dV_{0A} + \mu_{0A}\,dn_A \tag{8.9}$$

hence, for "practical" use, the following definitions are convenient:

$$E_s = E - E_{0A}, \qquad V_s = V - V_{0A}, \qquad S_s = S - S_{0A}, \qquad H_s = E_s + PV_s. \tag{8.10}$$

In general, the condensed phase has a very small pressure dependence, so that $H_s \approx E_s$ and

$$\bar{H}_1 = \left(\frac{\partial H}{\partial n_1}\right)_{n_A,P,T} = \left(\frac{\partial H_s}{\partial n_1}\right)_{n_A,P,T} \approx \left(\frac{\partial H_s}{\partial n_1}\right)_{n_A,T} \approx \left(\frac{\partial E_s}{\partial n_1}\right)_{n_A,T};$$

$$\bar{S}_1 \simeq \left(\frac{\partial S_s}{\partial n_1}\right)_{n_A,T}. \tag{8.11}$$

However, the total or integral thermodynamic functions of the condensed phase H, S, etc., are retained, because the corresponding molar integral quantities H^m, S^m, (i.e. H/n_1 and S/n_1), have direct statistical-mechanical significance. It is stressed that the isosteric heat derived from isotherms at different temperatures and constant values of Γ is a differential enthalpy given by

$$RT^2\left(\frac{\partial \ln P}{\partial T}\right)_\Gamma = H_g^m - \left(\frac{\partial H}{\partial n_1}\right)_{n_A,T} = q_{st}. \tag{8.12}$$

8.2. CALORIMETRIC HEATS

Calorimetrically determined heats of chemisorption, q, are differential heats, i.e. $q = \partial Q/\partial n_1$, where Q is the integral heat, $Q = \int_0^{n_1} q\,dn$. The precise meaning of q depends on the method of measurement, so that the system and surroundings must be unambiguously defined and the conditions under which the measurements are made must be carefully specified.

8.2.1. Differential heat of adsorption [3]

For an isothermal determination, the entire calorimetric system is immersed in a constant temperature bath and the heat of adsorption is the total heat transferred to the bath. For a constant-volume system, V_g, V, n_A are constants and, if the measurements is such that no external ($P\,dV$) work is done, then

$$dQ_{V_g, V, n_A, T} = d(E_g - E) = E_g^m \, dn_1 - \left(\frac{\partial E}{\partial n_1}\right)_{V_g, V, n_A, T} dn_1, \qquad (8.13)$$

or

$$q_d = \left(\frac{\partial Q}{\partial n_1}\right)_{V_g, V, n_A, T} = E_g^m - \left(\frac{\partial E}{\partial n_1}\right)_{V_g, V, n_A, T}, \qquad (8.14)$$

where q_d is the differential heat of adsorption. Since

$$q_{st} = H_g^m - \bar{H}_1 = E_g^m + PV_g - \left(\frac{\partial(E + PV)}{\partial n_1}\right)_{n_A, T}, \qquad (8.15)$$

then

$$q_{st} = q_d + RT \qquad (8.16)$$

for a perfect gas, since the term $\partial(PV)/\partial n_1$ is negligible.

8.2.2. Isothermal heat of adsorption [3]

The heat of adsorption is measured isothermally and reversibly and includes a $P\,dV$ work term. Moreover, since the gas pressure decreases from an initial value (no adsorption) to the lower equilibrium pressure (after adsorption), an additional term (the heat of compression) must be included. The heat transferred to the calorimeter is

$$Q = (E_g + PV_g) - (E + PV), \qquad (8.17)$$

or

$$dQ_{n_A, T} = dH_g + V_g \, dP = (dE + P\,dV + V\,dP); \qquad (8.18)$$

and

$$dE = T dS - P\,dV + \mu_1 \, dn_1.$$

Hence

$$dQ_{n_A, T} = dH_g + V_g \, dP - T\,dS + P\,dV - \mu_1 \, dn_1 - V\,dP$$
$$- H_g^m \, dn_1 - T\bar{S}_1 \, dn_1 - \mu_1 \, dn_1 + V_g \, dP, \qquad (8.20)$$

since $V\,dP$ is negligible, or

$$\left(\frac{\partial Q}{\partial n_1}\right)_{n_A, T} = H_g^m - T\bar{S}_1 - \mu_1 + V_g\left(\frac{\partial P}{\partial n_1}\right)_{n_A, T}. \qquad (8.21)$$

At equilibrium,

$$\mu_1 = \mu_g = H_g^m - TS_g^m; \tag{8.22}$$

hence,

$$q_i = \left(\frac{\partial Q}{\partial n_1}\right)_{n_A, T} = T(S_g^m - \bar{S}_1) + V_g\left(\frac{\partial P}{\partial n_1}\right)_{n_A, T} = q_{st} + V_g\left(\frac{\partial P}{\partial n_1}\right)_{n_A, T}. \tag{8.23}$$

The term $V_g(\partial P/\partial n_1)_{n_A, T}$ is the heat of compression, and can be evaluated from the slope of the isotherm. It is usually negligibly small compared with the large value of q_{st} in chemisorption, so that

$$q_i \simeq q_{st}. \tag{8.24}$$

8.2.3. Adiabatic heat of adsorption [4]†

In adiabatic calorimetry, the process takes place at constant total entropy, S_t.

The total internal energy E_t is

$$E_t = E_g + E + E_c, \tag{8.25}$$

where E_c is the internal energy of the calorimeter and adsorbent (in the absence of adsorbate), E is that of the adsorbed phase and E_g that of the gas phase.

The adiabatic heat of adsorption q_a is defined through the equation,

$$q_a = (C_c + n_g C_P^g + n_1 C_P^1)(\mathrm{d}T/\mathrm{d}n_1)_{S_t}, \tag{8.26}$$

where C_P^g is the molar heat capacity of the gas and C_P^1 that of the condensed phase, both at constant P and n_A; and C_c is the heat capacity of calorimeter and adsorbent. For constant total entropy, S_t,

$$\mathrm{d}E_c + \mathrm{d}E_g + \mathrm{d}E + P\,\mathrm{d}V + P\,\mathrm{d}V_g = 0, \tag{8.27}$$

which may be transformed to

$$C_c\,\mathrm{d}T + (T\,\mathrm{d}S_g - P\,\mathrm{d}V_g + \mu_g\,\mathrm{d}n_g) + (T\,\mathrm{d}S - P\,\mathrm{d}V + \mu_1\,\mathrm{d}n_1) + P\,\mathrm{d}V + P\,\mathrm{d}V_g = 0. \tag{8.28}$$

† For correction to Kingston and Aston's paper [4], see footnote on p. 75 of Young and Crowell [5].

At equilibrium, $\mu_g = \mu_1$; and $dn_g = -dn_1$. Hence, at constant total entropy,

$$C_c\,dT + T\,dS_g + T\,dS = 0 \tag{8.29}$$

The entropy of the gas may be written as $f(P, T, n_g)$, so that

$$dS_g = \left(\frac{\partial S_g}{\partial T}\right)_{P,n_g} dT + \left(\frac{\partial S_g}{\partial P}\right)_{T,n_g} dP + \left(\frac{\partial S_g}{\partial n_g}\right)_{T,P} dn_g. \tag{8.30}$$

Similarly, $S = f(P, T, n)$, so that

$$dS = \left(\frac{\partial S}{\partial T}\right)_{P,n_1} dT + \left(\frac{\partial S}{\partial P}\right)_{T,n_1} dP + \left(\frac{\partial S}{\partial n_1}\right)_{T,P} dn_1. \tag{8.31}$$

From the Maxwell relationships,

$$\left(\frac{\partial S}{\partial P}\right)_{T,n_1} = -\left(\frac{\partial V}{\partial T}\right)_{P,n_1} \simeq 0 \text{ (since } V \text{ is essentially constant)}, \tag{8.32}$$

and

$$\left(\frac{\partial S_g}{\partial P}\right)_{T,n_g} = -\left(\frac{\partial V_g}{\partial T}\right)_{P,n_g} = -\frac{V_g}{T} \text{ (for a perfect gas)}. \tag{8.33}$$

The heat capacities of the gaseous adsorbate and of the condensed phase at constant pressure are

$$C_P^g = \frac{T}{n_g}\left(\frac{\partial S_g}{\partial T}\right)_{P,n_g} \quad \text{and} \quad C_P^1 = \frac{T}{n_1}\left(\frac{\partial S}{\partial T}\right)_{P,n_1}. \tag{8.34}$$

Hence,

$$dS_g = \frac{n_g C_P^g\,dT}{T} - \frac{V_g\,dP}{T} - S_g^m\,dn_1 \tag{8.35}$$

and

$$dS = \frac{n_1 C_P^1\,dT}{T} + \bar{S}_1\,dn_1. \tag{8.36}$$

Combination of Eqs (8.35), (8.36) and (8.28) leads to

$$C_c\,dT + n_g C_P^g\,dT + n_1 C_P^1\,dT - V_g\,dP + T\bar{S}_1\,dn_1 - TS_g^m\,dn_1 = 0. \tag{8.37}$$

But

$$q_A = (C_c\,dT + n_g\,C_g^P\,dT + n_1 C_P^1\,dT)/dn_1 \tag{8.38}$$

and

$$q_{st} = TS_g^m - \overline{T}\overline{S}_1; \tag{8.39}$$

hence,

$$q_a = q_{st} + V_g\left(\frac{\partial P}{\partial n_1}\right)_{S_t}. \tag{8.40}$$

At constant n_A (and S_t), $n_1 = f(P, T)$ or $P = f(n_1, T)$; hence,

$$(dP)_{S_t} = \left(\frac{\partial P}{\partial T}\right)_{n_1, n_A}(dT)_{S_t} + \left(\frac{\partial P}{\partial n_1}\right)_{T, n_A}(dn_1)_{S_t}, \tag{8.41}$$

or

$$\left(\frac{\partial P}{\partial n_1}\right)_{S_t} = \left(\frac{\partial P}{\partial T}\right)_{n_1, n_A}\left(\frac{\partial T}{\partial n_1}\right)_{S_t} + \left(\frac{\partial P}{\partial n_1}\right)_{T, n_A}. \tag{8.42}$$

The adiabatic heat q_a is, therefore, related to the isosteric heat through Eq. (43):

$$q_a = q_{st} + V_g\left(\frac{\partial P}{\partial n_1}\right)_{T, n_A} + V_g\left(\frac{\partial P}{\partial T}\right)_{n_1, n_A}\left(\frac{\partial T}{\partial n_1}\right)_{S_t}. \tag{8.43}$$

8.3. ENTROPY OF THE SURFACE PHASE [6]

The entropy of the two-component surface phase (n_1 moles adsorbate, n_A moles adsorbent) is S and the integral molar entropy $S^m = S/n_1$. The quantity obtained experimentally is the differential molar entropy $\bar{S}_1$, where

$$\bar{S}_1 = (\partial S/\partial n_1)_{T, P, n_A} = n_1(\partial S^m/\partial n_1)_{T, P, n_A} + S^m_{n_1}. \tag{8.44}$$

The isosteric heat of adsorption $q_{st} = H_s^m - \overline{H}_1$, where H_g^m is the molar enthalpy of the gas, and $\overline{H}_1$ is the differential molar enthalpy of the surface phase for n_1 moles of adsorbate. The differential molar entropy is $S_g^m - \bar{S}_1$.

For the reversible isothermal transfer of n_1 moles of adsorbate from the gas phase at 1 atm pressure to the surface phase, the change in the differential molar free energy ΔG is

$$\Delta G = -(H_g^m - \overline{H}_1) + (S_g^m - \bar{S}_1)\,T = -q_{st} + (S_g^m - \bar{S}_1)\,T, \tag{8.45}$$

where $\Delta G = RT \ln P$ and P is the equilibrium pressure. The system is maintained at a constant temperature T K; and n_1 moles of adsorbate have been adsorbed. The value of ΔG is calculated from the equilibrium pressure. The isosteric heat q_{st} is evaluated from isosteres experimentally determined over a range of temperatures. The integral molar entropy S_g^m of the gaseous adsorbate at 1 atm pressure at a temperature T is calculated from statistical mechanics when not available from thermodynamic tables. Using Eq. (8.45), the "experimental" differential molar enthalpy $\bar{S}_1$ is evaluated.

However, the quantity of direct statistical-mechanical significance is the integral molar entropy $S_{n_1}^m$ obtained by integrating $\bar{S}_1$ from $n_1 = 0$ to $n_1 = n_1$, i.e.

$$S_{n_1}^m = (1/n_1) \int_0^{n_1} \bar{S}_1 \, dn_1. \tag{8.46}$$

Hence, from Eqs (8.45) and (8.46),

$$S_{n_1}^m = S_g^m - (R/n_1) \int_0^{n_1} \ln P \, dn_1 - (1/n_1 T) \int_0^{n_1} q_{st} \, dn_1. \tag{8.47}$$

After graphical integrations of the plot of $\ln P$ as a function of n_1 between the limits 0 and n_1 from the isotherm at T K, and of the plot of q_{st} as a function of n_1 between the same limits, $S_{n_1}^m$ can be evaluated. These $S_{n_1}^m$ values refer to the two-component phase and can only be identified as the integral molar entropies of the adsorbed molecules after making some physically unrealistic assumptions, viz.: (i) the perturbation of the surface metal atoms by chemisorption is negligible (i.e. the adsorbent is completely inert); and (ii) there are no interactions within the adsorbed layer. Nevertheless, "experimental" $S_{n_1}^m$ values have been used in an attempt to differentiate between mobile and immobile adspecies.

8.3.1. Entropy as a criterion of mobility [6]

In forming an immobile adspecies, the three translational degrees of freedom of the free molecule are transformed to three vibrational modes of the admolecule. The vibration frequency ν_z perpendicular to the surface can be estimated using the chemisorption bond energy to obtain the force constant of the vibration, assumed to be simple harmonic motion. The heat of chemisorption of carbon monoxide on tungsten varies with the crystal plane, but an average value [7, 8] is $\sim 300 \text{ kJ mol}^{-1}$ corresponding to a vibration frequency ν_z of $2 \times 10^{12} \text{ s}^{-1}$. The vibrational entropy contribution [9] is

$$S_z = R(h\nu_z/kT)[(\exp(h\nu_z//kT) - 1)^{-1} - \ln(1 - \exp(h\nu_z/kT))] \tag{8.48}$$

or $17 \text{ J mol}^{-1} \text{ K}^{-1}$ at 300 K.

The adspecies also vibrates in the (x, y) surface plane with frequencies ν_x and ν_y. It is confined to a potential well of barrier height equal to the activation energy for migration [10], which is about 80 kJ mol^{-1}. Using the same procedure as for $\nu_z, \nu_x = \nu_y = ca\ 1 \times 10^{12}\ \mathrm{s}^{-1}$ and their total entropy contribution is 42 J mol^{-1} K^{-1} at 300 K.

There is also a configurational entropy S_c term. For an adsorbed layer comprising N adsorbed molecules distributed amongst M sites ($M > N$),

$$S_c = kT \ln[M!/N!(M - N)!]. \tag{8.49}$$

Application of Stirling's theorem with $\theta = N/M$, gives

$$S_c = R[\ln \theta + [(1 - \theta)/\theta] \ln(1 - \theta)] \tag{8.50}$$

and, for $\theta = 0{\cdot}5$, $S_c = 13$ J mol^{-1} K^{-1}. The total entropy contribution is, therefore, 72 J mol^{-1} K^{-1}.

For a freely mobile adspecies, $S_c = 0$, and the partition function of the adsorbed layer is $[(2\pi mkT)\, eA]^N/h^2N$, assuming that the layer approximates to a two-dimensional gas. The average area available to an adsorbed CO molecule, given by $a = A/N$, at $\theta = 0{\cdot}5$, is $\sim 15 \times 10^{-16}\ \mathrm{cm}^2$ and the total entropy contribution [9] S_m is

$$S_m \approx R \ln(MTa) + 276 \approx 67\ \mathrm{J\ mol^{-1}\ K^{-1}} \tag{8.51}$$

at 300 K, compared with 72 J mol^{-1} K for an immobile adspecies.

The contribution of rotational degrees of freedom of the adspecies is now considered. The usual assumptions are that an immobile adspecies has no rotational modes and a mobile one has retained those possessed by the free molecule. However, chemisorption is localized and the mobile adspecies is bonded to an adsorption site for most of its life and spends little time in translational motion across the surface . Its rotational entropy, therefore, closely approximates that of an immobile adspecies and, consequently, the difference between the rotational entropy contributions of a mobile and immobile adspecies is very small. It is, thus, evident that on the basis of the values, 72 for immobility and 67 J mol^{-1} K^{-1} for mobility, no distinction can be made. Moreover, the "experimental" value,† at $\theta = 0{\cdot}7$ and 280 K, is 42 J mol^{-1} K^{-1}; it has been quoted as evidence [11] of substantial mobility, whereas a field-emission microscopic investigation visibly demonstrates that migration is, in fact, highly restricted [12, 13]!

† See Chapter 9.

The use of "experimental" $S^{m}_{n_1}$ values as a criterion for mobility or immobility cannot, therefore, be recommended. The main reasons are: (i) they relate to the two-component surface phase and not to the adspecies; (ii) interaction between adspecies can be neglected only for highly dilute layers, whereas practical determinations have been carried out for $\theta \geqslant 0{\cdot}4$; (iii) the "theoretical" $S^{m}_{n_1}$ values refer to a uniform surface and are derived using highly approximate partition functions; and (iv) the entropy difference between a mobile and immobile layer is expected to be small. In particular, any calculation purporting to give the variation of the fractional number of mobile admolecules [14] as a function of coverage is valueless.

REFERENCES

1. T. L. Hill, *J. Chem. Phys.*, 1949, **17**, 507; *Trans. Faraday Soc.*, 1951, **47**, 376.
2. T. L. Hill, *Adv. Catalysis*, 1952, **4**, 211.
3. T. L. Hill, *J. Chem. Phys.*, 1949, **17**, 520.
4. G. L. Kingston and J. G. Aston, *J. Am. Chem. Soc.*, 1951, **73**, 1929.
5. D. M. Young and A. D. Crowell, "Physical Adsorption of Gases", Butterworths, London, 1962.
6. C. Kemball, *Adv. Catalysis*, 1950, **2**, 233.
7. T. A. Delchar and G. Ehrlich, *J. Chem. Phys.*, 1965, **42**, 2686.
8. D. L. Adams and L. H. Germer, *Surface Sci.*, 1971, **27**, 21.
9. C. Kemball, *Proc. Roy. Soc. A*, 1946, **187**, 73.
10. G. Ehrlich and F. G. Hudda, *J. Chem. Phys.*, 1961, **35**, 1421.
11. E. K. Rideal and B. M. Trapnell, *Proc. Roy. Soc. A*, 1951, **205**, 409.
12. R. Klein, *J. Chem. Phys.*, 1959, **31**, 1306.
13. R. Gomer, *Disc. Faraday Soc.*, 1959, **28**, 23.
14. E. K. Rideal and F. G. Sweett, *Proc. Roy. Soc. A*, 1960, **257**, 291.

9

Statistical Thermodynamics of Adsorption

Two types of monolayers are distinguished: (i) a mobile monolayer in which the adsorbed molecule is tightly bound to the surface of the adsorbent in the normal direction, but has complete freedom of movement over the surface (two-dimensional gas); and (ii) a localized monolayer in which each adsorbed molecule is bonded to definite points on the solid surface (adsorption sites); the adsorbed molecule has three degrees of vibrational freedom at the site and has lost all its translational motion (lattice gas) [1, 2].

9.1. THE IDEAL LOCALIZED MONOLAYER

The surface of the adsorbent comprises M equivalent, distinguishable, and independent sites arranged in a regular two-dimensional array; at equilibrium, N of these sites are occupied by molecules of adsorbate bound one per site. The adsorbed molecules are in equilibrium with the gas phase at a pressure P, the whole system being at a constant temperature T. It is assumed that there are no interactions between adsorbed molecules. The minimum energy required to evaporate, or desorb, an adsorbed molecule from its lowest energy state at a site in the monolayer (taken as energy zero) is denoted by χ; the partition function for the internal degrees of the molecule, including the vibrations relative to its mean position at the site, is q_i. The partition function of a single adsorbed molecule is q_A, the energy zero being the lowest internal state of the isolated gas molecule at rest at infinite separation from the surface. Both q_i and q_A are functions of temperature and

$$q_A = q_i \exp(\chi/kT) \tag{9.1}$$

Of the M sites, N are occupied by adsorbed molecules, and $M - N$ are unoccupied ($M \geqslant N$). Although the sites are equivalent, they are distinguishable as they form a two-dimensional lattice; hence, the configuration degeneracy is given by

$$M!/N!(M - N)! \tag{9.2}$$

The complete partition function for the assembly of N adsorbed molecules is, therefore,

$$Q_A(N, T) = M!q_A(T)^N/N!(M - N)!, \tag{9.3}$$

and the Helmholtz free energy, F_A, of the adsorbed layer is

$$F_A = -kT \ln Q_A(N, T) = -kT \ln\left\{\frac{M!q_A(T)^N}{N!(M - N)!}\right\}.$$

The use of Stirling's theorem ($\ln x! = x \ln x - x$) provides the alternative expression,

$$F_A = -kT\{M \ln M - N \ln N - (M - N) \ln (M - N) + N \ln q_A\}, \tag{9.5}$$

from which the chemical potential μ_A of the adsorbed molecules may be derived as

$$\mu_A = kT\left(\frac{\partial \ln F_A}{\partial N}\right)_{T,M} = -kT\left\{\ln\left(\frac{N}{M - N}\right) - \ln q_A\right\}. \tag{9.6}$$

The fractional coverage of the M sites by N adsorbed molecules is $\theta = N/M$, so that

$$\mu_A = kT \ln\left\{\frac{\theta}{(1 - \theta)q_A}\right\}. \tag{9.7}$$

Since the adsorbed phase is in equilibrium with the gas phase at pressure P and temperature T, then $\mu_A = \mu_g$, where μ_g is the chemical potential of the gas-phase molecules, given by

$$\mu_g = kT \ln\left\{\frac{Ph^3}{kT(2\pi mkT)^{\frac{3}{2}}q_g(T)}\right\}. \tag{9.8}$$

where h is Planck's constant. The term $q_g(T)$ represents the partition function for internal degrees of freedom of the gaseous molecule. Substitution of

$$P_0 = kT\left\{\frac{(2\pi mkT)^{\frac{3}{2}}}{h^3}\, q_g(T)\right\}, \tag{9.9}$$

and insertion of the equilibrium condition leads to

$$\frac{\theta}{(1-\theta)q_A} = \frac{P}{P_0} \tag{9.10}$$

or

$$\theta(P, T) = q_A(T)P/(P_0 + q_A(T)P). \tag{9.11}$$

Equation (9.11) is the Langmuir adsorption isotherm, which therefore relates to a uniform, chemically inert surface, with no interaction between adsorbed molecules, and single occupancy of adsorption sites.

9.1.1. Dissociative adsorption of diatomic molecules

With some diatomic molecules (N_2, O_2, H_2, etc.), chemisorption on transition metals is dissociative and the adsorbed species is the atomic state. Equation (9.7) is still valid, so that with the coverage θ' expressed in terms of sites occupied by adatoms and with q'_A denoting the partition function of an adatom,

$$\mu'_A = kT\ln\left\{\frac{\theta'}{(1-\theta')q'_A}\right\}. \tag{9.12}$$

Similarly [cf. Eqs (9.8) and (9.9)],

$$\mu_g = kT\ln(P/P_0), \tag{9.13}$$

where P is the pressure of the gas-phase molecules. The condition for equilibrium, however, is now given by

$$\mu_g = 2\mu'_A. \tag{9.14}$$

Hence,

$$\theta'/(1-\theta')q'_A = (P/P_0)^{\frac{1}{2}}, \tag{9.15}$$

or

$$\theta' = q'_A(T)P^{\frac{1}{2}}/(P_0^{\frac{1}{2}} + q'_A(T)P^{\frac{1}{2}}). \tag{9.16}$$

9.2. THE LOCALIZED MONOLAYER WITH INTERACTION BETWEEN ADSORBED MOLECULES

9.2.1. Bragg–Williams lattice theory [3]

It is assumed that the only interaction energy between adsorbed molecules is that ω between pairs of nearest neighbours. The pairs are independent and do not interact with each other. The minimum energy to desorb an adsorbed molecule in its lowest energy from a site for which all neighbouring sites are unoccupied is denoted by χ° and its partition function by q_A°. Each site is surrounded by Z nearest neighbour sites, e.g. $Z = 4$ for a square lattice array; the total energy of interaction is the sum of the contributions of these pairs of nearest neighbours. The monolayer comprises M sites of which N are occupied and $M - N$ are empty. There are N_{11} pairs of neighbouring sites both occupied, N_{00} pairs of neighbouring sites both empty, and N_{10} pairs of neighbouring sites of which one is occupied and the other is empty. The assumption is a completely random distribution of adsorption molecules in the monolayer. This is an approximation, since it is evident that configurations of lower free energy occur more frequently than those of higher energy.

Since each occupied site has Z nearest neighbours, then for complete randomness, each site has a probability N/M of being occupied. For any give N, the average number of neighbours is ZN/M and, since there are N occupied sites, $N_{11} = \frac{1}{2}ZN^2/M$, the numerical factor $\frac{1}{2}$ being included to ensure that both members of a pair are not counted twice.

The complete partition function for the layer of adsorbed molecules is, therefore,

$$Q(N, N_{11}, T) = \frac{M!(q_A^\circ)^N \exp(-N_{11}\omega/kT)}{N!(M - N)!} \tag{9.17}$$

where $N_{11}\omega$ is the total interaction energy, and

$$F_A = -kT\{M \ln M - N \ln N - (M - N)\ln(M - N) + N \ln q_A^\circ + (ZN^2\omega/2MkT)\}, \tag{9.18}$$

since $N_{11} = ZN^2/2M$.

The chemical potential of the adsorbed molecules is

$$\mu_A = \left(\frac{\partial F_A}{\partial N}\right)_{T,M} = \ln\left\{\frac{\theta}{(1 - \theta)q_A^\circ}\right\} + \frac{Z\theta\omega}{kT}. \tag{9.19}$$

On equating $\mu_A = \mu_g$, the chemical potential of the gas phase, with $\mu_g = kT\ln(P/P_0)$,

$$\frac{P}{P_0} = \frac{\theta}{(1-\theta)q_A^\circ}\exp(Z\theta\omega/kT), \tag{9.20}$$

where $q_A^\circ = q_A \exp(\chi_0/kT)$ and θ is the fractional occupancy of sites. This equation reduces to Eq. (9.10), the Langmuir adsorption isotherm, as $\omega \to 0$. The contribution of the interaction energy E_i to the total energy is

$$E_i = N_{11}\omega = ZM\theta^2\omega/2. \tag{9.21}$$

9.2.2. Quasi-chemical approximation

At equilibrium with the adsorbed layer,

$$2N_{01} \rightleftarrows N_{11} + N_{00} \tag{9.22}$$

and a factor $\exp(-\omega/kT)$ takes care of the change in energy when an N_{11} pair is formed from two N_{01} pairs. The relationship between N_{11}, N_{00}, N_{10} and N_{01} must satisfy certain conditions. The total number of pairs of sites is $ZM/2$ and each pair of sites can be occupied in four different ways as N_{11}, N_{01}, N_{10} and N_{00} pairs; hence, the relationships are

$$ZN/2 = N_{11} + N_{01}/2 \text{ for occupied sites,} \tag{9.23}$$

and

$$Z(M-N)/2 = N_{00} + N_{01}/2 \text{ for empty sites.} \tag{9.24}$$

The equilibrium constant K of the reaction given in (9.22) may now be written in terms of the partition functions of the various pairs, q_{00}, q_{11} and q_{10}, i.e.

$$K = \frac{q_{00}q_{11}}{(q_{01})^2}\exp(-\omega/kT). \tag{9.25}$$

Since $q_{00} = 1$, $q_{11} = \exp(-\omega/kT)$, $q_{01} = 2$ (configuration degeneracy arises from the two equivalent configurations N_{01} and N_{10}),

$$K = \exp(-\omega/kT)/4. \tag{9.26}$$

Equation (9.26) may be now reformulated as

$$(N_{11})(N_{00}) = \frac{(N_{01})^2}{4}\exp(-\omega/kT), \tag{9.27}$$

and, by use of Eqs (9.23) and (9.24), Eq. (9.27) may be conveniently written as

$$2N_{11}(ZM - 2ZN + 2N_{11}) = (ZN - 2N_{11})^2 \exp(-\omega/kT). \quad (9.28)$$

On substitution of $\theta = N/M$ and $x = 2N_{11}/ZM$, Eq. (9.29) is obtained:

$$x(1 - 2\theta + x) = (\theta - x)^2 \exp(-\omega/kT). \quad (9.29)$$

The solution of the quadratic (9.29) is

$$x = \theta + \frac{\{1 - 4\theta(1 - \theta)(1 - \exp(-\omega/kT))\}^{\frac{1}{2}}}{2(1 - \exp(-\omega/kT))}, \quad (9.30)$$

or

$$N_{11} = \frac{ZM\theta^2}{2} - \frac{\theta(1 - \theta)ZM}{\{1 - 4\theta(1 - \theta)(1 - \exp(-\omega/kT))\}^{\frac{1}{2}} + 1}. \quad (9.31)$$

Replacement of the term in braces by β, i.e.

$$\beta = \{1 - 4\theta(1 - \theta)(1 - \exp(-\omega/kT))\}^{\frac{1}{2}} \quad (9.32)$$

gives

$$N_{11} = ZM\left(\frac{\theta}{2} - \frac{\theta(1 - \theta)}{\beta + 1}\right), \quad (9.33)$$

instead of $N_{11} = \frac{1}{2}ZN^2/M = \frac{1}{2}Z\theta^2 M$ in the Bragg–Williams approximation. Following the previous procedure [Eqs (9.17) to (9.20)], Eq. (9.34) is obtained after some lengthy algebra:

$$\frac{P}{P_0} = \frac{\theta}{(1 - \theta)q_A^\circ}\left(\frac{2 - 2\theta}{\beta + 1 - 2\theta}\right)^Z. \quad (9.34)$$

The contribution of interaction energy E_i to the total energy is

$$E_i = N_{11}\omega = ZM\omega\left\{\frac{\theta}{2} - \frac{\theta(1 - \theta)}{(\beta + 1)}\right\}. \quad (9.35)$$

Equation (9.35) reduces to the Bragg–Williams approximation $E_i = Z\omega\theta^2 M/2$ after substituting for β by Eq. (9.32) and expanding in powers of ω/kT with neglect of higher powers than the first. The quasi-chemical approximation is significantly better than the Bragg–Williams one and is equivalent to the Bethe method [5].

9.2.3. Energy of desorption

The total energy of desorption is derived from the free energy of the adsorbed layer by means of the thermodynamic equation

$$E_{\mathrm{d}} = -T^2 \frac{\partial(F_{\mathrm{A}}/T)}{\partial T}. \tag{9.36}$$

However, terms involving $\partial(\ln q(T))/\partial T$ cannot be evaluated accurately, and approximations must be introduced. In particular, the internal degrees of freedom of adsorbed and gas-phase molecules are assumed to be the same, and the three translational degrees of freedom of the gas-phase molecule are transformed to three harmonic vibrational degrees for the adsorbed molecule. The energy of desorption E_{d} of N molecules is then given by

$$F_{\mathrm{d}} = N(\chi - \tfrac{3}{2}kT) \tag{9.37}$$

and

$$E_{\mathrm{d}} = N(\chi - \tfrac{3}{2}kT) - N_{11}\omega \tag{9.38}$$

when interactions between adsorbed molecules are included.

In the Bragg–Williams approximation,

$$E^{\mathrm{d}} = N(\chi_0 - \tfrac{3}{2}kT) - Z\theta^2\omega M/2, \tag{9.39}$$

and

$$E_{\mathrm{d}} = N(\chi_0 - \tfrac{3}{2}kT) - MZx\omega/2 \tag{9.40}$$

in the quasi-chemical approximation, with x given by Eq. (9.30).

9.3. LOCALIZED MONOLAYERS ON A NON-UNIFORM SURFACE WITH NO INTERACTIONS [6–11]

The total number M of sites on the surface comprises a number $M_1, M_2, \ldots, M_n$ of sites having different desorption energies $\chi_1, \chi_2, \ldots, \chi_n$, and the different partition functions for adsorbed molecules on the various sites are denoted by $q_1(T), q_2(T), \ldots, q_n(T)$.

The complete partition function for the monolayer is then

$$\prod_i \frac{M_i! q_i(T)^{N_i}}{(M_i - N_i)! N_i!}, \qquad i = 1, 2, \ldots, n, \tag{9.41}$$

where $N_1, N_2, \ldots, N_n$ are the number of the different sites occupied by adsorbed molecules at equilibrium with the gas phase at a pressure P and temperature T.

The free energy of the adsorbed layer F_A is, therefore,

$$F_A = -kT \sum_i [M_i \ln M_i - N_i \ln N_i - (M_i - N_i) \ln(M_i - N_i) + N_i \ln q_i(T)] \tag{9.42}$$

and the chemical potential μ_i of molecules adsorbed on sites i is

$$\mu_i = kT \ln\left\{\frac{N_i}{(M_i - N_i)q_i(T)}\right\} = kT \ln\left\{\frac{\theta_i}{(1 - \theta_i)q_i(T)}\right\}. \tag{9.43}$$

At equilibrium,

$$\mu_1 = \mu_2 = \ldots = \mu_n = \mu_g = kT \ln\left\{\frac{P}{kT}\frac{h^3}{(2\pi mkT)^{\frac{3}{2}}} q_g(T)\right\} \tag{9.44}$$

or

$$\frac{\theta_i}{(1 - \theta_i)q_i(T)} = \frac{P}{P_0} \qquad \text{for } i = 1, 2, \ldots, n. \tag{9.45}$$

The fractional coverage of the total surface is Θ, where with $c_i = N_i / \sum_i N_i$,

$$\Theta = \sum_i c_i \theta_i \tag{9.46}$$

or

$$\Theta = \sum_i \frac{c_i q_i(T) P}{P^\circ + q_i(T)P}. \tag{9.47}$$

Assuming a continuous range of χ values and that the partition function for the internal degrees of freedom of the adsorbed molecule is independent of the magnitude of χ, i.e. independent of the particular site to which it is attached, Eq. (9.47) may be written in the form

$$\Theta = \int \frac{N_\chi \, d\chi}{(a/P) \exp(-\chi/kT) + 1}, \tag{9.48}$$

where

$$\int N_\chi \, d\chi = 1. \tag{9.49}$$

9.3.1. The Freundlich equation

The integration of Eq. (9.48) requires an analytical expression for the distribution function N_χ. For an exponential distribution [6–8],

$$N_\chi = b \exp(-\chi/\chi_m), \tag{9.50}$$

where b and χ_m are constants. To facilitate integration, the limits of the integrations in Eqs (9.48) and (9.49) are placed at $-\infty$ and $+\infty$, although these appear to include a large number of sites with repulsive adsorption energies, but may mean that the actual distribution function extends beyond the weakest sites that are ever filled and the strongest sites that are ever emptied. Substituting Eq. (9.50) into Eq. (9.48) and denoting

$$(a/P)\exp(-\chi/kT) = \phi,$$

then

$$\begin{aligned}\Theta &= (P/a)^{kT/\chi_m}\, bkT \int_{-\infty}^{\infty} \phi\, \frac{kT/\chi_m - 1}{1+\phi}\, d\phi, \\ &= (P/a)^{kT/\chi_m} bkT\pi \operatorname{cosec}(\pi kT/\chi_m), \end{aligned}\tag{9.51}$$

which, for small values of $\pi kT/\chi_m$, becomes

$$\Theta = (P/a)^{kT/\chi_m} b\chi_m, \tag{9.52}$$

corresponding to the Freundlich isotherm.

The isosteric heat q_{st} is

$$q_{st} = k\left(\frac{\partial \ln P}{\partial(1/T)}\right)_\theta = -\chi_m \ln \Theta + \chi_m \ln(b\chi_m). \tag{9.53}$$

For $(P/a) = 1$, Eq. (9.52) shows that $\Theta = b\chi_m$; and from Eq. (9.53), $q_{st} = 0$. The $\ln P$ against $\ln \Theta$ plots are linear with slopes linearly related to the temperature and, on extrapolation, intersect at a single point where $q_{st} = 0$, corresponding to a pressure P_0. For $P/a > 1$, $\Theta > b\chi_m$ [from Eq. (9.52)] and q_{st} is negative [from Eq. (9.53)]. Thus, $b\chi_m$ is the maximum adsorption and the Θ scale is, therefore, defined by placing $\Theta = 1$ at this value. Hence,

$$\Theta = (P/a)^{kT/\chi_m}, \quad \text{or} \quad \ln(P/P_0) = (\chi_m/kT)\ln \Theta, \tag{9.54}$$

and

$$q_{st} = -\chi_m \ln \Theta. \tag{9.55}$$

9.3.2. The Temkin equation

Another simple distribution of adsorption energies is given by

$$N_\chi = \text{const.} = C, \tag{9.56}$$

i.e. the surface comprises groups with the same number of adsorption sites, but for which χ varies linearly up to a maximum value χ_m. Hence,

$$\Theta = \int_0^{\chi_m} \frac{C\,d\chi}{1 + (a/P)\exp(-\chi/kT)}, \tag{9.57}$$

or

$$\Theta = kTC \ln\left\{\frac{1 + (P/a)\exp(\chi_m/kT)}{P/a + 1}\right\}. \tag{9.58}$$

For $P/a \ll 1$ and $(P/a)\exp(\chi_m/kT) \gg 1$,

$$\Theta = CkT \ln\{(P/a)\exp(\chi_m/kT)\}, \tag{9.59}$$

which is a form of the Temkin equation in which Θ is a linear function of $\ln P$ at constant temperature and

$$q_{st} = \chi_m(1 - \Theta/\chi_m C), \tag{9.60}$$

i.e. the isosteric heat varies linearly with coverage and χ_m is the value of χ at $\Theta = 0$.

9.4. MOBILE LAYERS

Chemisorbed molecules are located at the minima of the oscillating potential energy function at the metal surface and are, therefore, localized. On completion of the primary chemisorbed layer, the amplitude of this function is greatly reduced because chemisorption forces are short-ranged. Nevertheless, at higher pressures, additional adsorption takes place on this layer and these adspecies have considerable mobility. The statistical-mechanical treatment of a mobile layer is, therefore, outlined.

9.4.1. The ideal layer

The ideal system is free translational motion of adsorbed molecules over an equipotential surface (two-dimensional ideal gas) in the (x, y) plane with vibrations of the molecules in the z direction normal to the surface. The molecular partition function in the (x, y) plane is, therefore,

$$q_{xy} = \frac{(2\pi mkT)A}{h^2}, \tag{9.61}$$

where A is the total area of the adsorbent surface available to the adsorbed molecules; the total molecular partition function is

$$q = q_{xy}q_z \exp(\chi/kT), \tag{9.62}$$

where q_z is the partition function of the internal degrees of freedom of the adsorbed molecule and of the vibration normal to the surface.

The assembly partition function of the adsorbed phase for this non-localized system is

$$Q = q^N/N!. \tag{9.63}$$

Hence,

$$F_A = -kT(\ln(q^N/N!))_{T,A}; \tag{9.64}$$

$$\mu_A = -kT\ln(q/N); \tag{9.65}$$

and

$$\mu_g = kT\ln(P/P_0). \tag{9.66}$$

At equilibrium,

$$\frac{(2\pi mkT)}{h^2}\frac{A}{N} = \frac{P}{P_0} \tag{9.67}$$

or

$$\theta = \frac{N}{M} = \left[\frac{a(2\pi mkT)}{h^2} q_z \exp\frac{\chi}{kT}\right]\frac{P}{P_0}, \tag{9.68}$$

where $Ma = A$, a being the area associated with each of the M sites.

Extension to higher coverages requires correction for the finite size of the adsorbed molecules. The partition function for the adsorbed layer is then expressed in the form

$$Q = \frac{1}{N!}\left[\frac{2\pi mkT}{h^2}(A - N\beta)\right]^N, \tag{9.69}$$

where β is a constant related to the molecular size and is the two-dimensional analogue of van der Waals' b constant for the three-dimensional gas. Denoting

$2\pi mkT/h^2$ by q', then

$$\mu_A = \left(\frac{\partial F_A}{\partial N}\right)_T = -\frac{\partial}{\partial N}\left\{(kT \ln[q'(A - N\beta)])^N \frac{1}{N!}\right.$$

$$= -kT\left\{\ln[q'(A - N\beta)/N] - \frac{N\beta}{(A - N\beta)}\right\}. \quad (9.70)$$

If N_s is the number of adsorbed molecules in the completed monlayer, then $A = N_s\beta$, and $\theta = N/N_s$, so that, at equilibrium,

$$\mu_A = kT\left\{\ln\frac{\theta}{(1-\theta)}q'\beta + \frac{\theta}{1-\theta}\right\} = \mu_g = kT(\ln P/P_0). \quad (9.71)$$

Hence

$$P = P_0\left\{\frac{\theta h^2}{(1-\theta)(2\pi mkT)\beta}\right\}\exp\left(\frac{\theta}{1-\theta}\right) \quad (9.72)$$

$$= \text{const.}\left(\frac{\theta}{1-\theta}\right)\exp\left(\frac{\theta}{(1-\theta)}\right). \quad (9.73)$$

As $\theta \to 1$, $\exp(\theta/(1-\theta))$ varies more rapidly than does $\theta/(1-\theta)$, and the saturation value is approached more quickly than for the localized Langmuir model.

9.5. MOBILE LAYERS WITH INTERACTIONS

The interaction term is usually repulsive between adsorbed molecules. Because of the assumed independence of the adsorbed molecules, the assembly partition function is, as before,

$$Q = q_A^N/N!, \quad (9.74)$$

but the area A of the surface must be replaced by a "free area" A_f (because of the excluded area arising from the finite size of the molecule); a Boltzmann factor $\exp(-\phi/2kT)$ is inserted to take care of the intermolecular potential field. The energy ϕ is a function of N/A and represents the interaction energy between any one molecule and all the other adsorbed molecules (the numerical factor being required since the pair interaction is shared between two molecules in the total interaction energy). Hence, the partition function of the adsorbed molecule is given by

$$q' = \frac{(A_f\, e^{-\phi/2kT})}{h^2}(2\pi mkT). \tag{9.75}$$

For a random distribution, there are no molecules in the neighbourhood of a particular molecule (at $r = 0$) between $r = 0$ and r_c (assuming a hard-sphere core model); from r_c to $r = \infty$ the density is constant. Between r and $r + dr$, there are $(2\pi r\, dr)N/A$ other molecules. The interaction energy between each of these molecules and the central molecule at $r = 0$ is ∞, and for $r > r_c$ equals $\varepsilon(r_c/r)^3$, assuming a dipole repulsion interaction energy varying as $1/r^3$. Hence,

$$\phi = \int_{r_c}^{\infty} \varepsilon(r_c/r)^3\,(N/A)2\pi r\, dr = 2\pi(N/A)\varepsilon r_c^2. \tag{9.75}$$

It is necessary to correct for the excluded area $N\beta$. Hence,

$$Q = [(A - N\beta)q']^N \exp\left(-\frac{\pi N^2 \varepsilon r_c^2}{AkT}\right)^N \frac{1}{N!} \tag{9.77}$$

$$F_A = -kT\left\{\ln[(A - N\beta)q']^N + \left[\frac{\pi N^2 \varepsilon r^2}{AkT^\delta}\right]^N - N\ln N + N\right\} \tag{9.78}$$

and

$$\mu_A = -kT\left\{\ln[q'(A - N\beta)/N] + \frac{N\beta}{A - N\beta} + \frac{2\pi N\varepsilon r_c^2}{AkT}\right\} = \ln\left(\frac{P}{P_0}\right). \tag{9.79}$$

Therefore,

$$\frac{P}{P_0} = \frac{\theta}{(1-\theta)q'\beta}\exp\left(\frac{\theta}{1-\theta}\right)\exp\left(\frac{2\pi N\varepsilon r_c^2}{AkT}\right) \tag{9.80}$$

$$= \frac{\theta}{(1-\theta)q'\beta}\exp\left[\frac{\theta}{1-\theta} + \frac{2\pi N_s \varepsilon r_c^2 \theta}{AkT}\right],$$

and

$$P = \frac{P_0}{q'\beta}\left(\frac{\theta}{1-\theta}\right)\exp\left[\frac{\theta}{1-\theta} + \alpha\theta\right], \tag{9.81}$$

where $\alpha = 2\pi N_s \varepsilon r_c^2/AkT$.

Equation (9.81) has a similar form to that of Eq. (9.73), except for the term $\alpha\theta$. Alternative treatments involve different expressions for the interaction energy, e.g. the two-dimensional analogy of an imperfect gas requires the introduction of the function $(1 - e^{-\varepsilon/kT})\,2\pi r\,dr$ to represent the interaction energy.

The extension to include departures from ideal two-dimensional gas behaviour has been outlined because there is often a high concentration of molecules in a mobile second layer (approaching that of liquid in a complete monolayer), with consequent effect on the desorption profile in the initial stages of evolution from a heavily-covered layer.

REFERENCES

1. R. H. Fowler and E. A. Guggenheim, "Statistical Thermodynamics", Cambridge University Press, Cambridge, 1939, Chap. X.
2. T. L. Hill, "Introduction to Statistical Thermodynamics", Addison-Wesley. Reading, Mass., 1960, Chaps 7 and 14.
3. W. L. Bragg and E. J. Williams, *Proc. Roy. Soc. A*, 1934, **145**, 699.
4. T. L. Hill, *Adv. Catalysis* 1952, **4**, 211.
5. H. Bethe, *Proc. Roy. Soc. A*, 1935, **150**, 552.
6. G. Halsey and H. S. Taylor, *J. Chem. Phys.*, 1947, **15**, 624.
7. R. Sips, *J. Chem. Phys.*, 1948, **16**, 490.
8. G. Halsey, *J. Chem. Phys.*, 1948, **16**, 931.
9. S. Z. Roginski, "Adsorption and Catalysis on Heterogeneous Surfaces", Akad. Nauk SSSR, Moscow, 1949.
10. F. C. Tompkins, *Trans. Faraday Soc.*, 1950, **46**, 569.
11. J. G. Tolpin, G. S. John and E. Field, *Adv. Catalysis*, 1953, **5**, 217.

10

Electronic Theory of Metals

In order to discuss the chemisorption of gases on metals, some knowledge of the electronic properties of metals is essential. Although the surface properties are clearly more relevant to considerations of the characteristic of the gas–metal chemisorption bond,† the loss of three-dimensional symmetry at the surface immensely complicates the mathematics, and one approach has been to examine the properties of the bulk and from these to infer those of the surface. One of the more fundamental approaches is the band theory of metals. Attention is here confined to a simple one-dimensional model, since extension to the three-dimensional case does not introduce any new fundamental concepts.

10.1. BAND THEORY (FREE-ELECTRON MODEL) [1–4]

Many properties of metals can be rationalized in terms of the free-electron model, the essential feature of which is the free mobility of the valence electrons of the metal atoms within the bulk metal. Interaction of these conduction electrons with the ion cores and repulsion between electrons are neglected. However, obedience to the Pauli exclusion principle is necessary, i.e. no two electrons of the same atom can have all of their individual quantum numbers equal.

For an electron in free space, the Schrödinger equation is

$$-\frac{\hbar^2}{2m}\nabla^2\psi = E\psi, \tag{10.1}$$

† See Chapter 11, to which this chapter provides an elementary introduction on the free-electron theory of metals.

where m is the mass and E the total (constant) energy of the electron and ψ its wave function; $\hbar^2 = h/2\pi$, where h is Planck's constant. An obvious solution is

$$\psi = \exp(i\mathbf{k} \,.\, \mathbf{r}), \tag{10.2}$$

so that

$$\nabla^2\psi = -k^2\psi \tag{10.3}$$

and

$$E = \hbar^2 k^2/2m. \tag{10.3}$$

The wave vector **k** is given by

$$k = |\mathbf{k}| = 2\pi/\lambda \tag{10.4}$$

in which λ is the wavelength of the wave associated with a free electron of energy $E = \frac{1}{2}mv^2$; and $\lambda = h/mv$ from the de Broglie equation. Hence,

$$E = \frac{\hbar^2}{2m}(2\pi/\lambda)^2 = \frac{1}{2m}(h/\lambda)^2. \tag{10.5}$$

For a one-dimensional system in which the electron is confined to a line of length l, the boundary conditions are

$$\psi(0) = \psi(l) = 0.$$

Hence, the allowed solutions of the wave equation are

$$\psi_n = (2/l)^{\frac{1}{2}} \sin(n\pi x/l), \tag{10.6}$$

in which $n = 0, 1, 2, \ldots$ and

$$E_n = (\hbar\pi n)^2/2ml^2. \tag{10.7}$$

The electron may, therefore, be represented by a plane wave of uniform probability density given by $\psi\psi^* = l^{-1}$, where ψ^* is the conjugate of ψ.

Within a monatomic one-dimensional metal of lattice constant a, the Bragg relationship $2a \sin\theta = n\lambda$ must be obeyed. Because of the periodic structure, Bragg reflections occur when $\mathbf{k} = \pm n\pi a$. Within the corresponding energy range, solutions of the wave equation do not exist, i.e. there is an energy gap in the distribution of electron states (see Fig. 10.1). The forbidden energy band in **k** space between $-\pi/a$ and $+\pi/a$ is called the first Brillouin

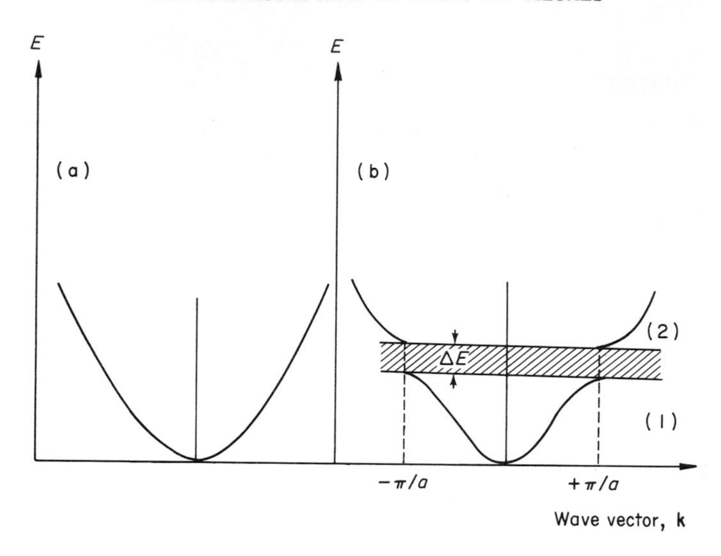

FIG. 10.1. Plot of energy E as a function of the wave vector **k** for: (a) a free electron, (b) a linear lattice constant a. The first Bragg reflections occur at $k = \pm\pi/a$; the region between is called the first Brillouin zone. The difference in the potential energy of the wave functions $(\psi_2 - \psi_1)$ is ΔE, the width of the forbidden band separating the first (1) and second (2) allowed bands.

zone [5]. At this k value, two identical waves travelling in opposite directions give a standing wave described by two independent wave functions,

$$\psi_1 \sim \sin(\pi x/a) \quad \text{and} \quad \psi_2 \sim \cos(\pi x/a). \tag{10.8}$$

Charge is preferentially concentrated for the former midway *between* the ion-cores, and *at* these cores for the latter, whereas for a plane wave there is a uniform distribution of electron charge over the length l (see Fig. 10.2). The average potential energy for these three conditions is, therefore, in order: $\psi_2 <$ plane wave $< \psi_1$, the difference $(\psi_2 - \psi_1)$ being the width of the energy gap.

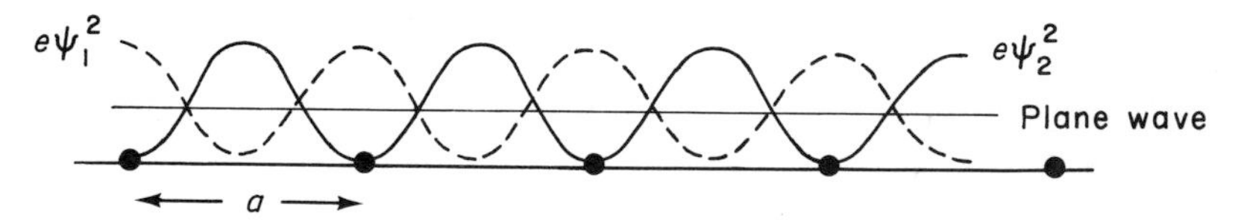

FIG. 10.2. Probability distribution of charge density for $|\psi_1|^2 = \sin^2 \pi x/a$ and $|\psi_2|^2 = \cos^2 \pi x/a$. The positive ion cores (●) are separated by a distance equal to the lattice constant a.

10.2 BLOCH FUNCTIONS [6]

A quantitative assessment of energy states of an electron in the periodic field of a metal lattice is complex. Solutions of the Schrödinger equation that include a periodic potential energy function $V(\mathbf{r})$ have been given by Bloch. The wave function of a free electron $[V(\mathbf{r}) = \text{const.}]$, viz.

$$\psi_k = \exp(i\mathbf{k}\,.\,\mathbf{r}), \tag{10.9}$$

is modified by inclusion of a modulating function $U_k(\mathbf{r})$ having the periodicity of the lattice to give

$$\psi_k = U_k(\mathbf{r}) \exp(i\mathbf{k}\,.\,\mathbf{r}). \tag{10.10}$$

This equation represents a plane wave modulated by the periodic ion-core potentials; its form, therefore, differs from that of the simple sinsoidal wave. However, as for the electron in free space, its energy $E_\mathbf{k}$ depends quadratically on the wave vector $\mathbf{k}$, i.e.

$$E_\mathbf{k} = (\hbar^2/2m_e)k^2. \tag{10.11}$$

The major difference is that the rest mass of the electron in the energy expression for a free electron is replaced by an effective mass m_e; m_e is defined as the mass that must be assigned to the free electron to make its velocity increment under an applied impulse equal to that which is actually attained by a valence electron in the metal under the same impulse.

One method of obtaining acceptable functions of $U_k(\mathbf{r})$ is to use the method of linear combination of atomic orbitals, and then to evaluate numerical magnitudes according to the so-called tight-binding approximation. This method is briefly outlined.

10.3 LINEAR COMBINATION OF ATOMIC ORBITALS (LCAO) [7]

The atomic orbitals are combined to give molecular orbitals. The wave function of the molecular orbital ψ is

$$\psi = \sum_n c_n \phi_n, \tag{10.12}$$

where ϕ_n are atomic orbitals. The wave equation is

$$H\psi = E\psi \tag{10.13}$$

where E is the orbital energy and H is the one-electron Hamiltonian operator. After multiplication by ψ, Eq. (10.2) is transformed to

$$E = \int \psi H \psi \, d\tau \Big/ \int \psi^2 \, d\tau \tag{10.14}$$

in which $d\tau$ is an element of volume. Combination of Eqs (10.12) and (10.14) gives

$$E = \int \sum_n c_n \phi_n H \sum_m c_m \phi_m \, d\tau \Big/ \int \sum_n c_n \phi_n \sum_m c_m \phi_m \, d\tau. \tag{10.15}$$

In the variational method, the energy is minimized with respect to each coefficient c_n, i.e. $(\partial E/\partial c_n) = 0$, to give a set of secular equations,

$$\sum_m (H_{nm} - E_{nm})c_m = 0, \qquad n = 1, 2, 3, \ldots. \tag{10.16}$$

The solutions of Eq. (10.16) are given by the determinant

$$|H_{nm} - ES_{nm}| = 0, \tag{10.17}$$

in which $H_{nm} = \int \phi_n H \phi_m \, d\tau$ is the resonance integral when $n \neq m$; or the coulomb integral H_{mm} when $n = m$; and $S_{nm} = \int \phi_n \phi_m \, d\tau$ is the overlap integral. Various procedures are used to provide numerical magnitudes of H_{nm} and S_{nm}, the most common being the tight-binding approximation in which the assumption is that

$$S_{nm} = 0 \quad \text{for} \quad n \neq m, \qquad S_{nm} = 1 \quad \text{when} \quad n = m.$$

In the extended Hückel molecular orbital method, both H_{nm} are evaluated using semi-empirical procedures.

Using this LCAO method, $U_k(\mathbf{r})$ in Eq. (10.10) can be expressed in the form

$$U_k(\mathbf{r}) = \sum_n [\exp i\mathbf{k}.(\mathbf{R}_n - \mathbf{r})]\phi(\mathbf{r} - \mathbf{R}_n), \tag{10.18}$$

where $\phi(\mathbf{r})$ is the atomic orbital of the free atom, and $\mathbf{R}_n$ is the lattice vector of the nth atom in the metal crystal. The solution of this equation shows that the atomic orbitals of the free atoms give rise to a group, or band, of allowed energies, the magnitudes of which are quasi-continuous.

10.4 ENERGY BANDS

The formation of energy bands in metals can be visualized by successively adding single metal atoms to one specific atom A having a single *s* atomic orbital ϕ_A. A second atom B having an identical atomic orbital, but labelled ϕ_B, is brought towards atom A. The wave functions ϕ_A, ϕ_B finally overlap to form a molecular orbital ψ given by the linear combination of the atomic orbitals, viz.

$$\psi = c_A\phi_A \pm c_B\phi_B,$$

where c_A, c_B are constants which have been evaluated by minimization of the orbital energy E. The energy of the $(c_A\phi_A - c_B\phi_B)$ state is higher, and that of the $(c_A\phi_A + c_B\phi_B)$ lower, than the energy of the isolated non-interacting atoms (see Fig. 10.3). On addition of a third atom, three energy states are formed; and by continuation of the addition, an extra energy level is generated for each atom that is incorporated; simultaneously, all the previous energy levels are slightly modified.

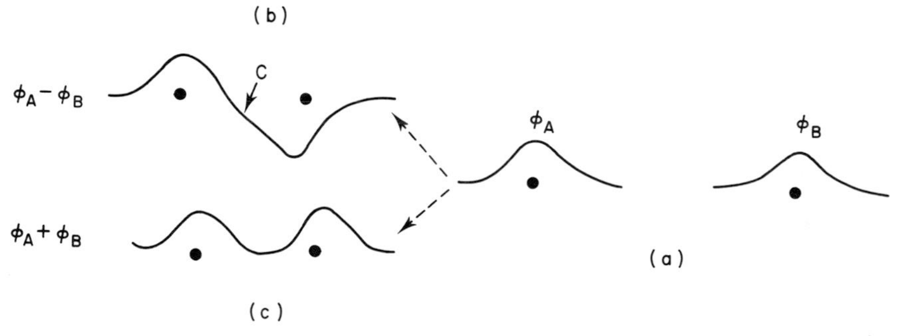

FIG. 10.3. Wave functions of identical atoms A and B at large separation (a). At small separation, the wave functions ϕ_A and ϕ_B overlap to give (b) the lower (ground state $(\phi_A + \phi_B)$ and (c) the higher (excited) state $(\phi_A - \phi_B)$; at C, the probability density is zero.

For an array of N atoms, the number of molecular orbitals equals the number of atomic orbitals associated with the originally isolated and non-interacting atoms, i.e. N energy levels are formed from the single *s* atomic orbitals of the N atoms (see Fig. 10.4). The energy levels of the molecular orbitals form a quasi-continuous set of allowed levels to give a band of energies extending from below and above the discrete atomic energy level. The width of the band is larger the greater the total interaction between the atoms at their equilibrium positions in the final array that constitutes the

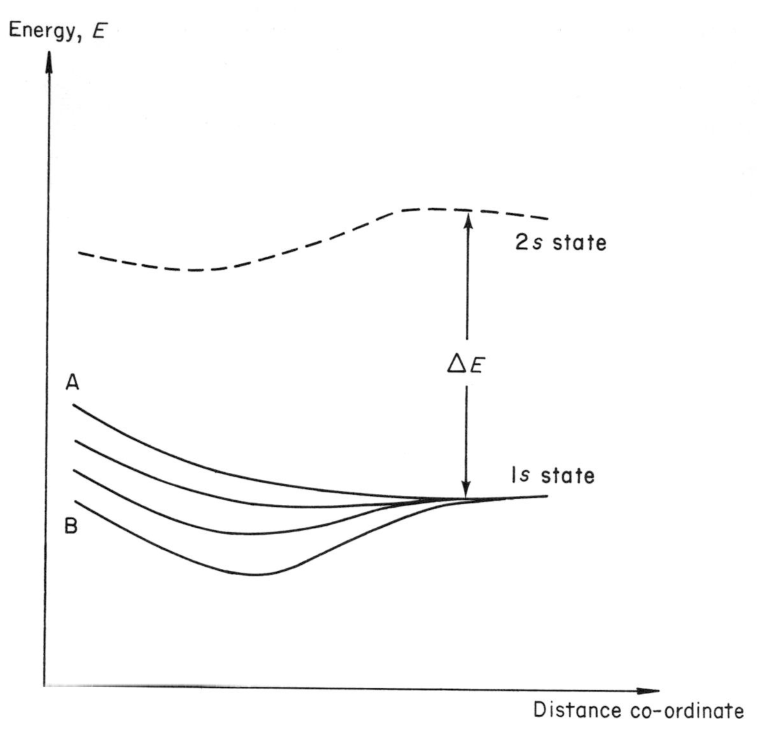

FIG. 10.4. Plot showing the splitting of the 1*s* energy levels for a linear array of four atoms as their distance apart is decreased to form the quasi-continuous energy band AB. The forbidden band of width ΔE separates the two allowed bands formed from the 1*s* and 2*s* energy states.

bulk metal. Should the free atom have more than one atomic orbital, each orbital forms its own energy band centred around the corresponding orbital energy of the non-interacting atoms.

10.4.1. Overlap of energy bands

There is often a marked difference in the chemisorption properties of transition and non-transition metals; our present interest is in the former group. In the first transition series (Sc to Ni), the orbital energies in the *d* states of the free atoms do not greatly differ from those of the *s* states, and, therefore, in the corresponding metals the 3*d* band strongly overlaps the 4*s* band. The 3*d* wave functions decrease more quickly with increasing distance from the nucleus than do the 3*s* wave functions and the internal overlap of the *d* atomic orbitals within the band is less. Consequently, the 3*d* atomic states are not broadened as much as the 4*s* atomic states, and the *d* bandwidth is about 3 eV compared with 6 to 7 for the 4*s* band. But, since the 3*d* atomic state is fivefold degenerate, it can accommodate up to ten electrons, and the density of states in the *d* band can be nearly 20 times greater than that in the *s* band. The energy distribution curve for the 3*d* band is, therefore, tall and narrow and that of the 4*s* band broad and short (see Fig. 10.5).

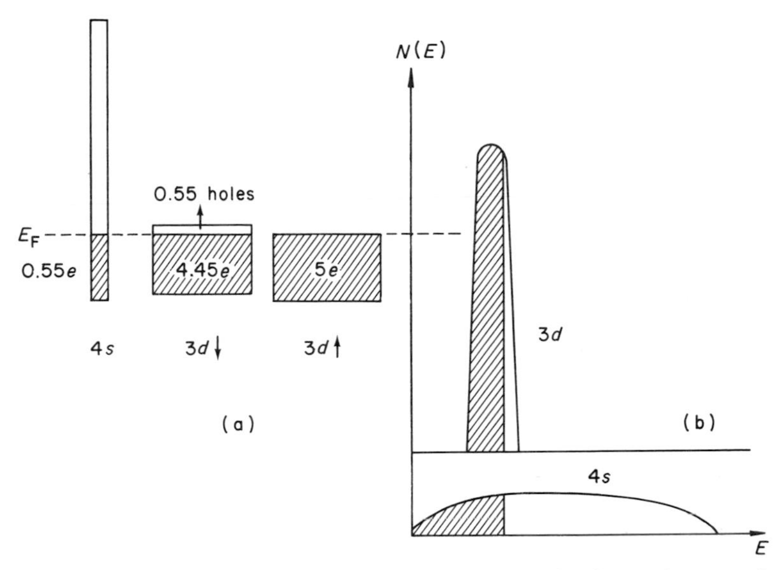

FIG. 10.5. (a) Relationship of the 4*s* and 3*d* bands on nickel. The 3*d* band comprises two substates 3*d*↓ and 3*d*↑ with opposite electron spins. The 3*d*↑ sub-band is fully occupied but the 3*d*↓ has 0·54 holes having transferred 0·55 electrons to the 4*s* band. (b) Electron density of states *N*(*E*) plotted as a function of the energy *E* in the 3*d* and 4*s* bands (schematic representation).

10.4.2. *d*-Band vacancies

Nickel has eight 3*d* and two 4*s* electrons in the atomic state. In the metal the corresponding bands overlap and the ten electrons are distributed between the bands; consequently the *d* band is not fully occupied, i.e. it contains unpaired electrons which are largely responsible for the ferromagnetism of nickel (a small contribution from the polarization of the *s* band and from the orbital motions of the electrons can be corrected for). The corrected saturation moment, in Bohr magnetons, is numerically equal to the number of unpaired electrons in the *d* band, since for nickel the spin-up band is full. The result is that there are 0·55 unpaired *d* electrons (or holes in the *d* band) and 0·55 electron in the *s* band, i.e. the electronic structure is denoted by 3*d* 9·45 4*s* 0·55.

10.4.3. Fermi surface

The band structure and density of occupied states at 0 K for nickel is illustrated in its simplest form in Fig. 10.5. For f.c.c. and b.c.c. metals, each band or Brillouin zone can accommodate two electrons for each atom of the crystal. At 0 K, the valence electrons occupy the lowest set of energy levels consistent with the Pauli exclusion principle; the highest occupied level at 0 K is called the Fermi level and the surface (in **k** space) is called the Fermi surface.

10.4.4. Limitations of simple band theory

The simple band theory takes no account of the spatial properties of the orbitals, nor of hybridization of *s*, *p*, *d* orbitals and band broadening, all of which modify the density of states as functions of energy. The density at the Fermi surface of transition metals may, however, be derived from specific heat and magnetic data. Nevertheless, although such physical properties as the electrical conductivity and magnetic susceptibility of bulk metals can be successfully explained, its application to surface problems involves additional major difficulties.

REFERENCES

1. N. F. Mott and H. Jones, "Theory of Properties of Metals and Alloys", Oxford University Press, London, 1926.
2. C. Kittell, "Introduction to Solid-state Physics", John Wiley and Sons, Inc., New York, 1956, Chap. 10.
3. S. Raines, "The Wave Mechanics of Electron on Metals", North Holland Publ. Co., Amsterdam, 1963.
4. N. F. Mott, *Adv. Phys.*, 1964, **13**, 325.
5. L. Brillouin, "Wave Propagation in Periodic Structures", McGraw-Hill, New York, 1946.
6. F. Bloch, *Z. Phys.*, 1928, **52**, 555.
7. C. A. Coulson, "Valence", Clarendon Press, Oxford, 1952.

11

Electronic Properties of Metal Surfaces

The theoretical understanding of the electronic properties of bulk metals has been greatly extended over the last few decades, but, due largely to the complications involved by loss of three-dimensional symmetry, knowledge of the electronic characteristics of their surfaces has progressed comparatively slowly except over the last decade.

Two general concepts of metals have been developed, one being largely restricted to ion-core–electron interactions, and the other to electron–electron interactions. Recently, however, self-consistent calculations that include both types of interactions have been performed.

11.1. THE BULK LATTICE MODEL

In the first of these theoretical approaches, the bulk lattice potential extends to the surface, where it is directly linked with the vacuum potential (Fig. 11.1), and the interest is largely centred on the possible existence of electronic states localized in the surface region (surface states). Such calculations are of a semi-quantitative character because no account is taken of the effect of the strong perturbation of the bulk properties at the surface. Nevertheless, they predict the presence of localized surface states with energies outside the range of band energies of the solid. The main features of the method can be exemplified following an LCAO treatment by Grimley [1]. The solid is considered to be a one-dimensional chain of $(N + 1)$ atoms, all of which have similar

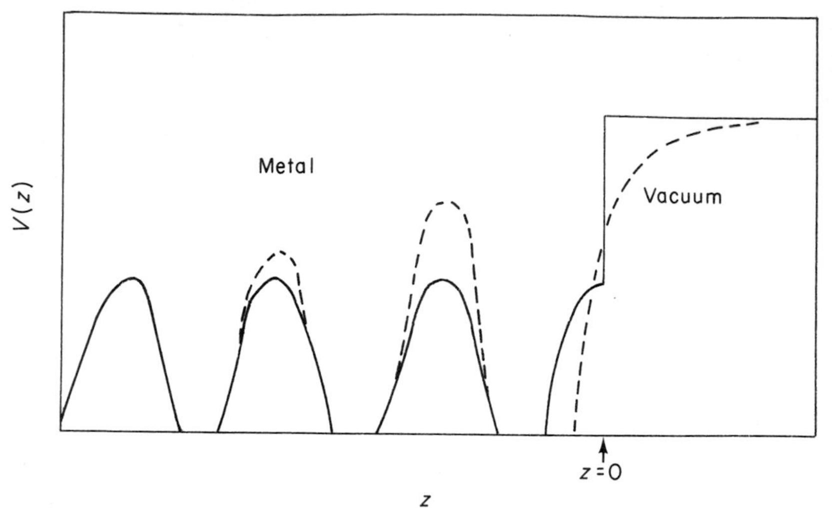

FIG. 11.1. Pictorial representation of potential energy $V(z)$ as a function of distance co-ordinate z perpendicular to the plane of the surface (at $z = 0$) for bulk lattice approximation (full line) and self-consistent lattice approximation (dashed line).

s-like atomic orbitals $\phi(r, m)$. The wave function for the chain is then given by

$$\psi = \sum_{m} c_m \phi(r, m), \tag{11.1}$$

in which c_m is the coefficient of the mth wave function, with the surface (or end) atoms having coefficients c_0 and c_N. Overlap between atomic orbitals is neglected. Substitution of Eq. (11.1) into the Schrödinger equation, $H\psi = E\psi$, gives a set of linear equations:

$$(E - H_{mm})c_m = \sum_{n \neq m} c_n H_{mn} \tag{11.2}$$

in which

$$H_{mn} = \int \phi(r, m)\, H\phi(r, n)\, \mathrm{d}r, \tag{11.3}$$

E denotes the energies, and H the one-electron Hamiltonian operator. The tight-binding or Hückel approximation is applied, i.e.

$$\begin{aligned} H_{mm} &= \alpha, & m &\neq 0, \\ H_{00} &= \alpha' & m &= 0, \\ H_{m, m \pm 1} &= \beta, \end{aligned}$$

all other matrix elements of H being neglected. The Coulomb integral α has a different value α' at the end atom ($m = 0$), so that Eq. (11.2) can be written as

$$(E - \alpha)c_m = \beta(c_{m+1} + c_{m-1}) \qquad \text{for } m \neq 0, \tag{11.4}$$

and

$$(E - \alpha')c_0 = \beta c_1 \qquad \text{for } m = 0. \tag{11.5}$$

For large N, c_N can be taken to be zero, and

$$c_m = \sin(N - m)\theta \qquad \text{for all } m.$$

On substitution for the coefficients of the wave functions in Eq. (11.4),

$$E = \alpha + 2\beta \cos \theta \tag{11.6}$$

in which the θ are the energy-eigenvalues. In order that the boundary conditions of Eq. (11.5) be simultaneously satisfied,

$$Z + \cos \theta + \sin \theta \cot N\theta = 0, \tag{11.7}$$

in which

$$Z = (\alpha - \alpha')/\beta.$$

Equation (11.7) has (at least) $(N - 1)$ roots with energy-eigenvalues in the range,

$$\alpha + 2\beta < E < \alpha - 2\beta; \tag{11.8}$$

consequently, a band of width 4β of non-localized orbitals is generated from a single state of the isolated state.

For $Z > (1 + N^{-1})$, the remaining Nth root has the form

$$\theta_N = i\zeta, \qquad \text{or } (\pi + i\zeta), \tag{11.9}$$

with ζ real and positive. For $\theta_N = i\zeta$ and large N, $Z - -\exp \zeta$, and this state exists when $Z < -1$. Its wave function is

$$c_m = c_0 \exp(-m\zeta) \tag{11.10}$$

and its energy is

$$E = \alpha + 2\beta \cosh \zeta, \tag{11.11}$$

with a magnitude lower than the lowest energy in the band. For $\theta_N = (\pi + i\zeta)$ and large N, $Z = \exp\zeta$, so that $Z > 1$. The wave function and energy are

$$c_m = c_0(-1)^m \exp(m\zeta) \tag{11.12}$$

and

$$E = \alpha - 2\beta \cosh \zeta, \tag{11.13}$$

the energy being higher than the highest energy of the band of non-localized states.

Since the atomic orbitals have the symmetry of s states, then $\beta < 0$. Consequently, when $Z = (\alpha - \alpha')/\beta$ becomes less than zero, the energy of the non-localized states decreases slowly (of order N^{-1}); at $Z = 1$, the energy of the localized state is less than that of the band energy and becomes an atomic orbital centred on the end-atom ($m = 0$) to form a surface state. Similarly, as Z is increased above unity, an atomic orbital is formed on the end atom with a higher energy state than that of the band. In both cases, the amplitude of the $(N - 1)$ non-localized states decreases to zero at the end atom.

Generalization to the surface of a three-dimensional solid is straightforward. For $Z > 1$, the single surface state of the end atom of the linear chain becomes a band of surface states of N^2 levels of width 8β. This latter overlaps the normal band states of the solid except when Z is large; then the two bands separate to give a two-dimensional array of atoms having electronic properties described by the band of surface states.

However, a metal atom usually has multi-band states; and the solution of the one-dimensional model requires more highly sophisticated mathematical techniques. Moreover, the tight-binding parameters of the bulk material are substantially altered by the perturbations arising from the free surface, so much so that, from a theoretical standpoint, the existence of surface states could be considered uncertain. Alternatively, CNDO (complete neglect of differential overlap) methods have been used [11, 12]; they make use of the atomic properties of orbitals and of the substrate atoms, but the results obtained are still only of a semi-quantitative character. In general, when these bulk lattice calculations have been employed to evaluate surface energies and work functions, they are of inadequate accuracy.

11.1.1. The surface bond with a chemisorbed atom [1–10]

The LCAO treatment for a one-dimensional chain of N atoms having s-like atomic orbitals with a chemisorbed atom λ interacting with the end-atom ($m = 0$) follows closely the procedure given in absence of a foreign atom.

The one-electron Hamiltonian operator now applies, however, to the chain plus the chemisorbed atom. Overlap between atomic orbitals is again neglected and the wave function of the complete system is still described by the equation

$$\psi = \sum_m c_m \phi(r, m). \tag{11.1}$$

But, in order to characterize the interaction of the chemisorbed atom, additional tight-binding approximations must be introduced, viz.

$$\begin{aligned} H(\lambda, \lambda) &= \alpha'', \\ H(0, \lambda) = H(\lambda, 0) &= \beta'. \end{aligned} \tag{11.14}$$

The system of linear equations

$$(E - \alpha)c_m = \beta(c_{m+1} + c_{m-1}), \qquad m \neq 0, \tag{11.15}$$

has, therefore, to be solved with the boundary conditions,

$$(E - \alpha')c_\lambda = \beta c_1 + \beta' c_\lambda, \tag{11.16}$$

and

$$(E - \alpha'')c_\lambda = \beta' c_0. \tag{11.17}$$

For N large, the solutions are

$$c_m = \sin(N - m)\theta, \qquad \text{for } m \neq \lambda. \tag{11.18}$$

$$c_\lambda = (\eta \sin N\theta)/(Z' + Z \cos \theta), \qquad \text{for } m = \lambda, \tag{11.19}$$

in which

$$Z = (\alpha - \alpha')/\beta, \qquad Z' = (\alpha - \alpha'')\beta, \qquad \eta = \beta'/\beta. \tag{11.20}$$

The energy levels are

$$(E - \alpha)/2\beta = E' = \cos \theta, \tag{11.21}$$

with a range $-1 < E' < 1$ and with θ as one of $(N - 1)$ roots of the equation

$$(Z + \cos \theta + \sin \cot N\theta)(Z' + 2 \cos \theta) = \eta^2. \tag{11.22}$$

Its associated wave functions are non-localized and within the energy range of the normal band of bulk states; but the remaining two roots of the total of $(N+1)$ roots can have the form $i\zeta$ and $(\pi + i\zeta)$, with ζ real and positive, and then localized states centred around the chemisorbed and end atoms can exist with energies outside the normal band.

For E' positive,

$$(Z + \exp\zeta)(Z' + 2\cosh\zeta) = \eta^2 \tag{11.23}$$

so that

$$c_m = \exp(-m\zeta) \qquad \text{for } m \neq \lambda, \tag{11.24}$$

$$c_\lambda = \eta/Z' + 2\cosh\zeta) \qquad \text{for } m = \lambda, \tag{11.25}$$

and

$$E' = \cosh\zeta. \tag{11.26}$$

For E' negative,

$$(Z - \exp\zeta)(Z' - 2\cosh\zeta) = \eta^2, \tag{11.27}$$

with

$$c_m = (-1)^m \exp(-m\zeta) \qquad \text{for } m \neq \lambda, \tag{11.28}$$

$$c_\lambda = \eta/(Z' - 2\cosh\zeta) \qquad \text{for } m = \lambda, \tag{11.29}$$

and

$$E' = -\cosh\zeta. \tag{11.30}$$

Localized states, therefore, occur when either Eq. (11.23) or Eq. (11.27) has real roots ζ; the existence of such states depends on the extent of interaction between the chemisorbed and end-atom, i.e. on the values of Z, Z' and η.

If the interaction is confined to one orbital on the chemisorbed atom and one band of bulk orbitals, a maximum of two localized states are formed when the perturbation of the solid by the chemisorbed atom is confined to the end-atom ($m = 0$). But bonds of multiple bond-orders can be formed if there is more than one orbital of the chemisorbed atom, or more than one band of solid states, of if the perturbation extends into the solid ($m > 0$).

For a three-dimensional model, with the surface completely covered by (N_m) chemisorbed atoms, the energy levels are

$$E = \alpha + 2\beta(\cos\theta_1 + \cos\theta_2 + \cos\theta_3) \tag{11.31}$$

for a cubic lattice. For the (100) plane, θ_1 is any root of the equation

$$(Z + \cos\theta_1 + \sin\theta_1 \cot N_m\theta_1)\ (Z' + 2(1-\eta')$$
$$\times (\cos\theta_2 + \cos\theta_3) + 2\cos\theta_1) = \eta^2, \tag{11.32}$$

with

$$\theta_2 = k_2\pi/N_m, \qquad \theta_3 = k_3\pi/N_m, \qquad k_2, k_3 = 1, 2, \ldots, N_m. \quad (11.33)$$

The additional interaction parameter $\eta' = \beta''/\beta$ is necessary so as to include the resonance integral β'' between nearest neighbour adatoms. The two discrete localized states of the one-dimensional chain are now transformed to two bands of surface states associated with the adsorbed layer on the solid surface; each band contains N_m^2 levels, i.e. corresponding to the number N_m of adatoms on the surface. Such bands can overlap or be separated from the normal band of bulk states by a suitable choice of the magnitude of Z and Z'. When $\beta'' < \beta$, some states can be absent from the band of N_m^2 levels and become non-localized in the normal bulk band; localized states can have energies both above and below the bulk band range of energies.

11.1.2. Discussion of the bulk lattice model

The LCAO treatment of the electronic characteristics of a clean metal surface and of the surface bond of a foreign atom with this surface leads to some important general conclusions, despite the semi-quantitative nature of this approach. The main interest in the free surface is the existence of free valences or dangling bonds, a concept which can be more clearly defined in terms of the presence of surface energy levels or bands which are outside the non-localized energies within the solid. The tunnelling effect in quantum mechanics indicates that an electron confined to a potential well has a definite probability of tunnelling through the potential barriers that classically confine the electron. The wave function of the electron, which in terms of a one-dimensional free-electron model is $\psi = \sin kx$ inside the lattice, should join smoothly at $x = 0$ to a decaying, approximately exponential, wave function outside in vacuum; the electron, therefore, has a finite probability ψ^2 of being outside the geometric surface. A highly simplified (delta function) model in which the potential is represented by a periodic square well with infinitely high barriers and infinitely narrow barriers between the wells was developed by Tamm [13]. His calculations revealed the existence of wave functions localized at the surface with energies outside those within the bulk, there being one surface level for each energy gap between the allowed energy bands. This type of surface state is referred to as a Tamm state.

In the LCAO treatment, using the tight binding approximation, the existence of localized surface states depends on the value of the parameter $Z = (\alpha - \alpha')/\beta$, where $H_{mm} = \alpha$, $H_{00} = \alpha'$ and $H_{m,m+1} = \beta$, the surface coulomb integral H_{00} being different from the bulk value H_{mm} to represent the distortion of the potential at the surface, and the resonance integral

$H_{m, m \pm 1}$ being confined to nearest neighbours. Within the energy limits

$$\alpha + 2\beta < E < \alpha - 2\beta,$$

only non-localized states (surface and bulk) within a quasi-continuous band exist. But for $Z < -1$, and N large, the Nth root has an energy $E = \alpha + 2\beta \cosh \xi$, where ξ is real and positive, below the lowest energy in the quasi-continuous band. It represents a localized (**P**) state ($m = 0$). For $Z < +1$, and N large, there is a localized (**N**) state of energy $E = \alpha - 2\beta \cosh \xi$ which is above the highest level of the band. These surface states can also be obtained by using a different value of β for the surface atoms. As $Z \to -\infty$, the lower is the surface energy state and the more discrete is the atomic orbital ϕ_0 on the end atom.

For the three-dimensional case, the discrete atomic orbitals ϕ_0 form a band of surface states, but only for large Z, and, since these arise from one

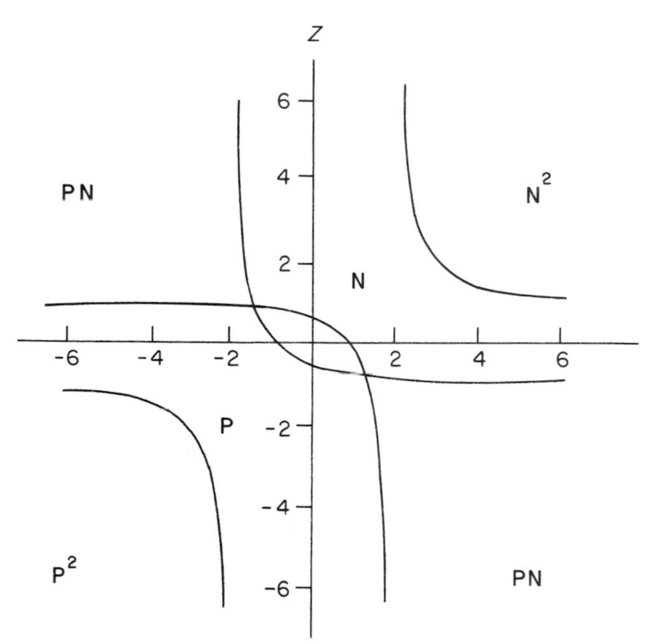

FIG. 11.2. Localized states for an adatom at a metal surface. The occurrence of localized states for a specific value of η^2 is defined by the equation

$$(Z \pm 1)(Z' \pm 2) = \eta^2 = (\beta''/\beta)^2,$$

where β = the resonance integral $H_{m, m+1}$ and β'' = the resonance integral between nearest-neighbour adatoms; $Z = (\alpha - \alpha')/\beta$. $Z' = (\alpha - \alpha'')/\beta$, in which $H_{mm} = \alpha$, $H_{00} = \alpha'$ and $H(\lambda, \lambda) = \alpha''$.

For $\eta^2 = 1$, the number of localized states in the various regions are given in parentheses:

$\mathbf{P}^2$(2); $\mathbf{N}^2$(2); **P**(1), **N**(1); **PN**, **P**(1), **N**(1) atomic orbital, they are Tamm states. For smaller Z values, localization is less complete and these states may extend to a significant extent into the bulk.

In the one-dimensional surface-bond treatment, Eq. (11.22) for large N indicates that localized states exist in regions of the ZZ' plane for a given η when (see Fig. 11.2), $(Z \pm 1)(Z' \pm 2) = \eta^2$. For $\eta^2 = 1$, in the regions $\mathbf{P}^2$ and $\mathbf{N}^2$, there are two localized states, but only one in the regions **P** and **N**; there are one **P** and one **N** state in the two **PN** regions. The bonding states depend on the sign of β, the resonance integral, which is negative for s atomic orbitals and positive for p orbitals of the solid. Table 11.1 gives a classification of possible bonding states [1–10].

TABLE 11.1

Adatom orbital	Solid atomic orbital	Bonding state
s	s, d_x^2	$\mathbf{P}_\sigma$
s	p_z	$\mathbf{N}_\sigma$
p_x	p_x, d_{zx}	$\mathbf{P}_\pi$
p_y	$p_y d_{yz}$	$\mathbf{P}_\pi$
p_z	p_x	$\mathbf{N}_\sigma$
p_z	d_z	$\mathbf{P}_\sigma$

The σ states are symmetrical about the linear chain axis and π states have a nodal plane passing through the axis; z is the axis along the chain.

Extension to the three-dimensional case complicates the mathematics, but the general conclusions are not greatly different [11, 12]. The problem is simplified when there is a monolayer of adatoms at the surface. It is necessary to include three Coulomb integrals with different α, α', α'' values, respectively, for atoms in the bulk, surface atoms and adatoms, and three corresponding resonance integrals β (bulk), β' (surface atom, adatom) and β'' (between nearest neighbours in the adsorbed layer), and also a new parameter $\eta' = \beta''/\beta$. For the special case $\eta' = 1$, the resonance integral between adsorbed atoms equals that between bulk atoms, and the solution then closely resembles that of the one-dimensional model, except that the discrete states become bands; otherwise no new features arise. When $\eta' \neq 1$, Z' has different limits $Z' \pm 4(1 - \eta')$ and the pattern shown is altered in detail, e.g. incomplete bands of **P** and **N** states with less than the N^2 levels can now arise.

The great weakness of the bulk lattice model is that a detailed knowledge of the magnitudes of the tight-binding parameters, α, β and analogous quantities is not available, except possibly some bulk values. For the surface region, however, the shift of charge can strongly affect the bulk values, and even approximate values of the surface parameters cannot be estimated. The tight-binding approximation for metals is a poor one, and is particularly so for transition metals, but the LCAO treatment does provide some broad insight into some of the characteristics of surface bonding.

11.2. THE JELLIUM MODEL

This model is an extension of the simple free-electron model; only the valence electrons of the metal contribute to the electron charge density, and the ion-cores (the nuclei plus inner-shell electrons) are represented by a smeared-out positive charge of density equal to that of the free electrons. The electron charge distribution within the bulk of the metal is uniform, but the distribution spreads outwards beyond the geometric boundary of the surface in order to decrease the kinetic energy of the free electrons; the process is limited by the increase in the electronic potential energy (or the electrostatic potential) and simultaneous smaller increases due to exchange and correlation terms (Fig. 11.3). In the bulk, the attractive exchange-correlation potential arises

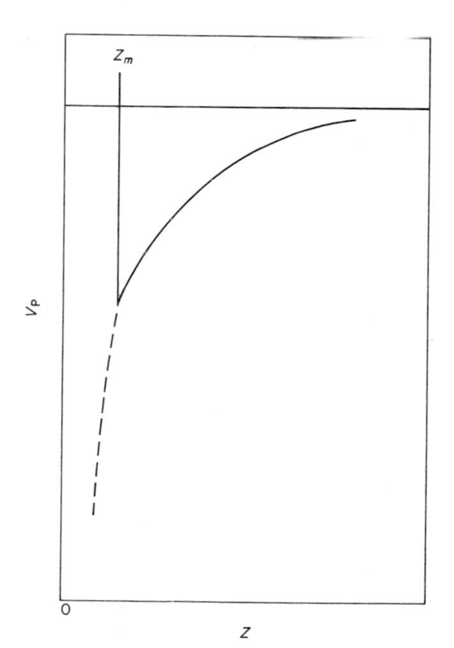

FIG. 11.3. Representation of pseudo-potential (full line) compared with core potential (dashed line) for $z > z_m$, $V_P = -e^2/z$ for a single-valence metal atom.

from the presence of a spherical cavity of reduced charge density surrounding each electron; this potential is responsible for the classical image term. It decreases in an approximately exponential manner with distance outside the surface, as does the electrostatic potential, i.e. the electron density decays exponentially at the surface. A double-charge layer is, therefore, set up with its negative charge outermost.

In the simplest jellium model, the ground-state energy of such a system, E_n, is a unique function of the electron number density $n(\mathbf{r})$. Its value is obtained by minimization of this energy with respect to variations of $n(\mathbf{r})$, while conserving the total number of electrons, N, where $N = \int n(\mathbf{r})\,d\mathbf{r}$, and $\mathbf{r}$ is the distance normal to the surface. The positive charge density is given by $n_+(\mathbf{r}) = \bar{n}$ for $\mathbf{r} < 0$, and $n_+(\mathbf{r}) = 0$ for $\mathbf{r} > 0$, and the electron density by $\lim n(\mathbf{r}) = \bar{n}$ as $\mathbf{r} \to -\infty$ and $\lim n(\mathbf{r}) = 0$ as $\mathbf{r} \to +\infty$. The minimization condition may be written as

$$(\delta/\delta n)\,[E(n) - \mu N] = (\delta/\delta n)\left[E(n) - \mu \int n(\mathbf{r})\,d\mathbf{r}\right] = 0, \qquad (11.34)$$

where μ is a Lagrange multiplier which is identified with the chemical potential. The energy E(n) of the interacting electron system comprises three terms: (i) the energy of the electrons in the external field $[V(\mathbf{r})]$ due to the positive charge density $n(\mathbf{r})$, i.e. $\int V(\mathbf{r})n(\mathbf{r})\,d\mathbf{r}$; (ii) electron–electron interaction terms given by

$$\frac{1}{2}\int \frac{n(\mathbf{r})n(\mathbf{r}')\,d\mathbf{r}\,d\mathbf{r}'}{(\mathbf{r} - \mathbf{r}')};$$

and (iii) a bulk term $G(n)$ which denotes the sum of kinetic, exchange and correlation terms; the chemical potential μ relative to the inner potential is, therefore, given by $\delta G(n)/\delta n(\mathbf{r})$. The first two two terms can be expressed in terms of an electrostatic potential V_E, so that

$$E(n) - \int V_E(\mathbf{r})\,n(\mathbf{r})\,d\mathbf{r} + G_n. \qquad (11.35)$$

This potential must obey the Poisson equation, i.e.

$$\nabla^2 V_E = -4\pi[n(\mathbf{r}) - n_+(\mathbf{r})] \qquad (11.36)$$

in order to preserve self-consistency. Various approximations have been used to evaluate $G(n)$ [14]. The electron charge density is then expressed as an exponential function of distance z with respect to a plane midway between the plates of the surface double layer, i.e.

$$\left.\begin{aligned} n(z) &= \bar{n} - \tfrac{1}{2}\bar{n}\,e^{\beta z} \qquad &\text{for } z < 0,\\ n(z) &= \tfrac{1}{2}\bar{n}\,e^{-\beta z} \qquad &\text{for } z > 0. \end{aligned}\right\} \tag{11.37}$$

The value of β is obtained by combining Eq. (11.34) with Eq. (11.37), inserting the calculated values of $G(n)$, and minimizing the energy $E(n)$ with respect to β. For various *sp* metals having different values of $\bar{n}$, β is approximately constant (1·2 to 1·3); this magnitude indicates a rapid decay of the electron density from the metal surface into the vacuum [19].

The jellium model allows an assessment of the work function and of the electron charge density at the surface. It has also been developed in a more sophisticated form to evaluate the binding energy of a point charge as a function of distance from the surface [20]. However, this model is invalid for *d* band metals since the band structure arising from the discrete ion potentials becomes a major factor (see later).

11.3. THE PSEUDO-POTENTIAL METHOD [21–25]

In the pseudopotential method, the crystal lattice potential is included and a self-consistent calculation of electron properties can be made. Bulk band structures can be satisfactorily elucidated and the method has also been applied with some success to surface properties. The main features of this approach are outlined (see Fig. 11.4).

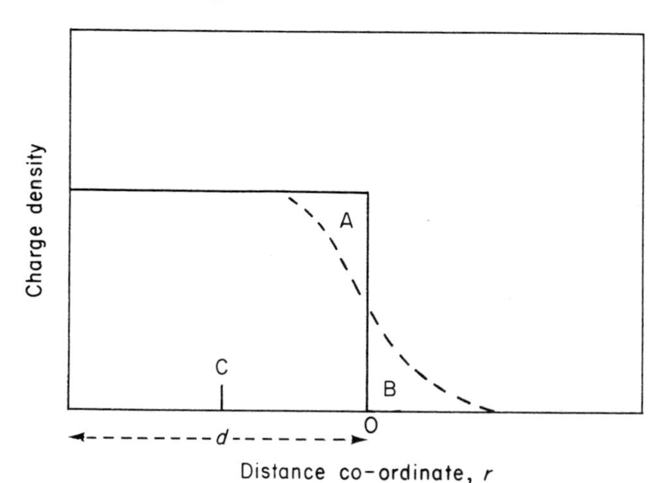

FIG. 11.4. Jellium model. A, positive charge density $n_+(r) = \bar{n}$ for $r < 0$ and $n_+(r) = 0$ for $r > 0$. B, electron charge density $n(r) \to 0$ as $r \to \infty$ and $\bar{n}$ as $r \to -\infty$. C, position of nuclei for surface plane.

The Bloch wave functions of the valence electrons are expressed as linear combinations of orthogonalized waves given by

$$W_{\mathbf{k}}(\mathbf{r}) = \exp(i\mathbf{k}\,.\,\mathbf{r}) - \sum_{\alpha} \varphi_{\alpha}(\mathbf{r}) \int \varphi_{\alpha}(\mathbf{r}') \exp i\mathbf{k}\,.\,\mathbf{r}'\, \mathrm{d}\mathbf{r}', \tag{11.38}$$

where $\mathbf{k}$ is the electron wave vector and φ_{α} represents the core states. The free-electron band states are then given by

$$\psi_{\mathbf{k}}(\mathbf{r}) = \sum_{\mathbf{G}} \alpha_{\mathbf{k}}(\mathbf{G})\, W_{\mathbf{k}+\mathbf{G}}(\mathbf{r}), \tag{11.39}$$

where $\mathbf{G}$ are reciprocal lattice vectors.

Insertion of $\psi_{\mathbf{k}}(\mathbf{r})$ into the Schrödinger equation gives

$$-\tfrac{1}{2}\Delta^2\psi_{\mathbf{k}}(\mathbf{r}) + V_{\mathrm{P}}\psi_{\mathbf{k}} = E_{\mathbf{k}}\psi_{\mathbf{k}}(\mathbf{r}), \tag{11.40}$$

where $\psi_{\mathbf{k}}(\mathbf{r}) = \sum_{\mathbf{G}} \alpha_{\mathbf{k}}(\mathbf{G}) \exp i(\mathbf{k} + \mathbf{G})\,.\,\mathbf{r}$, corresponding to a linear combination of plane waves.

The pseudo-potential V_{P} is

$$V_{\mathrm{P}}\psi_{\mathbf{k}} = V(\mathbf{r})\,\psi_{\mathbf{k}}(\mathbf{r}) + \sum_{\alpha} (E_{\mathbf{k}} - E_{\alpha})\,\varphi_{\alpha}(\mathbf{r}) \int \varphi_{\alpha}(\mathbf{r}')\,\psi_{\mathbf{k}}(\mathbf{r}')\,\mathrm{d}\mathbf{r}', \tag{11.41}$$

the first term being the attractive lattice potential. The second is a repulsive term arising from the orthogonalization; it is energy-dependent and non-localized.

The pseudo-potential method can be used to evaluate the electronic properties of surfaces of *sp* metal. The bulk pseudo-potential wave functions

$$\psi_{\mathbf{k}}(\mathbf{r}) = \sum_{\mathbf{G}} \alpha_{\mathbf{k}}(\mathbf{G}) \exp(i(\mathbf{k} + \mathbf{G})\,.\,\mathbf{r}) \tag{11.42}$$

are smoothly joined to the wave functions in the surface region (comprising a small number of atomic layers) and the vacuum outside. These have the form

$$\psi_{\mathbf{r}} = \sum_{\mathbf{G}} U(z) \exp[i(\mathbf{k} + \mathbf{G})\,.\,\mathbf{r}] \tag{11.43}$$

in the direction of the surface plane, where $U(z)$ decays into vacuum. The eigen-energies are obtained by substitution into the Schrödinger equation in which the potential energy is given by

$$V(\mathbf{r}) = \varphi(\mathbf{r}) + Fn^{\frac{1}{3}} + V_{\mathrm{P}}(\mathbf{r}), \tag{11.44}$$

where $V(\mathbf{r})$ is the electrostatic potential determined from Poisson's equation, $Fn^{\frac{1}{3}}$ is the exchange and correlation term with F as an adjustable parameter; $V_P(\mathbf{r})$ is the pseudo-potential. The surface energies and work function of Na(100) have been calculated in this manner [26].

For bulk transition metals the d-electron bands are treated as atomic-like orbitals which interact with sp pseudo-potentials [27]. This interaction causes the atomic d level to increase by several electron volts at the main level of the d bandwidth in the bulk and the bottom of the s band is depressed below that of the s atomic level. Extension to transition metal surfaces has also been attempted. It would appear that the main differences from the electronic structure in bulk is: (i) leakage of s electron charge into the vacuum; and (ii) a decrease in the number of d-band holes in the surface compared with that in bulk.

11.4. THE WORK FUNCTION

The electron work function Φ can be defined as the difference in energy of an electron in the bulk metal at the Fermi level and at the vacuum level outside the surface at a distance such that it has negligible interaction with the surface ($>10^{-4}$ cm). Similarly, the chemical potential μ of the electron in the bulk can be equated to the difference in its energy within a metal having an equal number of positive and negative changes and one having one electron less. At the surface, electrons spread into vacuum to smooth out the discontinuity in potential, and also spread across the surface to decrease the amplitude of the variation in potential over the surface. A dipole surface layer is thereby created, the metal surface potential $\Delta\chi$ being given by Poisson's equation,

$$\Delta\chi = 4\pi \int_{\text{surface}} z[n(z) - n_+(z)]\, \mathrm{d}z, \tag{11.45}$$

where z is the distance co-ordinate outwards and normal to the surface, and $n(z)$ and $n_+(z)$ are the average electron and positive charge densities in a plane at a distance z.

The work function, therefore, comprises a volume contribution equal to the work done against the chemical potential μ of the electron at the Fermi level, and a surface contribution comprising that done against the difference of electrostatic potential at the surface. It follows that

$$e\Phi = e\Delta\chi - \mu \tag{11.46}$$

and

$$\Delta\chi = \Phi_{\text{vacuum}} - \Phi_{\text{Fermi level}} = \Phi_{\text{outer}} - \Phi_{\text{inner}}. \tag{11.47}$$

11.4.1. Theoretical evaluation of the work function [28, 29]

The most successful calculation of the value of the work function of a metal has been carried out using the jellium model. The work function, or the minimum amount of energy required to remove an electron from the metal at 0 K, is given by

$$e\Phi = -e(\Phi_{\text{inner}} - \Phi_{\text{outer}}) - \mu. \tag{11.48}$$

When the work function, represented by $e\Phi_0$, is expressed relative to the electrostatic potential as zero, i.e. $\Phi_{\text{inner}} = 0$, then

$$e\Phi_0 = \Phi_{\text{outer}} - \mu, \tag{11.49}$$

in which $\mu = \langle \delta G(n)/\delta n(z) \rangle$ for the jellium model and

$$e\Phi_0 = 4\pi \int_{-\infty}^{+\infty} [n(z) - n_+(z)] z \, dz, \tag{11.50}$$

with $n_+(z) = \bar{n}$ for $z \leqslant 0$, and $n_+(z) = 0$ for $z > 0$. The necessary self-consistency between the electronic charge density and the potential arising from the spread of density into vacuum has already been ensured by making the potential obey the Poisson equation. As before, the electron charge density normal to the surface is represented by

$$\begin{aligned} n(z) &= \bar{n} - \tfrac{1}{2}\bar{n} \exp(\beta z) \qquad &\text{for } z \leqslant 0, \\ n(z) &= \tfrac{1}{2}\bar{n} \exp(-\beta z) \qquad &\text{for } z > 0, \end{aligned} \tag{11.51}$$

for which a value of $\beta = 1{\cdot}2$ to $1{\cdot}3$ has been calculated for a variety of *sp* metals.

The calculated work functions of *sp* metals obtained by this procedure are within 0·5 eV (or less) of the experimental values (see Fig. 11.5). The simple jellium model cannot be applied to *d*-band metals; moreover, the effect of different atomic densities at the various crystallographic planes of a metal has been neglected.

11.4.2. The effect of crystallinity [30]

The surface electrostatic barrier set up by the charge double-layer arises from the spreading of the electron distribution from the geometric surface into vacuum in an attempt to lower the kinetic energy of the electrons. The double layer is, therefore, negative on the vacuum side and electrons have to penetrate this barrier (~0·4 eV) in order to escape from the metal surface, i.e. they must overcome the image potential arising from the existence of equal but positive charges within the metal. The magnitude of this potential is largely determined by: (i) an outward spread of electrons, the nature of which is fairly insensitive to the atomic density of the surface; and (ii) a smoothing of the surface charges in order to minimize the localized effects of the positive ion-cores at the surface by increasing the charge density in the valleys between the

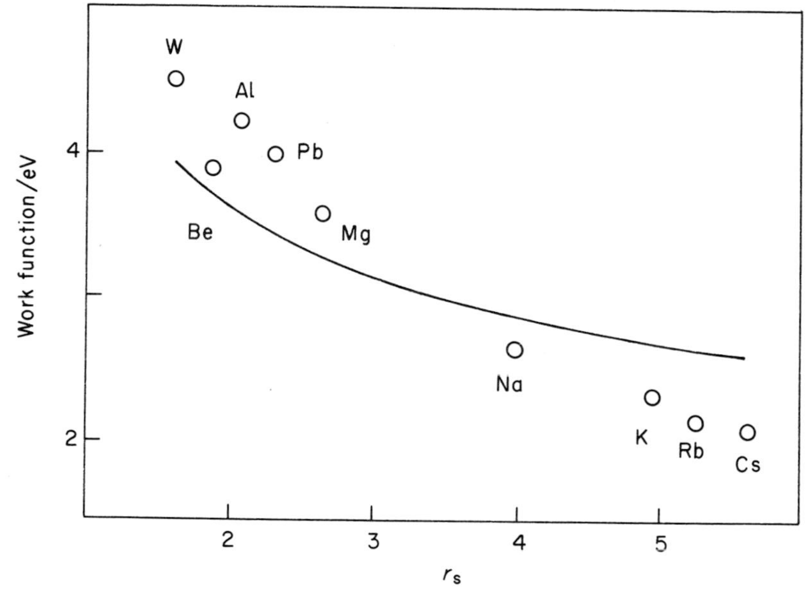

FIG. 11.5. Some theoretical and experimental values of the work function. —, Theoretical values (Smith [28]); ○, experimental values (Lang and Kohn [29]; $r_c = (\frac{1}{4}\pi n_+)^{\frac{1}{3}}$.

cores and decreasing it in the immediate vicinity of the protruding atoms. Its effect is opposite to that of the outward spreading of electrons and, hence, the work function is reduced; moreover, the degree of smoothing is very sensitive to the atomic density of the crystal plane. Consequently, the image potential is larger at the high-density low-index planes, and such planes have higher work functions. Thus, the most densely packed plane of tungsten W(110) has a work function of 5·3 eV compared with that (4·3 eV) of the lower-density W(310) plane.

To take account of the different atomic densities, the jellium model has been extended so as to include the effect of discrete ion-cores [20, 28, 29, 31, 32] by use of pseudo-potentials of the form

$$\begin{aligned} V_{\mathrm{P}}(\mathbf{r}) &= 0 \qquad \text{for } \mathbf{r} < r_{\mathrm{c}} \\ V_{\mathrm{P}}(\mathbf{r}) &= z/\mathbf{r} \qquad \text{for } \mathbf{r} > r_{\mathrm{c}} \end{aligned} \tag{11.52}$$

(where r_{c} is the radius of the ion-core), by application of first-order perturbation theory. The perturbing potential δV_{P} is the difference of the potential of this array of pseudo-potentials and that derived from the jellium model. It is related to the change of work function through Eq. (11.53),

$$\delta\Phi = \int \delta V_{\mathrm{P}}(\mathbf{r})\, n_{\mathrm{d}}(z)\, \mathrm{d}\mathbf{r}, \tag{11.53}$$

where $n_{\mathrm{d}}(z)$ is the difference in the electron number density brought about by removing one electron from the unperturbed jellium model. It follows that

$$n_{\mathrm{d}}(z)\, \mathrm{d}\mathbf{r} = -1 \tag{11.54}$$

and

$$\delta\Phi = -\int \delta V_{\mathrm{P}}(z)\, n_{\mathrm{d}}(z)\, \mathrm{d}z \Big/ \int n_{\mathrm{d}}(z)\, \mathrm{d}z, \tag{11.55}$$

in which $\delta V_{\mathrm{P}}(z) = \int \delta V(\mathbf{r})\, \mathrm{d}x\, \mathrm{d}y$.

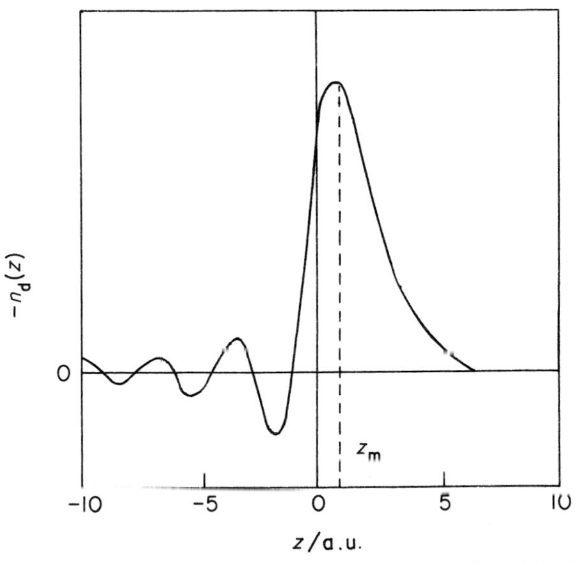

FIG. 11.6. Typical plot of screening charge density $n_{\mathrm{d}}(z)$ as a function of the distance co-ordinate z. It has a maximum at z_{m} to give the position of the image plane; $z = 0$ is the edge of the uniform positive background.

The term $n_d(z)$ may be represented as a surface charge density induced by a weak electric field perpendicular to the surface, and has a sharp maximum at a particular value, z_m, which, therefore, gives the location of the image plane (see Fig. 11.6), and can be evaluated using Eq. (11.56):

$$z_m = \int_{-\infty}^{+\infty} z n_d(z)\,dz \Big/ \int_{-\infty}^{+\infty} n_d(z)\,dz. \qquad (11.56)$$

With z_m known and $n_d(z)$ evaluated from the jellium model for a particular ion-core radius, the magnitude of $\delta\Phi$ can be calculated.

REFERENCES

1. T. B. Grimley, *Adv. Catalysis*, 1960, **12**, 1.
2. T. B. Grimley, *Proc. Phys. Soc.*, 1967, **90**, 751.
3. T. B. Grimley, *Proc. Phys. Soc.*, 1967, **92**, 776.
4. T. B. Grimley and S. M. Walker, *Surface Sci.*, 1969, **14**, 395.
5. D. M. Newns, *Phys. Rev.*, 1969, **178**, 1123.
6. D. M. Newns, *Phys. Rev.*, 1970, **24**, 1575.
7. D. M. Newns, *Phys. Rev. B*, 1970, **1**, 3304.
8. D. M. Newns, *Phys. Letters*, 1972, **38**, 341.
9. B. R. Cooper and A. J. Bennett, *Phys. Rev. B*, 1970, **1**, 4654.
10. J. W. Gadzuk, *Surface Sci.*, 1974, **43**, 44.
11. J. A. Pople, D. D. Santry and G. A. Segal, *J. Chem. Phys.*, 1965, **43**, 129.
12. A. J. Bennett, B. McCarroll and R. P. Messner, *Phys. Rev. B*, 1971, **3**, 1397.
13. I. Tamm, *Phys. Z. Sowjet*, 1932, **1**, 733.
14. P. Hohenburg and W. Kohn, *Phys. Rev. B*, 1964, **136**, 864.
15. W. Kohn and L. J. Sham, *Phys. Rev. A*, 1965, **140**, 1133.
16. N. D. Lang and W. Kohn, *Phys. Rev.*, 1970, **1**, 4555.
17. N. D. Lang and W. Kohn, *Phys. Rev.*, 1971, **3**, 1215.
18. N. D. Lang, *Solid State Phys.*, 1973, **28**, 225.
19. J. R. Smith, *Phys. Rev.*, 1969, **181**, 522.
20. J. R. Smith, S. C. Ying and W. Kohn, *Phys. Rev. Letters*, 1973, **30**, 610.
21. W. A. Harrison, "Pseudo-potentials in the Theory of Metals", W. A. Benjamin, New York, 1966.
22. V. Heine, *Surface Sci.*, 1964, **2**, 1.
23. V. Heine, *Phys. Rev.*, 1967, **153**, 673.
24. J. A. Applebaum and D. R. Hamann, *Phys. Rev. B*, 1972, **6**, 2166.
25. J. A. Applebaum and D. R. Hamann, *Phys. Rev. Letters*, 1973, **31**, 106.
26. N. D. Lang, *Solid State Comm.*, 1969, **7**, 1047.
27. K. Levin, A. Liebsch and K. H. Bennerman, *Phys. Rev. B*, 1973, **7**, 3066.
28. J. R. Smith, *Phys. Rev.*, 1969, **181**, 522.
29. N. D. Lang and W. Kohn, *Phys. Rev. B*, 1971, **3**, 1215; 1973, **8**, 6010.
30. R. Smoluchowski, *Phys. Rev.*, 1941, **60**, 661.
31. N. W. Ashcroft and D. C. Langreth, *Phys. Rev.*, 1967, **155**, 1682; **159**, 1967.
32. J. A. Appelbaum and D. R. Hamann, *Phys. Rev. B*, 1972, **6**, 2166.

12

Perturbation of Surface Electronic Properties by Chemisorption. I. The Work Function

12.1. INTRODUCTION

The formation of a chemisorption bond is accompanied by a change in the distribution of the local density of states of the valence electrons of the surface metal atoms. The nature and magnitude of the redistribution largely depend on the electronegativities of the adspecies and the metal atoms, but the symmetry characteristics and degree of occupancy of the metal orbitals are also relevant factors. This perturbation of surface atoms is often substantial and they may be displaced from their original locations.

The modification of the charge distribution by chemisorption is evidenced by the change of: (i) the work function of the metal; (ii) its electrical conductivity; and (iii) should bonding involve the pairing of unpaired spins of *d* electrons, its paramagnetic susceptibility. Accurate measurements of the work function of single crystal planes of extremely small area can be accomplished and, for *sp* metal surfaces, the theoretical background is sufficiently well advanced that *ab initio* calculations of its magnitude can be satisfactorily carried out.

In contrast, electrical conductivity and magnetic susceptibility investigations involve large surface-to-volume ratios and measurements have been largely confined to finely divided (supported) powders and evaporated films.

The interpretation of the results is controversial and imprecise, and many of the conclusions drawn are of doubtful validity. Nevertheless, in principle, information about gas–metal interactions should emerge following advances in our understanding of both the experimental and theoretical aspects of such investigations.

12.2. THE WORK FUNCTION (OR SURFACE POTENTIAL) OF ADSORBATES

The work function Φ of a metal has been defined in Chap. 11. The relationship between the various terms involved are summarized in Fig. 12.1. The value of Φ depends on the atomic density of the ion-cores at the crystal plane under examination and is markedly changed by the chemical nature and surface concentration of a chemisorbed overlayer.

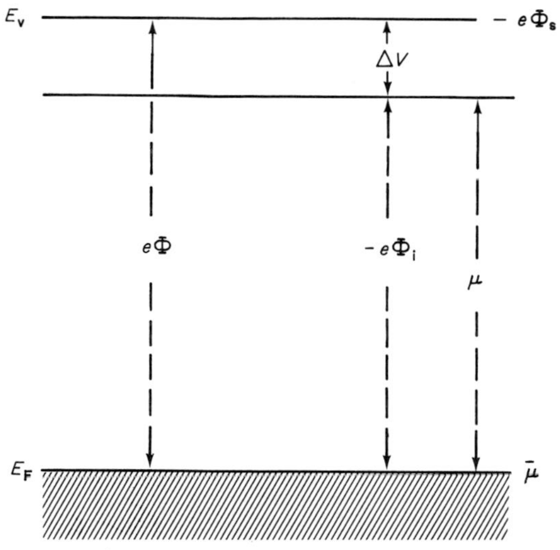

FIG. 12.1. The energy to transfer an electron from the Fermi (E_F) to the vacuum (E_v) level equals $e\Phi$, where Φ is the work function. At the surface, due to the outward spread of conduction electrons, a surface, or outer, electrostatic potential Φ_s is set up. The difference between Φ_s and Φ_i, the inner electrostatic potential of electrons inside the bulk metal, gives the potential difference ΔV across the double layer, $\Delta V = \Phi_s - \Phi_i$. The chemical potential of free electrons in the metal is μ and $\Phi = -\mu/e + (\Phi_i - \Phi_s) = -\bar{\mu}$, where $\bar{\mu}$ is the electrochemical potential and equals E_F. A usual reference state is $-e\Phi_s = E_v = 0$.

When chemisorption takes place, the chemisorbed molecule is polarized and each molecule may be considered as a dipole aligned perpendicularly to the surface. For a complete monolayer of adsorbate comprising n_m cm^{-2}

adsorbed molecules, the charge density is uniform over the two charge sheets of the double layer. The system represents a parallel-plate capacitor with its plates separated by a distance d between the centres of gravity of charge. The dipole moment μ_0 of the adsorbed molecule is defined as $\mu_0 = ed$, e being the electronic charge. No external field emanates from the double layer, but the work function of the metal, Φ, is changed by the presence of the chemisorbed layer. When the dipole has its negative charge outermost from the surface, the work function is increased. It is now customary to refer to the surface potential of the adsorbed species $\Delta\chi = -\Delta\Phi$ and, in the circumstance above, it carries a negative sign. From classical electrostatic theory, the potential between the plates of this parallel-plate capacitor (or the change in surface potential, $\Delta\chi$) is given by

$$\Delta\chi = 4\pi n_{\mathrm{m}}\mu_{\mathrm{m}}, \tag{12.1}$$

where μ_{m} is the dipole moment of each admolecule in the full monolayer, and includes the effect of mutual depolarization of adjacent molecules in the adsorbed layer, all of which have dipoles of the same sign. At lower fractional coverages θ, the charge distribution is discontinuous but can be assumed [1–4] to be uniformly smoothed out over the dipole sheets that form the double layer so that

$$-\Delta\Phi = \Delta\chi_{\theta} = 4\pi n_{\mathrm{m}}\theta\mu_0. \tag{12.2}$$

The importance of measurements of $\Delta\chi$ is the information that they provide about the charge transfer between adatoms and surface atoms and, therefore, the effect of chemisorption on the local density of electron states at the surface.

12.2.1. Experimental methods [5, 6]

In the *Kelvin or contact potential method* [7], two plates, comprising different metals having work functions Φ_1, Φ_2, form the plates of a parallel-plate capacitor of capacitance C; they are connected in series with a sensitive ammeter and a source of external potential V_{E} (Fig. 12.2). Equilibrium requires that the electrochemical potentials, $\bar{\mu}_1, \bar{\mu}_2$ of the free electrons of the two metals should be equal. The potentials just outside the surface of the metals are

$$V_1 = (-\Phi_1 + \bar{\mu}_1/e), \qquad V_2 = (-\Phi_2 + \bar{\mu}_2/e): \tag{12.3}$$

hence,

$$e(V_1 - V_2) = \Phi_2 - \Phi_1 = eV_{12}, \tag{12.4}$$

where V_{12} is the contact potential difference. Hence, V_{12} generates a current; this is reduced to zero by adjusting a backing-off potential V_E until $-V_E = V_{12}$. For chemisorption studies, one plate (2), the work function of which is unchanged by presence of the gaseous adsorbate, acts as an inert stable reference electrode. The $\Delta\chi$ value is given by the difference of V_E with and without chemisorption on plate (1).

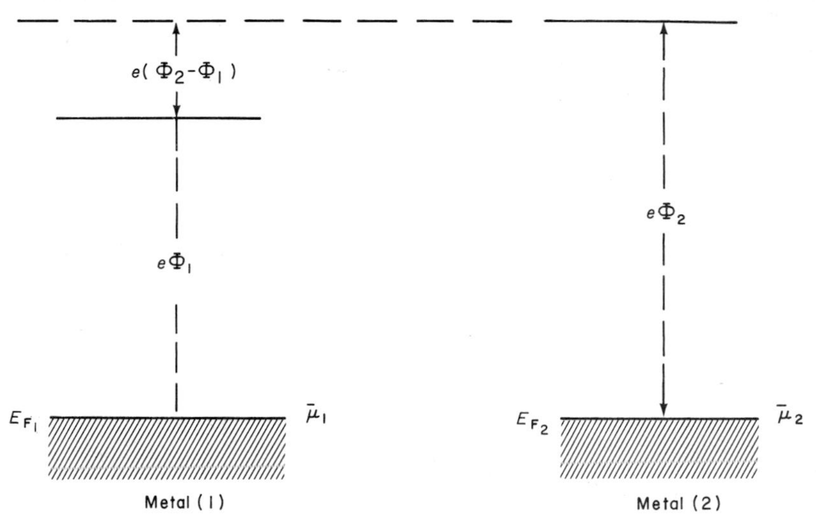

FIG. 12.2. Contact potential difference. Metals (1) and (2) have different values of E_F. When they are put in electrical contact, electron transfer takes place until $E_{F_1} = E_{F_2}$, and their electrochemical potentials are equalized, $\bar{\mu}_1 - \bar{\mu}_2 = -e(\Phi_1 - \Phi_2) = V_{12}$, the contact potential difference or Volta potential. The condition of zero current is given by $-V_{12} = V_E$, the opposing applied potential difference.

The *static capacitor method* [8, 9] is similar, except that the tendency for charge to flow is detected and a compensatory potential is automatically applied to cause cessation of flow. The essential requirement is the setting-up of the potential in a time shorter than the time constant of the RC circuit comprising the condenser capacitance C and the external resistance R ($\sim 10^{12}\,\Omega$) (see Fig. 12.3).

In the *vibrating condenser method* [10–13], plate (1) is mechanically vibrated, thereby varying the capacitance C (see Fig. 12.4). To maintain the charge on the plates at an essentially constant value, a high external resistance R is placed in circuit such that $RC > \tau$, the period of the mechanical vibration. An a.c. current is generated which is reduced to zero when $-V_E = V_{12}$. These capacitor methods are generally superior to diode or photoelectric methods.

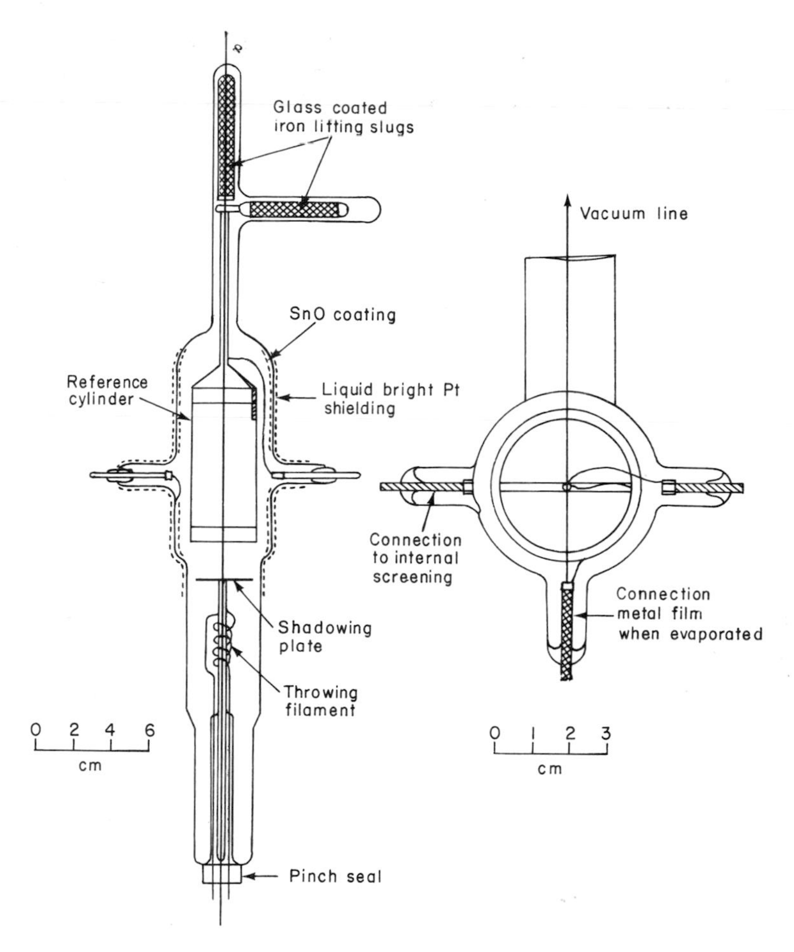

FIG. 12.3. Static capacitor for metal films. The capacitor comprises two concentric cylinders separated by a 2-mm gap. Film is deposited with the inner cylinder in the upper part of the cell. Measurements are taken with the inner cylinder resting on the shoulder of the lower part of the cell. The surface of the cylinder, coated with conducting SnO film, forms the reference electrode.

In the *diode method* [14–17], the apparatus is essentially a thermionic diode (Fig. 12.5). Thermal emission of electrons of low current density from a heated W-filament cathode builds up a space charge at the cathode surface, thereby forming an electrostatic barrier to electron flow to the anode on which chemisorption is effected. The current flow between anode and cathode is almost entirely controlled by the applied potential between the electrodes and the work function of the anode. Current-applied voltage plots are obtained in the presence and absence of gaseous adsorbate; two parallel characteristic curves result (Fig. 12.6). The displacement on the potential axis gives the change of work function of the anode. The cathode is maintained at a sufficiently high temperature for chemisorption to be absent and, hence, is an inert reference electrode. The limitations of the method are that the pressure must be less than 10^{-2} Pa to minimize ionization, and that electron impact desorption of weakly-bonded adsorbate should not occur.

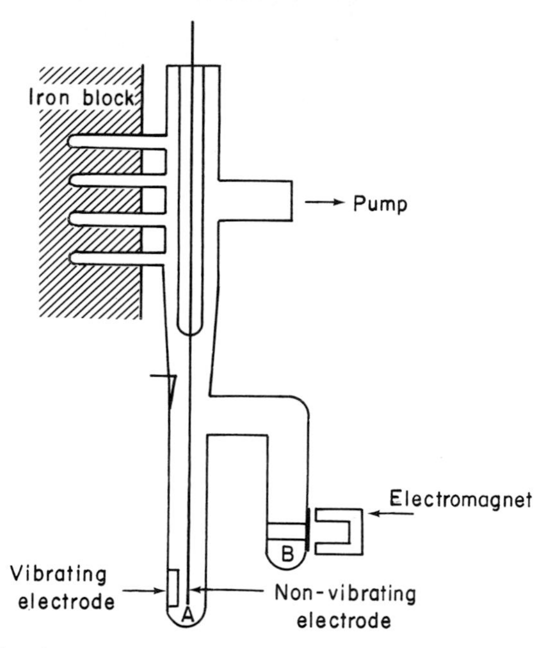

FIG. 12.4. One form of a vibrating condenser apparatus. The upper section is rigidly cemented to an iron block and carries the stationary electrode; movement of the vibrating electrode is produced by an a.c. current excitation of the electromagnet operating on an iron plate attached to the cell at B. Film is deposited on the inside of the cell at A, using an auxiliary filament (not shown), with shielding of the vibrating electrode.

In the *photoelectric method* [18–20], monochromatic light of energy $h\nu$ illuminates the metal surface and effects emission of electrons of energy $h(\nu - \nu_0)$, where ν_0 is the threshold frequency, $\nu_0 = e\Phi/h$. The photoelectric current is recorded for varying retarding potential V_E and is zero when $eV'_E = h(\nu - \nu_0)$, where V'_E is the intercept on the voltage axis. The $\Delta\chi$ caused by chemisorption is given by the voltage displacement of this intercept. Severe limitations of this method are: (i) the small photoelectric currents involved ($\sim 10^{-14}$ A); (ii) the fact that ν is in the far ultra-violet (for $\Phi > 5$ eV); and (iii) the occurrence of electrical instability due to static accumulation of charge and the development of thermoelectric voltages.

In *field-emission microscopy* [21, 22],† the basis of the method is the application of a high external electric field of 3 to 15 kV to a fine metal tip of radius of curvature $\sim 10^{-4}$ cm to effect cold emission of electrons from the

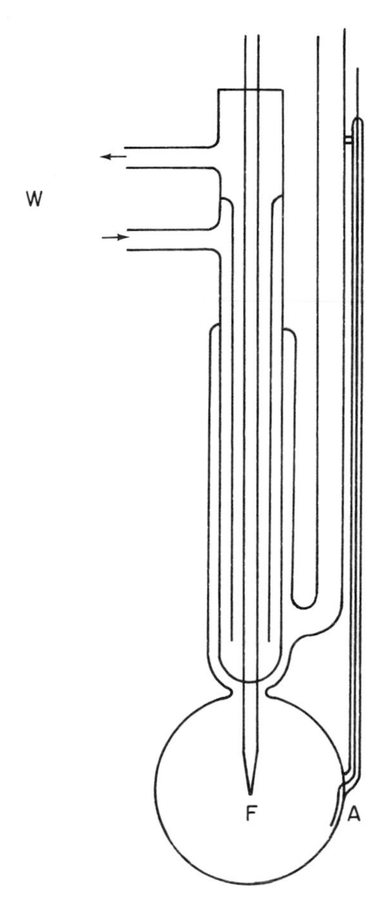

FIG. 12.5. The cathode is a tungsten filament F at the centre of a spherical vessel on the inside of which a film is condensed by complete evaporation of a metal loop on the filament. Connection to the anode film is by a tungsten wire A partially embedded inside the glass vessel. The cathode leads are thermostatted by water circulation (W).

Fermi level into vacuum. The total current density is given by the Fowler–Nordheim equation

$$j/V^2 = ab \exp[-6{\cdot}8 \times 10^7 (e\Phi)^{-\frac{3}{2}}/KV], \qquad (12.5)$$

† See also Chapter 19 and references therein.

where a is the emission area (cm^2), K a constant in cm^{-1} (approximately given by the reciprocal of the radius of curvature of the tip), Φ the work function of the metal cathode in eV, and V the applied voltage in volts. The constant b is given by $6{\cdot}2 \times 10^{16} K^2(E_F/e\Phi)^{\frac{1}{2}}/(E_F + e\Phi)$, where E_F is the energy at the Fermi level. Since Φ is essentially independent of the applied voltage, the plot of $\ln(j/V^2)$ against $1/V$ is linear and of slope $(e\Phi)^{\frac{3}{2}}/K$. Slopes (S_a, S_p)

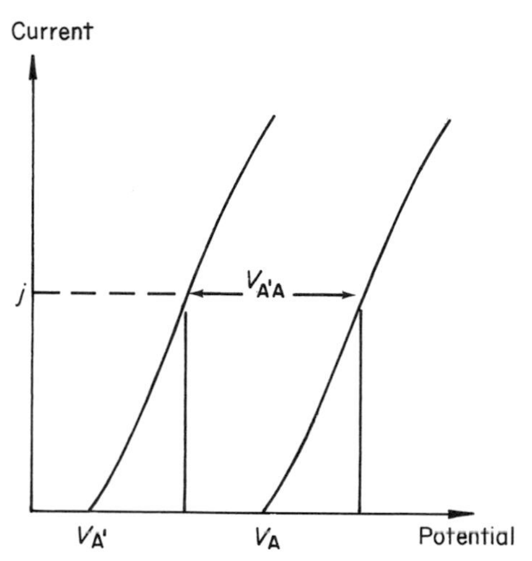

FIG. 12.6. Current/applied voltage curves in space-charge-limited diode method for clean (A) and covered (A′) anode surfaces. Displacement of the parallel characteristic plots $V_{A'A}$ equals the change of work function by chemisorption.

are determined in the absence and presence of the chemisorbed layer and $\Delta\chi$ is then evaluated as

$$-\Delta\chi = \Phi[(S_a/S_p)^{\frac{2}{3}} - 1]. \tag{12.6}$$

By using a probe-hole device, the simultaneous changes of χ on individual planes can be recorded. Unfortunately, only the total amount of chemisorbed material can be measured. Since the surface concentration may vary widely on the different planes, the individual amounts adsorbed on specific planes are unknown. Moreover, the measured currents may be as low as 10^{-9} A and an accuracy of one part in 10^3 is required. Similarly, applied voltages of several kV must be measured to within 1 part in 10^4. A major advantage is that the visual picture produced on the fluorescent screen by electron emission from the tip before adsorption gives an excellent criterion of cleanliness that is not available in the other procedures.

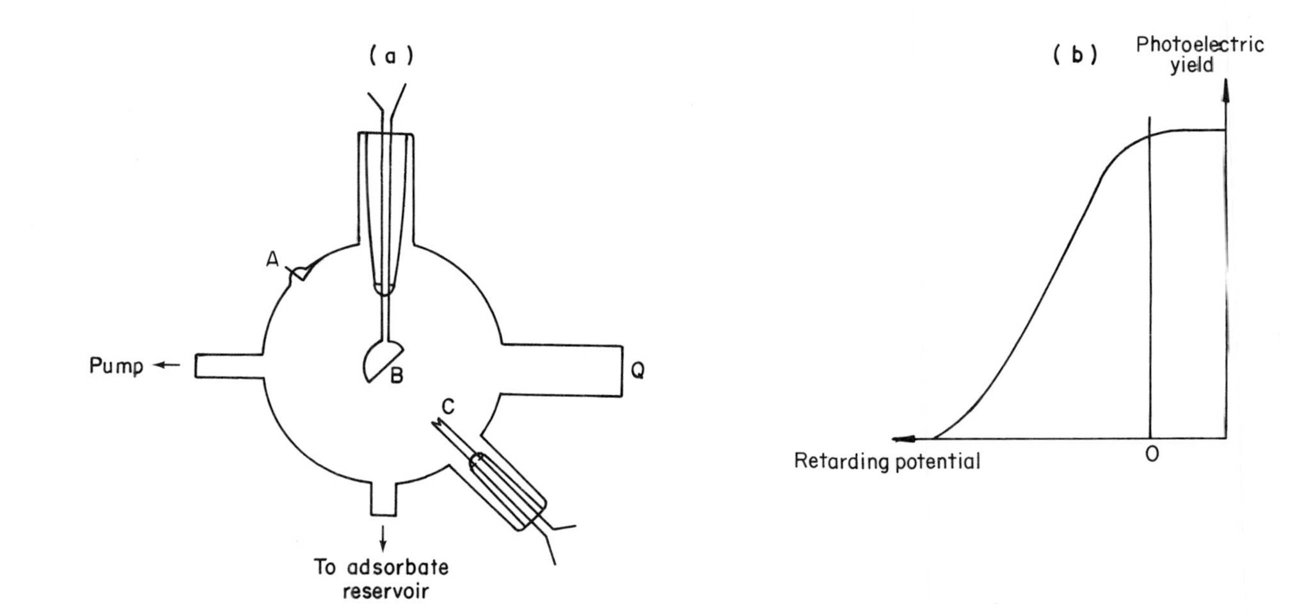

FIG. 12.7. (a) Photoelectric work function cell. Monochromatic ultra-violet light passes through a quartz window Q to illuminate the cathode B which is covered with a surface film previously evaporated from filament C. The anode A is a metal coating on the internal wall of the glass vessel.

(b) Photocurrent generated by electrons photoemitted from the cathode surface is plotted against the applied retarding potential V_E. The threshold frequency ν_0 is derived from the intercept V'_E for zero current ($eV'_E = h(\nu - \nu_0)$) and the incident frequency ν; the work function is then given by $h\nu_0/e$. The displacement of V'_E from the value obtained for the clean cathode gives $\Delta\chi$, the surface potential.

12.3. THEORETICAL EVALUATION OF THE CHANGE OF WORK FUNCTION BY FOREIGN METAL ADATOMS [23–25]

The chemical potential μ of electrons in the bulk metal, given by $\delta G(n)/\delta n(\mathbf{r})$ in Chapter 11, is virtually unaffected by the presence of an adsorbed layer because the surface region forms a negligible part of the volume of the metal adsorbent. The work function change is, therefore, brought about by modification of the surface electrostatic barrier. The static external potential $V(r)$ is simply the sum of the electrostatic potential of the jellium metal at an appropriate value of the ion-core radius and that of the adsorbate layer of density $\bar{n} = N/d$, where N is the number of foreign metal adatoms per unit area and d is the thickness of the layer (or the spacing between the closest-packed lattice planes) [23]. The new value of Φ can, therefore, be calculated using the same procedure as that for the clean surface. The method gives acceptable magnitudes for the change of work functions for alkali metal adatoms on *sp* metals (Fig. 11.5).

In another method that has been applied to sodium adatoms, hydrogen-like wave functions $\psi_{\mathrm{n,s}}$ are assigned to the adatoms, and the potential between the ion-core (or the atomic electron) and the metal is ascribed to the classical image charge. The one-electron Hamiltonian is then approximated to

$$H = \frac{-\hbar^2}{2m}\psi^2 - \frac{e^2}{4r_1} + \frac{e^2}{r_2} - \frac{e^2}{r_3}. \tag{12.7}$$

The first term is the kinetic operator, the second is the image potential of the electron, the third is the repulsive potential between the electron and the ion-core image, and the fourth is the attractive potential between the electron and the ion-core. The perturbing potential on the alkali-metal electron is

$$H_{\mathrm{a-m}} = \frac{-e^2}{4r_1} + \frac{e^2}{r_2}. \tag{12.8}$$

The energy shift due to interaction of the adatom and the metal is then

$$\Delta E = \frac{\langle \psi_{\mathrm{n,s}} | H_{\mathrm{a-m}} | \psi_{\mathrm{n,s}} \rangle}{\langle \psi_{\mathrm{n,s}} \psi_{\mathrm{n,s}} \rangle} \tag{12.9}$$

and the electronic level broadening is

$$\Gamma \approx \frac{2\pi}{h} \rho_{\mathrm{K}} \langle \psi_{\mathrm{n,s}} | H_{\mathrm{m-a}} | \psi_1 \rangle, \tag{12.10}$$

where ψ_1 is the single-electron wave function, ρ_K the density of states in the metal and

$$H_{m-a} - e^2/r_3. \tag{12.11}$$

The average charge q on the adatom is calculated from the values of ΔE and Γ, and the density distribution for the electronic energy levels of the adatom, which is usually taken as Lorentzian. The equilibrium distance d of the adatom from the effective surface is not calculable from first principles, but reasonably self-consistent values of ΔE, Γ and d can be selected to evaluate the surface dipole μ_0.

The semi-empirical equation

$$\mu_0 = (r_i + 1{\cdot}12\rho^{-\frac{1}{3}})q \tag{12.12}$$

where r_i is the radius of the ion in Å, ρ is the free electron density in the same units, and q the effective charge on the adatom, has been found to give reasonable agreement with experiment (see Fig. 11.5). As an example, the dipole moment calculated for a Na adatom on Ni(111) and Ni(100) is 8·9 eV compared with an experimental value $[\mu_0 = (\Delta\Phi/2\pi n)]_{\theta\to 0}$ for an ionic adsorbate] of 7·3 eV derived from the measured work function. The plot of Φ with increasing coverage, evaluated by both of the above procedures, displays a minimum at fractional monolayer coverage in agreement with experiment. The adatom–metal bond may be ionic, polar or metallic in character, depending on whether the broadening and shifting of the atomic valence level place it above, at or below the Fermi level. For sodium, this level lies above E_F and indicates that adatoms of this metal are partially ionized.

REFERENCES

1. R. Gomer, *J. Chem. Phys.*, 1953, **21**, 1869.
2. J. Topping, *Proc. Roy. Soc. A*, 1927, **114**, 67.
3. C. E. Carroll and J. W. May, *Surface Sci.*, 1972, **29**, 60.
4. E. R. Gyftopoulos and D. Levine, *J. Appl. Phys.*, 1962, **33**, 67.
5. J. C. Rivière, *in* "Solid State Surface", (M. Green, ed.), Marcel Decker, New York, 1969, Vol. 1, p. 179.
6. R. V. Culver and F. C. Tompkins, *Adv. Catalysis*, 1959, **11**, 104.
7. P. P. Craig and V. Radeka, *Rev. Scient. Instrum.*, 1970, **41**, 258.
8. T. A. Delchar, A. Eberhagen and F. C. Tompkins, *J. Scient. Instrum.*, 1963, **40**, 105.
9. T. L. Ashcroft, J. Riney and N. Hackerman, *Rev. Scient. Instrum.*, 1963, **34**, 5.
10. J. C. P. Mignolet, *Disc. Faraday Soc.*, 1950, **8**, 326.
11. C. T. Kirk and E. E. Huber, *Surface Sci.*, 1968, **9**, 217.
12. J. C. Rivière, *Brit. J. Appl. Phys.*, 1964, **15**, 1341.
13. T. A. Delchar and G. Ehrlich, *J. Chem. Phys.*, 1965, **42**, 2686.

14. A. G. Knapp, *Surface Sci.*, 1973, **34**, 289.
15. J. Pritchard, *Trans. Faraday Soc.*, 1963, **59**, 437.
16. J. R. Anderson, *J. Phys. Chem. Solids*, 1960, **16**, 291.
17. B. J. Hopkins and R. K. Ross, *Brit. J. Appl. Phys.*, 1965, **16**, 1319.
18. R. Suhrmann, J. M. Heras, L. V. de Heras and G. Wedler, *Ber. Bunsenges Phys. Chem.*, 1964, **68**, 511.
19. W. M. H. Sachtler and G. J. H. Dongelo, *J. Chim. Phys.*, 1957, **54**, 27.
20. J. Eisinger, *J. Chem. Phys.*, 1958, **29**, 1154.
21. R. Gomer, *J. Chem. Phys.*, 1953, **21**, 1869.
22. R. W. Strayer, W. Mackie and L. W. Swenson, *Surface Sci.*, 1973, **34**, 225 (retarding field method).
23. N. D. Lang, *Phys. Rev. B*, 1971, **4**, 4234.
24. J. R. Smith, *Phys. Rev.*, 1969, **181**, 522.
25. N. D. Lang and W. Kohn, *Phys. Rev. B*, 1971, **3**, 1215; 1973, **8**, 6010.

13

Perturbation of Surface Electronic Properties by Chemisorption. II. Magnetic Susceptibility†

13.1. INTRODUCTION

The magnetic induction B of a metal is related to the intensity of magnetization M and the strength of the electric field H, externally applied to the specimen, by the equation

$$B = H + 4\pi M. \tag{13.1}$$

The magnetic susceptibility χ_v per unit volume is defined as the ratio M/H. Atoms of transition metals have d electrons, some of which have unpaired spins. An unpaired electron has an atomic magnetic moment and, consequently, such metals exhibit paramagnetism. The tendency of atomic magnets to align themselves in the direction of an applied magnetic field is opposed by their thermal motion and, in the absence of the field, their orientation is random. The fraction of the total number of atomic moments

† See Selwood [1].

aligned in a field H at a temperature T is given by Langevin's equation

$$\frac{M}{M_s} = \coth\left(\frac{NH}{kT}\right) - \frac{kT}{H\tilde{\mu}}, \tag{13.2}$$

where M_s is the saturation magnetization at infinite field and 0 K (when all the atomic magnets are oriented in the direction of the field) and equals $N\tilde{\mu}$, where N is the number of magnets per unit volume, each having a magnetic moment $\tilde{\mu}$. The magnetization M, at a temperature T in an external field H, is usually small compared with M_s, and Eq. (13.1), for $M/M_s \ll 1$, is approximated to

$$M = N\tilde{\mu}^2 H/3kT, \tag{13.3}$$

so that

$$\chi_v = M/H = N\tilde{\mu}^2/3kT. \tag{13.4}$$

Atomic magnetic moments are expressed in terms of the Bohr magneton, $\beta = eh/4\pi c$, where e is the electronic charge, m is the electron mass, c is the velocity of light and h is Planck's constant; β equals $9{\cdot}27 \times 10^{-24}$ A m^2.

In addition, all substances exhibit diamagnetic susceptibility due to the Larmor precession of the motion of electrons in the applied magnetic field. It is negative in sign (i.e. the magnets are aligned to oppose the field) and independent of temperature and field strength. A plot of the total susceptibility at constant field as a function of reciprocal temperature therefore has an intercept corresponding to the diamagnetic component; the paramagnetic susceptibility is derived from the slope of this plot.

A few paramagnetic substances, such as iron, cobalt and nickel, attain an exceptionally high magnetization in very weak fields and (at moderate temperatures and below) rapidly reach their saturation magnetization with little increase of field intensity. This ferromagnetization decreases as the temperature is raised and falls abruptly to a very low value at a specific temperature T_c, termed the Curie temperature. Below this temperature ($\sim$1050 K for Fe, $\sim$900 K for Ni, etc.), a ferromagnetic metal comprises a large number of Weiss domains [2], or small elements of volume in which all the electron spins are parallel (even in absence of an applied field) due to the operation of exchange forces; but the alignment of the domains themselves is random. On application of an external field, the domains are oriented, and saturation magnetization corresponds to complete alignment in the direction of the field. However, above the Curie temperature, the co-operative effect within the domains is destroyed; the ferromagnetic metal then behaves as a typical paramagnetic element and its magnetization varies inversely as the temperature.

13.2. EXPERIMENTAL MEASUREMENTS

Measurements are conveniently made using an a.c. permeammeter [1]; this comprises a step-down transformer consisting of a few turns of platinum wire to form the secondary which is inserted in the core of a solenoid of several thousand turns. A second secondary is used to balance out normal induction in both secondaries. The sample of pelleted material is placed in one of the secondary coils and both secondaries are connected to a milli-voltmeter, the readings of which are directly proportional to the magnetization of the sample.

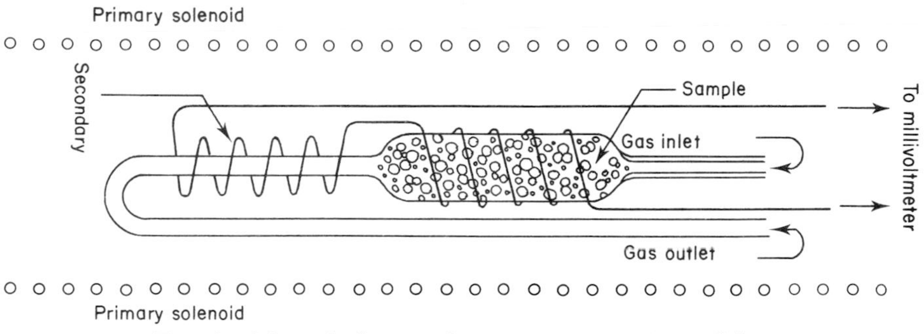

FIG. 13.1. Schematic diagram of a.c. permeameter (Selwood [1]).

A diagrammatic representation of the a.c. permeammeter is shown in Fig. 13.1. The primary solenoid consists of 5000 turns of copper wire energized by an 0·75-A 60-Hz a.c. which is stabilized by a voltage regulator. The two secondaries comprise coils of non-magnetic metal (Pt, or chromel A no. 26) of 40 turns arranged in opposition and the sample is placed in one of them. They are connected to the millivoltmeter for recording the magnetization of the sample. The low-frequency of 60 Hz ensures that the relaxation time is sufficiently short to allow the magnetization to attain its maximum value.

13.3. APPLICATION TO CHEMISORBED LAYERS

The change of saturation magnetization brought about by chemisorption is obtained from measurements below the Curie temperature. For saturation, the moments of the Weiss domains, and also of the atomic magnets of the surface metal atoms, are aligned in parallel. When the alignment of the surface moments is unaffected by chemisorption, the change of atomic moments can be directly derived from the measured change of saturation magnetization. But the formation of the chemisorption bond usually effects decoupling of the surface atomic moments since the thermal energy (10^{-21} J at 70 K) is larger than the orientation energy (10^{-22} J) of an isolated magnetic moment, although very much smaller than that required to orient a Weiss domain (i.e. $10^{-22}n$ J, where n is the large number of atoms in a domain). This decoupling, which is, therefore, strongly temperature dependent, effects a decrease of the saturation magnetization above 70 K, irrespective of any decrease or increase of the atomic moment. However, measurement of the change of magnetization at the temperature of liquid helium provides a direct measure of the change of surface atomic moments due to chemisorption.

In chemisorption investigations, the fractional change of magnetization $\Delta M_0/M_0$, is expressed in terms of $\tilde{\mu}$, the atomic moment, i.e.

$$\Delta M_0/M_0 = \Delta(N\tilde{\mu})/N\tilde{\mu}, \tag{13.5}$$

where M_0 is the saturation magnetization of the clean metal at 0 K and N is the number of metal atoms in the system. Chemisorption causes an apparent change in $\tilde{\mu}$, which is usually formulated in terms of ε, defined as the change brought about by bonding one adspecies to a single surface metal atom. Hence,

$$\varepsilon = (\Delta M_0/M_0)(N\tilde{\mu}/n_a), \tag{13.6}$$

in which n_a is the number of adspecies in the chemisorbed layer, and ε has units of the Bohr magneton β.

To obtain measurable changes of magnetization, a large surface-to-volume ratio of the metal is essential. By reducing the particle size to linear dimensions of less than 100 Å, about 10% of the metal atoms are present at the surface; each particle then comprises a single magnetic domain, which acts as a paramagnetic atom of high magnetic moment and is said to exhibit collective paramagnetism. Consequently, when the saturation magnetization is plotted as a function of H/T for different values of H and T, a single characteristic plot is obtained [see Eq. (13.3), and Fig. 13.2]. The atomic magnetic moment can then be calculated since the number of metal atoms within the applied field may be derived from the total mass and density of the metal. The use of low temperatures and high field means that a good extrapolation to give the true saturation magnetization M_0 at 0 K can be made. Calibration of the system is accomplished by using a sample of a pure metal of accurately known magnetization properties.

The experimental sample comprises a set of n particles each of volume v for particles of uniform size, the total volume V being given by nv. The sample must display collective paramagnetism, i.e. confirmation that the state of

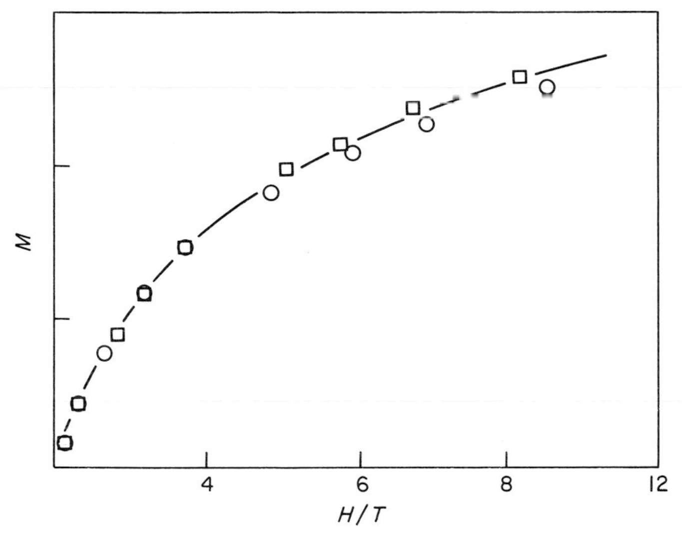

FIG. 13.2. Plot of magnetization M against H/T for a nickel–silica sample. ○, 77 K; □, 296 K. (After Dietz and Selwood [3].) Arbitrary units.

subdivision or particle size conforms to a relationship similar to Eq. (13.3) should be experimentally tested. The appropriate equation for a set of particles is

$$M = n\mu^2 H/3kT, \tag{13.7}$$

where n is the number of particles and μ the magnetic moment of a particle comprising a single Weiss domain of large paramagnetic moment. Equation (13.7) is usually expressed in terms of the spontaneous magnetization defined as $M_{sp} = \mu/v$, i.e.

$$M = n(M_{sp}v)^2 H/3kT. \tag{13.8}$$

When n_H atoms are chemisorbed, the particle magnetic moment is reduced to

$$\mu' = N_{sp}v - n_H\varepsilon\beta \tag{13.9}$$

where, as before, ε is the change in the number of Bohr magnetons brought about by one H adatom. The magnetization in presence of the adsorbed layer M' is, therefore,

$$M' = n(M_{sp}v - n_H\varepsilon\beta)^2 H/3kT, \tag{13.10}$$

and

$$\frac{\Delta M}{M} = \frac{-2n_{\mathrm{H}}\varepsilon\beta}{M_{\mathrm{sp}}v} + \left(\frac{n_{\mathrm{H}}\varepsilon\beta}{M_{\mathrm{sp}}v}\right)^2 \approx \frac{-2n_{\mathrm{H}}\varepsilon\beta}{M_{\mathrm{sp}}v}, \tag{13.11}$$

i.e. the fractional decrease of magnetization is directly proportional to the number of adatoms per metal particle and varies inversely as its volume. The numerical factor 2 applies when the particles are of uniform size, but varies for different distributions. It is, however, a constant for a particular distribution being 1·87 for a Maxwellian one. Hence, the slope of a plot of $\Delta M/M$ against n_{H} at a particular temperature varies for different distribution functions. In addition, M_{sp} decreases slightly with increase of temperature, so the slope is temperature dependent (see Fig. 13.3).

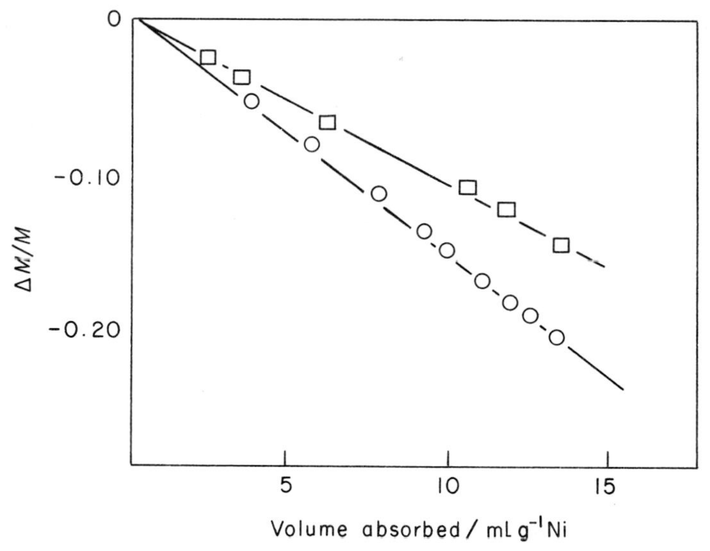

FIG. 13.3. Fractional change of magnetization $\Delta M/M$ as a function of the amount of hydrogen adsorbed per gram of nickel at two temperatures: ○——○, 480 K; □——□, 195 K. The ratio of the slopes is 0·71 compared with a value of 0·69 from the temperature coefficient of the spontaneous magnetization.

The most extensively investigated system has been the dissociative chemisorption of hydrogen on nickel, for which the saturation atomic moment is $0{\cdot}61\beta$. A series of experimental measurements of $\Delta M_0/M_0$ for various weights of nickel and surface concentrations of hydrogen atoms gives an average value of around 0·71 for ε. If the change of magnetization is ascribed solely to pairing of electron spins of the metal atoms by the hydrogen adatom, then the number of spins paired by one adatom is 0·71.

13.3.1. The H_2/Ni and CO/Ni systems

In general, the validity of the theoretical relationships given above is confirmed experimentally. The $\Delta M/M$ plot decreases linearly with the amount of hydrogen chemisorbed on the nickel particles [4], and the increase of slope with rise of temperature is in reasonable accord with the measured decrease of M_{sp} with increase of temperature (see Fig. 13.3). This agreement indicates that ε is independent of temperature; the linearity of the plot at constant temperature shows that ε is also independent of the extent of chemisorption over a wide range. The constant value of ε (or of the number of electron spins paired per adatom of hydrogen) suggests that only one type of chemisorption bond is formed, and that the chemisorption of hydrogen by nickel involves the pairing of electron spins. It is probable that the bond is formed by interaction of the 1*s* electron of the adatom with a vacant *d* orbital of a metal atom, but distinction between an ionic bond of the type H^+—Ni, or a covalent bond in which both the 1*s* electron of the hydrogen and a *d* electron of the metal contribute, is not possible.

With carbon monoxide as adsorbate, the fractional magnetization isotherm is initially linear with a slope of about one-half that of the corresponding hydrogen isotherm. Since hydrogen is dissociatively adsorbed, the difference of slopes suggests that one carbon monoxide molecule is bonded

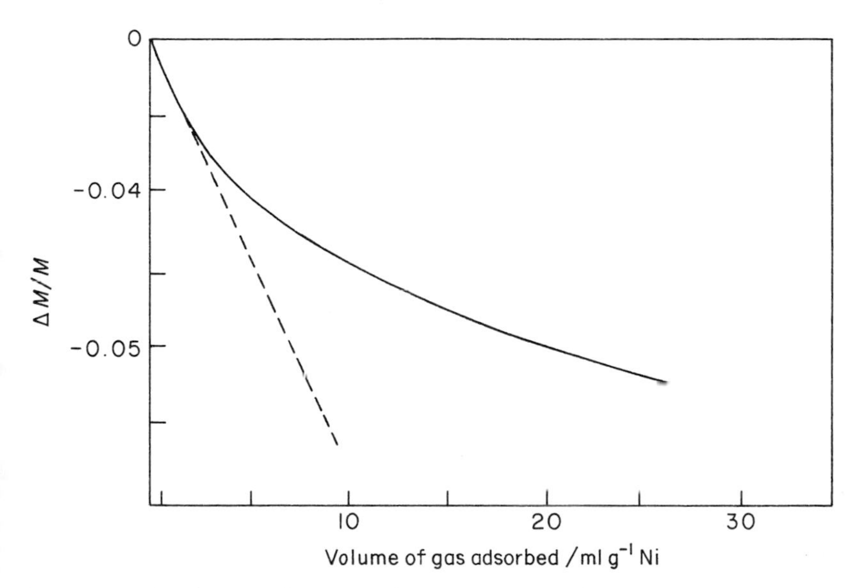

FIG. 13.4. Fractional change of magnetization $\Delta M/M$ as a function of the amount of gas adsorbed at room temperature per gram of sintered nickel supported on kieselguhr. ———, Carbon monoxide; -----, hydrogen atoms.

to a single surface metal atom to form the linear adspecies Ni=C=O at lower coverages. However, with increase of concentration of the chemisorbed layer, the slope decreases fairly sharply to a value consistent with two or more adsorbate molecules being bound to a single metal atom. This mode of attachment is weak, since, after evacuation at 150 °C, virtually all such chemisorbed molecules are desorbed, whereas those admolecules that have the linear structure remain on the surface. These conclusions are, however, doubtful since the shape of the magnetization isotherm depends on the particle size of the metal; for example, for large particles, the initial slope approaches that of the corresponding hydrogen isotherm but then continuously decreases after quite small amounts of gas have been adsorbed [5] (see Fig. 13.4).

13.4. BOND ORDER

The bond order of an adsorbate is defined as the number of metal atoms taking part in the bonding of one molecule of the adsorbed species. Its numerical magnitude is referenced to the bond order of hydrogen which is given a value of 2. Comparison is made between the slope of the fractional magnetization isotherm for the adspecies and that of hydrogen; the ratio of the two slopes divided by two gives the bond order. Its magnitude gives some indication of the possible dissociative nature of the chemisorption process. For example [1], the bond order of ethylene on nickel at 0 °C is two, consistent with the occurrence of associative adsorption, whereas the value of four for hydrogen sulphide implies that each hydrogen atom is bonded to a single nickel atom, with the S adatom attached to two adjacent metal atoms to form a bridge structure. Chemisorbed benzene has a bond order of six at around 100 °C, suggesting a structure in which the benzene ring is flat on the surface with simultaneous bonding of its six hydrogen atoms. The order increases to around 18 at about 200 °C: this increase has been interpreted as a dissociative chemisorption following an activated rupture of C—H and C—C bonds [1].

13.5. APPRAISAL OF RESULTS

Although experimental measurements are restricted to those ferromagnetic elements which display collective paramagnetization for particles in a sufficiently fine state of division, some general information has been gained about the electronic interactions involved in the formation of chemisorption bonds; in particular there is some evidence that the bonding can, in part, be related to pairing of unpaired spins arising from the presence of partially filled *d* orbitals in the metal. However, too much confidence cannot be placed on the more quantitative aspects of the results nor on their theoretical interpretation. Thus, the adsorption of krypton at $-78\,^{\circ}C$ effects a decrease of magnetization amounting to about one-fifth of that brought about by the chemisorption of hydrogen. Since the work function of a metal is markedly modified by the adsorption of krypton, a redistribution of the surface electron density takes place, but this polarization effect is not associated with the pairing of electron spins. It is evident that the value of magnetization studies must be viewed with caution until a more detailed and deeper understanding of the theoretical basis of the changes in magnetic susceptibility accompanying an adsorption process has been developed.

REFERENCES

1. P. W. Selwood, "Adsorption and Collective Paramagnetism", Academic Press, New York and London, 1962.
2. C. Kittel, "Introduction to Solid State Physics", 2nd ed, Wiley, New York, 1956, Chap. 15.
3. R. E. Dietz and P. W. Selwood, *J. Chem. Phys.*, 1961, **35**, 270.
4. J. A. Silvent and P. W. Selwood, *J. Amer. Chem. Soc.*, 1961, **83**, 1034.
5. I. Den Besten, P. G. Fox and P. W. Selwood, *J. Phys. Chem.*, 1962, **66**, 450.

14

Perturbation of Surface Electronic Properties by Chemisorption. III. Electrical Conductivity

14.1. INTRODUCTION

14.1.1. Bulk conductivity

Electrical conductivity is due to the movement of some (or all) of the valence electrons of the metal atoms under an externally-applied potential difference across the metal (usually in the form of a filament or wire). For a metal with an ideal periodic structure, there is no loss of kinetic energy of the electrons to the lattice. In practice, however, departures from ideality arise (e.g. owing to thermal vibrations of the ion cores about their equilibrium positions) and partial dissipation of kinetic energy always occurs. For a specific magnitude of the applied field, the drift velocity of the electrons therefore attains an average finite value $\bar{v}$.

The measured resistance R of a metal specimen is given by $R = (l/bt)\rho$, where l, b, t are, respectively, the length, breadth and thickness of the specimen, and ρ is its specific resistivity. The ratio (l/bt) is termed the geometric factor G. The specific conductivity, $\sigma = 1/\rho$, can be expressed in terms of the rest mass m_0 and charge e of the electron, n the number of conduction electrons per unit volume of the metal, and $\bar{\lambda}$ their mean free path, as

$$\sigma = ne^2\bar{\lambda}/m_0\bar{v} = ne^2\tau/m_0, \qquad (14.1)$$

in which $\tau = \bar{\lambda}/\bar{v}$ is the relaxation time. The electron mobility, or the mean velocity per unit electric field, is given by $e\tau/\bar{v}m_0$. For copper, $n = 8{\cdot}5 \times 10^{22}$ cm^{-5} (provided that each atom contributes one valence electron to the conduction band); other magnitudes are $\tau \sim 2 \times 10^{-14}$ s at room temperature, $\bar{\lambda} \sim 3 \times 10^{-6}$ cm (or about 100 times the lattice constant); the drift velocity is $\sim 4 \times 10^5$ cm s^{-1} compared with the intrinsic velocity of the electrons of 2×10^8 cm s^{-1} and the mobility is ~ 30 (cm s^{-1})/(V cm^{-1}).

Equation (14.1) does not take into account the interactions between the conduction electrons and the ion-cores, but this can be done by replacing m_0 by the effective electron mass m, defined as the mass that must be assigned to a free electron such that its velocity increment under an applied impulse becomes equal to that actually attained by the conduction electron in the metal under the same impulse.

14.1.2. Effect of the surface

The surface of a metal reflects the mobile conduction electrons; those that are specularly reflected undergo no loss of momentum, but diffuse scattering is accompanied by loss of some of their kinetic energy. Surface atoms have a lower co-ordination than those in the bulk and their mean square displacement perpendicular to the surface is 2 to 5 times larger than that of the lattice atoms. Consequently, for thin films, the surface loss predominates over the phonon scattering in the bulk.

14.2. EFFECT OF CHEMISORPTION [1, 2]

14.2.1. Filaments

14.2.1.1 *Reflection of electrons by the adsorbate*

Chemisorption brings about a redistribution of electrons in a surface region that extends to about three atomic layers into the bulk. The result is that the potential energy of the ion-cores of the surface metal atoms involved in forming the chemisorption bond is different from that of the ion-cores of the bulk metal atoms. The magnitude of the difference increases with the strength of the chemisorption bond, particularly when rehybridization of surface atoms occurs and some displacement in the location of these atoms takes place.

Since the valence electrons are also displaced towards or away from the surface metal atoms, there is little screening of the charge of their ion-cores within an atomic diameter, and the chemisorbed layer corresponds to a surface array of virtual charges. The effect is to increase the diffuse scattering and decrease the specular reflexion of the conduction electrons;

consequently, the resistivity of a metal filament is increased by the presence of a chemisorbed layer.

For a flat surface, constructive interference of the scattered electrons occurs for particular locations of the surface charges. When the distance between charges is equal or less than the wavelength of the conduction electrons (which is around 10 Å at the Fermi level), specular reflexion results; whereas for distances apart greater than the electron wavelength, diffuse scattering predominates. The resistivity change accompanying chemisorption, therefore, depends on the spatial distribution of the chemisorbed species. Moreover, since physisorption does not markedly affect the potential energy of the ion cores of the surface atoms, a weak chemisorption may be distinguished from physical adsorption, e.g. at low temperatures where the accompanying physisorption is large.

In addition, it is possible, in principle, to identify whether chemisorbed adspecies are immobile or mobile, and, if the latter, whether the interaction between them is attractive or repulsive in character. Thus, for immobile adspecies, the distribution is random and, at low coverages, the scattering charges are widely spaced; as the coverage is increased, the resistivity should first increase and then decrease as the distance apart of the adspecies decreases. At full coverage, when each surface metal atom is bonded to an

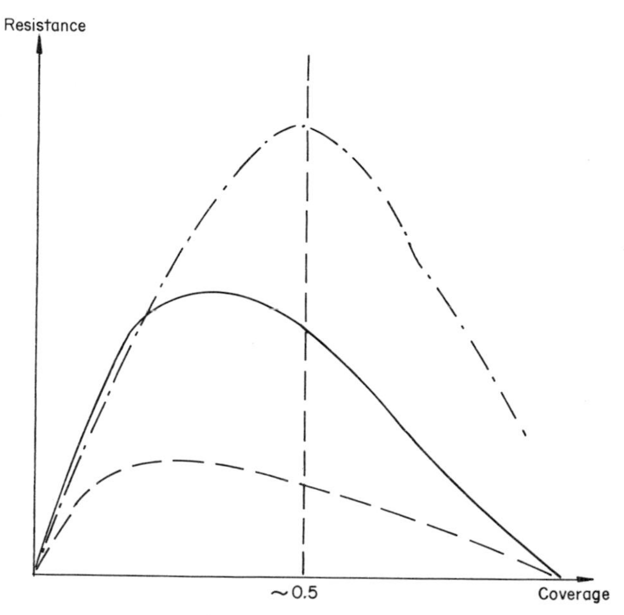

FIG. 14.1. Diagrammatic representation of change of resistance with coverage of chemisorbed adatoms. —·—, Mobile adatoms having repulsive interactions; ———, mobile adatoms having attractive interactions; – – – –, random distribution of immobile atoms.

adspecies, specular reflexion predominates and the resistivity may then approach that corresponding to the clean surface (see Fig. 14.1).

When the adspecies are mobile and their interactions are repulsive, there is some degree of ordering at low coverages which is, in the ideal case, complete at half-coverage. Consequently, with increasing coverage, the resistivity should increase to a sharp maximum and then decrease as the sites between the adspecies are filled up. On the other hand, with attractive interactions, the increase resistivity should be smaller with a broad maximum. In practice, most filaments have a roughness factor greater than unity, and differentiation between the three cases above is not a practical proposition.

14.2.1.2. *Change of surface resistivity*

In addition to the change of resistivity brought about by the increase in diffuse reflexion that accompanies chemisorption, formation of a two-dimensional adsorbate–metal complex of resistivity different from that of metal atoms takes place. Scattering is, therefore, brought about by both structural and thermal effects. The probabilities of scattering are, to a good approximation, additive, i.e.

$$\frac{1}{\tau_s} = \frac{1}{\tau_R} + \frac{1}{\tau_T}, \qquad (14.2)$$

in which the subscripts R and T refer to the structural and thermal contributions. Because of the reciprocal relationship between resistivity ρ and τ, $\rho_s = \rho_R + \rho_T$. The term ρ_R is usually called the residual resistivity.

It is reasonable to assume that Mattheison's rule (i.e. that the temperature coefficient of resistivity $d\rho/dT$ is independent of the defect and impurity concentration in the bulk metal) can be applied to the surface phase formed by chemisorption. In this case, its temperature coefficient of resistivity is independent of the extent of chemisorption, i.e. of the coverage.

Experimentally, the measured resistance R_M of the filament contains contributions from the surface phase (R_s) and that of the bulk metal (R_B). These two resistances are in parallel so that

$$\frac{1}{R_M} = \frac{1}{R_s} + \frac{1}{R_B}, \qquad (14.3)$$

or the two conductivities are additive,

$$\sigma_M = \sigma_s + \sigma_B. \qquad (14.4)$$

Now, the temperature coefficients for diffuse scattering of conduction electrons at the surface and the scattering of the electrons within the bulk by phonons of the lattice (which is responsible for the increase of bulk resistivity at higher temperatures) are roughly the same for most filaments. Hence, the residual resistivity can be derived from measurements of the resistance of the filament at two different temperatures, provided the extent of chemisorption is the same. Alternatively, the change of residual resistivity with increasing coverage at constant temperature can be evaluated. In this manner, the complication arising from variation of ρ_T with coverage at constant temperature can be eliminated.

14.2.2. Evaporated films

14.2.2.1. *Island films*

Very thin films, particularly those deposited on low-temperature substrates, have a discontinuous structure comprising an array of isolated crystallites of linear dimensions as small as 30 Å for metals of high cohesive energy. Conduction through the film can be brought about by: (i) electron tunnelling between the crystallites, provided that their distance apart is sufficiently small; (ii) thermionic; emission across larger distances, particularly at elevated temperatures; and (iii) conduction through the non-metallic substrate, usually Pyrex.

In contrast to the bulk metal, the resistance of island films decreases with rise of temperature and, hence, the process of conduction involves an activation energy. Electron tunnelling is essentially a non-activated process, but an activated one is simulated because the crystallites become charged during transport of electrons and the Coulombic interactions between the particles create an electrostatic energy barrier. Similarly, the thermionic emission of electrons requires an activation energy in order to eject an electron at the Fermi level out of the metal.

Chemisorption studies can be used to identify the dominant conduction mechanism, as, for example, in the adsorption of hydrogen by an island film of platinum on a Pyrex substrate. Conduction through the Pyrex substrate can be eliminated, since the chemisorption of hydrogen on Pyrex at room temperature is negligible and, therefore, no change of resistance can result. Similarly, electron tunnelling is not a dominant factor, because the electrostatic energy barrier arising from Coulombic interactions between the charged crystallites is not affected by a change of work function of the platinum when hydrogen is chemisorbed and the resistance should, therefore, remain unchanged. Experimentally it is found to increase [3–6]. The dominant conduction mechanism must, then, be thermionic emission and transport across the gaps between the crystallites.

This conclusion is confirmed by measurements of the change of work function [7]. When platinum chemisorbs hydrogen at 273 K, the work function is increased, i.e. the activation energy for thermionic emission and the resistivity are higher. At 77 K, as the ambient hydrogen pressure is increased, physisorption of molecular hydrogen decreases the work function from its maximum value which, however, may be recovered by evacuation, with consequent desorption of the molecularly adsorbed state. In accord with the thermionic emission mechanism, the measured resistance is found to increase initially, then to decrease as the hydrogen pressure is increased, and finally to return after evacuation to the maximum value.

In general, thermionic emission appears to be the dominant conduction process through island films of most metals. Nevertheless, there are other complicating factors arising from the fact that the mean free path of the conduction electrons is comparable with that of the dimensions of the crystallites; there is also electron scattering at grain boundaries, the number of which is sensitively dependent on the deposition temperature of the film and the cohesive energy of the metal. These effects, however, do not make a significant contribution for island films, but can do so for continuous ones.

14.2.2.2. *Continuous films*

Thick films are usually sufficiently well annealed so that the exothermicity of the chemisorption process does not bring about further sintering or movement of grain boundaries. However, films of metal of high cohesive energy may still have some porosity and electron transport across gaps can contribute to the conductivity of the film. For a truly continuous film, however, the resistance change that occurs on chemisorption can be accounted for by the increase scattering of mobile electrons at the surface and by the formation of a surface phase of different resistivity from that of the bulk metal. The extent of scattering largely depends on whether the film has a thickness comparable with the wavelength of the electrons at the Fermi level and on the grain boundary content of the film. Otherwise, continuous films may be treated as filaments.

14.2.3. Interpretation of results

In general, the results obtained using various gas/metal systems provide little definitive information. The interpretation of the experimental information is highly speculative because of complications due to the presence of different crystal planes, mostly of low index, the accessibility of the pore structure, the mobility of the adspecies, and the change of the geometric factor by chemisorption; in addition one must consider the effects of diffuse electron scattering and the conductivity of the surface phase, the magnitude of which is assumed to be comparable with that of the corresponding three-dimensional compound. The complex nature of the results and their interpretation is exemplified by the hydrogen/nickel [3–6] and carbon monoxide/tungsten systems [8].

14.2.3.1. *The H_2/Ni and CO/W systems*

A nickel film deposited at 273 K is largely continuous. The plot of $\Delta\sigma/\sigma_0$ as a function of coverage of chemisorbed hydrogen displays [3–6] a shallow minimum amounting to about 2% at around 0·1 fractional coverage. Since the adsorbed layer is mobile, with repulsive interactions between the adspecies, the results were ascribed in imprecise terms to diffuse scattering at the surface. On bombardment of the film with hydrogen atoms, there was a marked decrease of $\Delta\sigma/\sigma_0$; this was attributed to the formation of a surface hydride with a resistivity greater than that of the metal, together with an increase in the diffuse scattering of electrons by the surface phase.

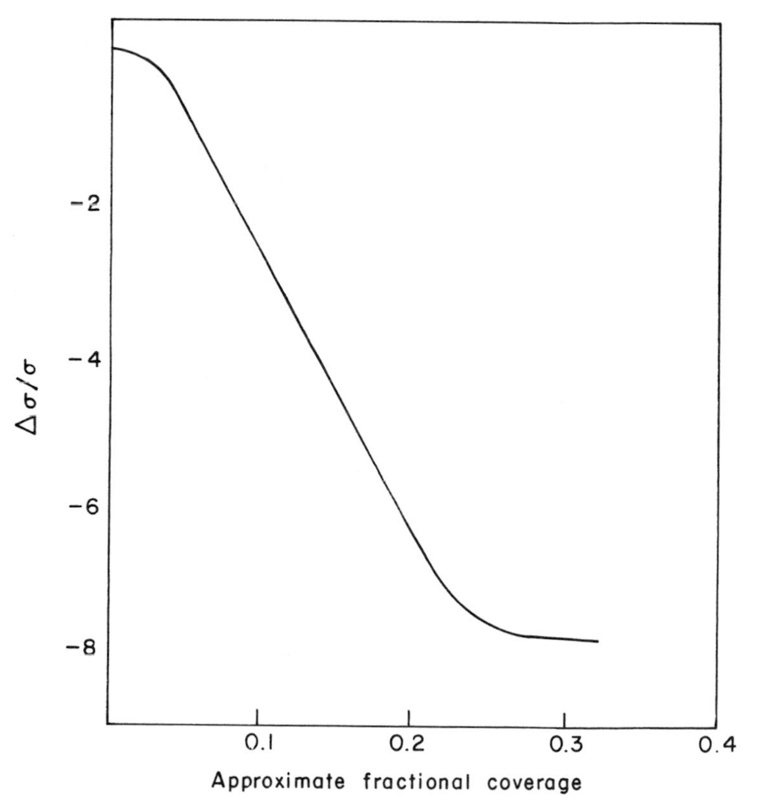

FIG. 14.2. Plot of relative percentage conductance of tungsten film at 273 K with increasing fractional coverage of adsorbed carbon monoxide.

Similar unexpected results were obtained for the carbon monoxide/tungsten-film system [8]. The relative conductance $\Delta\sigma/\sigma_0$ decreased almost linearly by about 9% as the fractional coverage was increased to around 0·3, and was largely independent of temperature (see Fig. 14.2). With further increase of coverage, $\Delta\sigma/\sigma_0$ remained approximately constant. These results were explained in terms of two different chemisorbed states, one of which had no effect on the conductance. Results on various other gas/metal systems are equally difficult to interpret in a satisfactory and consistent manner.

14.2.4. Appraisal of results

Despite the fact that experimental results obtained by various workers are often comparable, one cannot place much confidence even in the qualitative conclusions that have been reported. Indeed, although knowledge has been gained about the mechanism of conduction by evaporated films, little information about the nature of the chemisorption bond has been provided. Certainly, earlier work purporting to provide semi-quantitative conclusions about the number of electrons that are shared in forming the bond must be regarded with grave suspicion (see, for example, Ehrlich [9]). Highly accurate experimentation with single-crystal filaments, however, might provide results of interest and value.

REFERENCES

1. R. Suhrmann, *Adv. Catalysis*, 1955, **7**, 303.
2. J. W. Geus, *in* "Chemisorption and Reactions on Metallic Films" (J. R. Anderson, ed.), Academic Press, London and New York, 1971, Vol. 1, Chap. 5. This reference contains a comprehensive set of references.
3. W. M. H. Sachtler and G. J. H. Dorgelo, *Bull. Soc. Chim. Belg.* 1958, **67**, 465.
4. R. Suhrmann, Y. Mizushima, A. Hermann and G. Wedler, *Z. Phys. Chem., Frankfurt*, 1958, **17**, 350.
5. P. Zwietering, H. L. T. Koks and C. van Heerden, *J. Phys. Chem. Solids*, 1959, **11**, 18.
6. V. Ponec and Z. Knor, *Coll. Czech. Chem. Comm.*, 1960, **25**, 2913.
7. J. C. P. Mignolet, *J. Chim. Phys.*, 1957, **54**, 19.
8. J. W. Geus, H. L. T. Koks and P. Zeitering, *J. Catalysis*, 1963, **2**, 274.
9. G. Ehrlich, *J. Chem. Phys.*, 1961, **35**, 2165.

15

Low-energy Electron Diffraction (LEED)

Detailed information about the geometric arrangement of atoms in the outermost layer of a solid is essential in studies of the properties of the surface atoms; low-energy electron diffraction is almost exclusively used to investigate the structure of crystal surfaces.

15.1. DIFFRACTION OF ELECTRON WAVES [1–3]

An electron of mass m having a velocity v has an associated electron wave of wavelength λ given by the de Broglie equation, $\lambda = h/mv$, where h is Planck's constant. Hence, when an electron initially at rest of charge e is accelerated by an applied voltage V,

$$\lambda = h/(2mE)^{\frac{1}{2}} = h/(2meV)^{\frac{1}{2}} = (150{\cdot}4/V)^{\frac{1}{2}}$$

in which $E = \frac{1}{2}mv^2$, λ is in Å and V in volts. Thus, for $V = 10$ V, $\lambda = 4$ Å and at 100 V, λ is 1·2 Å.

These waves are rapidly attenuated as they penetrate the solid due to strong inelastic scattering and, for this reasons, LEED is a powerful surface technique for exploring the two-dimensional geometry of the outermost layer of a crystal.

At the surface of a metal crystal, the ion-cores form a periodic array of wave-scattering units. When the scattered waves from adjacent rows of lattice points have a path difference that is a multiple of λ. there is constructive interference along directions determined by Bragg's law, viz.

$$n\lambda = d \sin \theta \tag{15.1}$$

for normal incidence of the primary beam, where d is the lattice constant of the crystal plane or the distance between adjacent rows of ion-cores; n is the order of the diffraction and θ the angle between the diffraction beam and the normal to the surface (Fig. 15.1) [4]. Diffraction patterns are formed only from the elastic scattering component of the diffracted wave. In elastic

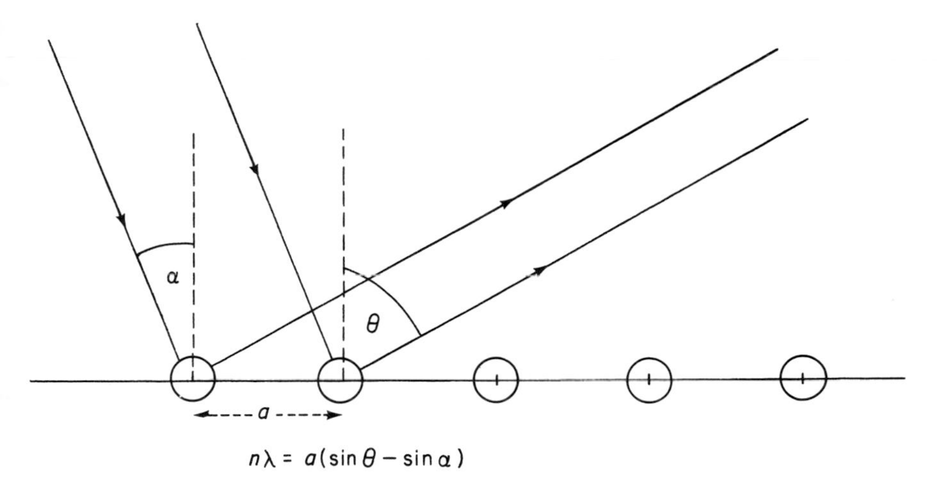

FIG. 15.1. Diffraction of electrons by a one-dimensional array of surface atoms. Electron wavelength $\lambda = h/mv = h/(2mE)^{\frac{1}{2}}$, where E = electron energy, m is electron mass. The angle of the incident beam with the normal is α and the angle of reflection is θ. Interference occurs when scattered waves from neighbouring lattice atoms have path differences that are multiples of λ and

$$n\lambda = a(\sin \theta - \sin \alpha).$$

For a first-order spectrum ($n = 1$) and $\alpha = 0°$, $\lambda = d_{hk} \sin \theta$, where h, k are the Miller indices (see Fig. 15.4).

scattering, the wave vectors of the incident and scattered electrons are the same and there is no loss of energy nor change of wavelength. However, the momentum vector is modified so that the angle of scattering of the diffracted beam is different from that of the reflected wave. By measurement of the angle θ for a given λ (or accelerating voltage), the direction and distance apart of the rows of ion-cores can be evaluated (Fig. 15.2).

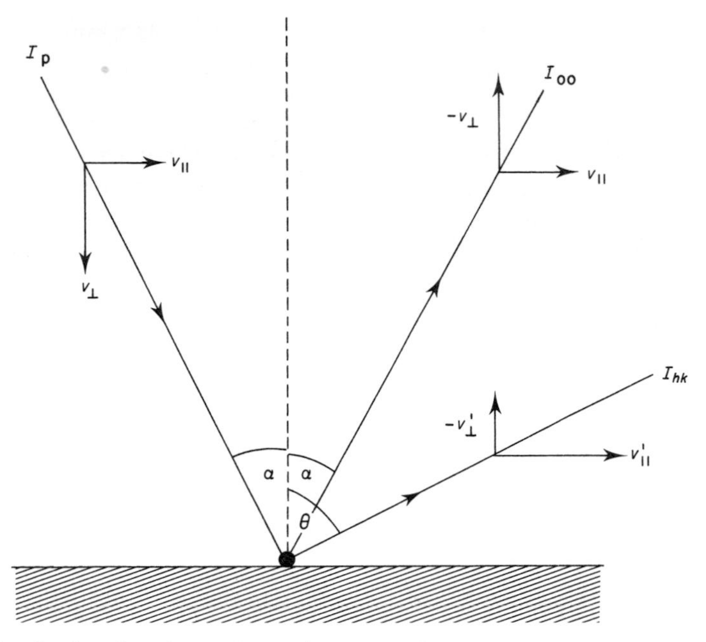

FIG. 15.2. Elastic reflection of a primary electron beam I_p to give specular beam I_{00} and Bragg beam I_{hk}, where h, k are Miller indices (see Fig. 15.4). Velocity vectors of the beam are: before $(v_{\|}, v_{\perp})$ and after $(-v'_{\|}, v'_{\perp})$ for I_{hk}, and $(-v_{\perp}, v_{\|})$ for I_{00}. Kinetic energy is conserved; note the changes of momenta.

15.2. THE LEED DIFFRACTOMETER [5–7]

The primary beam is generated by an electron gun, usually a heated oxide-on-metal filament, which forms the cathode (see Fig. 15.3). The electrons are accelerated towards the surface of the sample by applying an external voltage of 20 to 500 V across an anode and the gun. The electron wavelength must be equal to or less than the lattice constant (2 to 3 Å) of the crystal to obtain a diffraction pattern. The voltage applied (100 to 200 V) is normally a compromise value; higher voltages generate smaller wavelengths, which are advantageous, but these penetrate more deeply into the surface and increase multiple scattering, whereas the diffraction pattern of interest is that formed by electrons elastically back-scattered from the outermost surface layer.

In practice, a small fraction of the incident electrons passes through a small hole in the anode, and the emergent beam is then focused electrostatically or magnetically to form an image of ~1 mm in diameter at the surface of the sample held at ground potential. Of the electrons back-scattered from the surface, a large number has undergone inelastic scattering with consequent

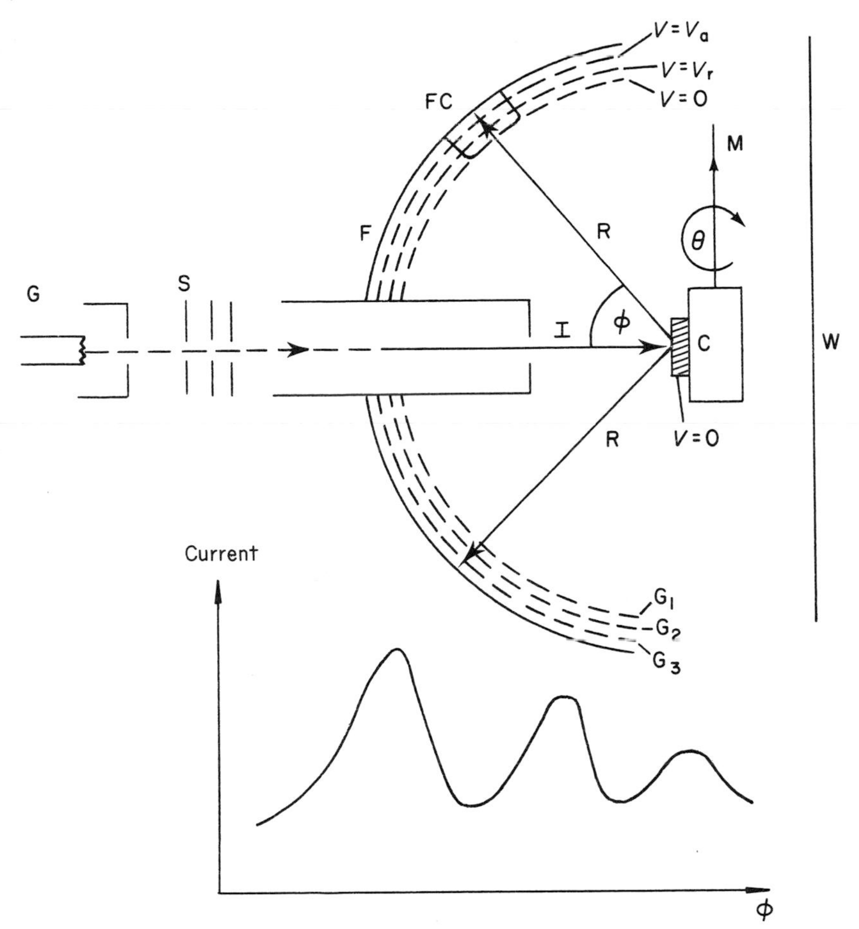

FIG. 15.3. Diagrammatic representation of a low-energy electron diffraction tube. G, electron gun, source of a low-energy primary electron beam; S, slits to select a narrow incident beam I, normal to the surface of the crystal, C; R, reflected electron beam; F, hemispherical fluorescent screen; G_i, concentric grids, G_1 is the innermost grid, maintained at the same potential ($V = 0$) as the sample (crystal C) to provide field-free space for the incident and reflected electrons, G_2 is a grid having a retarding potential V_r to screen out inelastically scattered electrons, G_3 is a grid for post-acceleration of reflected electrons ($V_a = 5$ kV) to excite fluorescence; M, crystal manipulator for moving sample to position for cleaning by electron or ion-bombardment, and for rotating crystal through an angle θ; W, window for viewing the screen through a transparent grid system and for photographing and photometry of screen pattern. The intensity of the image is linearly related to the beam intensity. Spherical symmetry of system gives a uniform diffuse background.

For measurement of intensities, a Faraday cage FC, movable over a hemispherical arc for varying ϕ, is employed.

(i) All beams at a given angle of refraction ϕ are collected for varying θ; absolute intensities are measured for one beam at a time.

(ii) Other angles of diffraction are collected by cage moved in an arc (varying ϕ) and the current is obtained as $f(\phi)$, as sketched in the lower part of the figure.

loss of energy. Since these electrons do not contribute to the diffraction pattern, they are removed by applying a retarding potential of suitable voltage, and only the higher-velocity electrons, which have been elastically scattered (about 5% of the original beam), pass through. They are then accelerated by a positive potential towards a hemispherical fluorescent screen where they produce a visible diffraction pattern.

For quantitative intensity measurements, a diffractometer detector replaces the screen. The essential feature is a movable Faraday cup or a channeltron, which selects the elastically scattered electrons over a small solid angle. Both sample and detector can be rotated to obtain variation of the incident and diffraction angles. For precise measurements the highest surface perfection is required; an ultra-high vacuum is essential.

The energy spread of electrons received by the detector arises mostly from the distribution of their thermal energies. For a cathode at 1000 K, the range is $\sim \pm 0{\cdot}2$ eV about the mean value. In addition, since the amplitude of the diffracted beam is not greatly different from that of the incident beam, the capture cross-section of the diffracted electrons is high ($\sim 10^{-14}$ mm^2), particularly so since their kinetic energy is small. Consequently, there is a high probability that a diffracted electron will suffer more than one scattering collision.

15.3. SHARPNESS OF DIFFRACTION SPOTS [8]

For a monoenergetic, parallel, electron beam and a perfect surface, a diffraction spot would be infinitely sharp. However, deviations from parallelism, the finite size of the electron source, the energy spread within the electron beam and the finite width of the aperture, ensure that any spot has finite size. The size of the surface region for which LEED information may be obtained, i.e. the diameter L of the coherence zone, is around 100 Å. If the mean diameter d of a surface region with perfect periodicity is less than L, then the pattern contains additional features such as streaking, splitting of spots, etc., that arise from the superposition of the amplitude of scattered waves from different regions. But for $d > L$, the pattern reflects the superposition of intensities of diffracted waves from regions of uniform periodicity. Should different domains of macroscopic dimensions be present, these can also be explored by scanning the surface with the primary beam.

In practice, the definition, or sharpness of the diffraction, depends on: (i) the degree of surface order over distances exceeding about 100 Å. Imperfections are responsible for an increase of background intensity and broadening of the diffracted beam. (ii) The extent of penetration of the incident electrons into the solid. The deeper the penetration, or the higher the energy of the primary beam, the larger is the proportion of inelastically scattered electrons in the diffracted beam; the intensity of the pattern is, therefore, less, both in an absolute sense and also relative to the background intensity. (iii) The thermal spread of the incident electrons and the excitation of lattice phonons lead to an additional increase in the half-width of the diffraction peak and a decrease in the height of the peak maximum. (iv) Multiple scattering is responsible for some re-diffraction of the diffracted beam and contributes to the observed intensity in a complex manner.

15.4. ANALYSIS OF DIFFRACTION PATTERNS [9, 10]

Reciprocal or Miller indices are used in analysing diffraction patterns since they provide a description of the atomic rows in terms of lines passing through

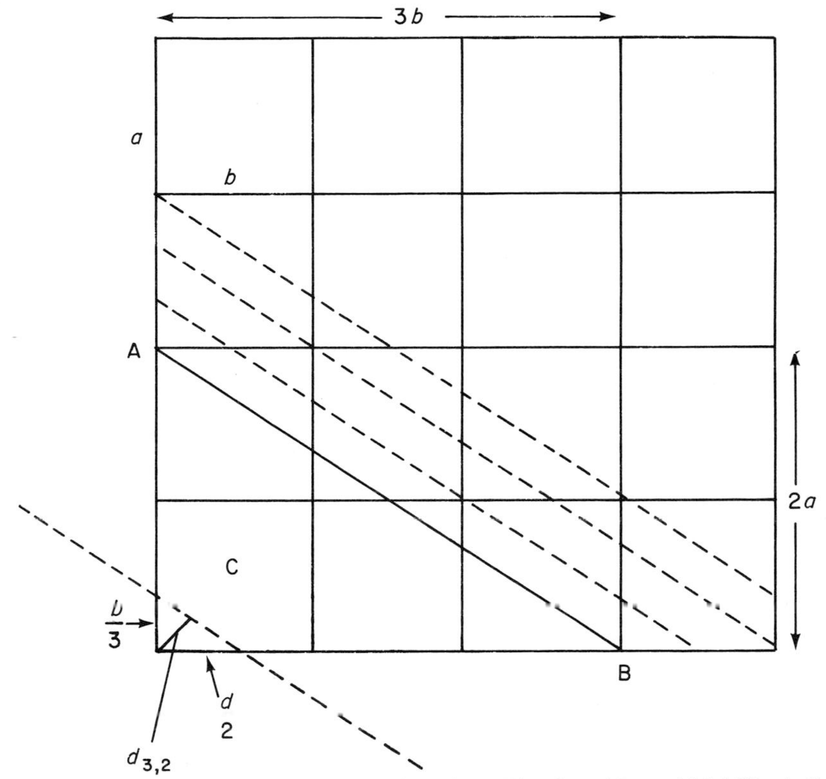

FIG. 15.4. Definition of Miller indices in two-dimensions. The plane AB, to which Miller indices are being assigned, intercepts the unit mesh at $2a$, $3b$. Dashed lines represent planes separated by d_{hk} with similar Miller indices hk. The unit cell C is intercepted at $a/3$ and $b/2$, where

$$d_{hk} = d_{3,2} = [(a/3)^2 + (b/2)^2]^{-\frac{1}{2}} = [(a/h)^2 + (b/k)^2]^{-\frac{1}{2}}.$$

a unit cell. An outline of the procedure for a two-dimensional array is, therefore, indicated. For a (primitive) rectangular net with lattice constants a and b at right-angles, the orientation of any line can be delineated by integral multiples of a and b, say, $2a$, $3b$ (Fig. 15.4). The lowest common denominator (6) of the reciprocals of these numbers is obtained, and the numerators of their quotients, 3 and 2, respectively, are called the Miller indices. The unit cell at the origin of the mesh is intercepted by the given line at $a/3$ and $b/2$ so that the perpendicular distance of this line to the origin from geometric considerations is

$$d_{hk} = \left(\frac{h^2}{a^2} + \frac{k^2}{b^2}\right)^{-\frac{1}{2}} \quad \text{or} \quad \frac{1}{d_{hk}^2} = \frac{h^2}{a^2} + \frac{k^2}{b^2}, \tag{15.2}$$

where h, k (= 3 and 2 here) are Miller indices, and d is the spacing between adjacent lines of a set described by the Miller indices hk passing through the surface. If the unit mesh is generated by the unit mesh vectors **a** and **b** subtending an angle γ (instead of 90°), then

$$\frac{\sin\gamma^2}{d_{hk}} = \frac{h^2}{a^2} + \frac{k^2}{b^2} - \left(\frac{2hk}{ab}\right)\cos\gamma. \tag{15.3}$$

There are five (Bravais) lattices in two-dimensions, viz. square ($a = b$,

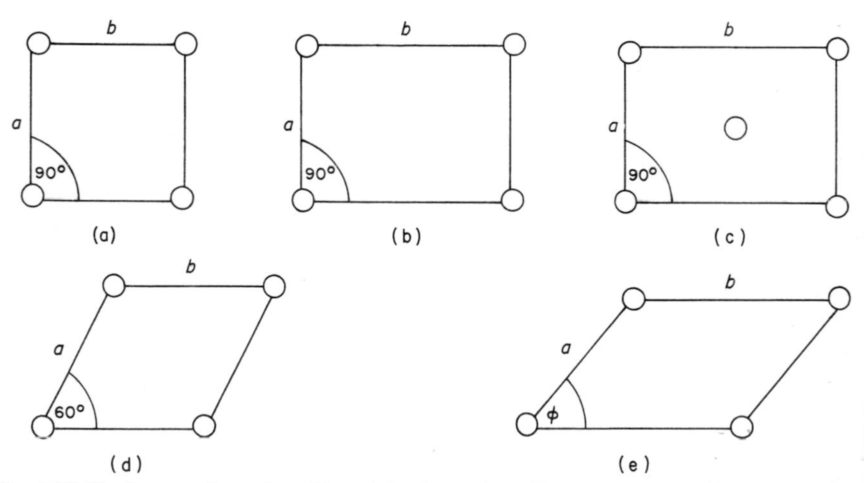

FIG. 15.5. The five two-dimensional Bravais lattices. a, b are distances between adjacent scattering surface atoms; ϕ is the angle subtended by the vectors **a**, **b**.
(a) Square: $a = b$, $\phi = 90°$; fourfold rotational symmetry.
(b) Primitive rectangular: $a \neq b$, $\phi = 90°$; twofold symmetry.
(c) Centred rectangular: $a \neq b$, $\phi = 90°$; twofold symmetry.
(d) Hexagonal: $a = b$, $\phi = 60°$; sixfold symmetry.
(e) Oblique: $a \neq b$, $\phi \neq 90°$, twofold symmetry.

$\gamma = 90°$), hexagonal ($a = b, \gamma = 120°$), rectangular primitive ($a \neq b, \gamma = 90°$), rectangular centred ($a \neq b, \gamma \neq 90°$), and oblique ($a \neq b, \gamma \neq 90°$) (Fig. 15.5).

Diffraction angles θ for different angles of incidence α (both measured from the normal to the surface) are given by

$$n\lambda = d_{hk} \sin \theta + d_{hk} \sin \alpha. \tag{15.4}$$

Since α, in practice, approaches normal incidence,

$$\sin \theta = \frac{n\lambda}{d_{hk}} = \frac{n}{d_{hk}} \left(\frac{150 \cdot 4}{V} \right)^{\frac{1}{2}}. \tag{15.5}$$

Hence, from the measured value θ, d_{hk} can be evaluated for a particular value of the accelerating voltage or electron wavelength. It is evident that, as V is increased, the first diffraction beam appears when $n = 1$ and $\theta = 90°$; furthermore, for a given voltage, larger unit nets produce more beams and their first diffraction maxima are closer to the normal.

15.5. THEORETICAL MODELS FOR ELECTRON SCATTERING

15.5.1. Kinematical theory [1, 2]

This theory, in its simplest form, takes account of the diffracted waves produced by the incident wave in the outermost layer of the solid, i.e. only single scattering processes are considered. Since multiple scattering contributes to the observed intensity, quantitative comparison of intensities are not meaningful, although information about steps and facetting, etc., can be extract

The general treatment is to describe the incident electron and the diffraction waves by wave vectors $\mathbf{k}_0$ and $\mathbf{k}$, each being represented as a plane wave. The amplitude of the scattered wave is obtained by summing over the amplitudes of waves scattered by all atoms. A product of two factors, F, G are involved. The F factor describes the scattering amplitude of the atoms, and its summation depends on the scattering angle and the electron wavelength. It varies slowly with angle and mainly determines the intensity ($I = |F|^2 |G|^2$) but, because of neglect of multiple scattering, it is not particularly meaningful. The G factor, or interference function, comprises a summation over completely equivalent unit meshes and is given by

$$G = \sum_{n_1 n_2} \exp[-i(\mathbf{k} - \mathbf{k}_0) \cdot (n_1 \mathbf{a}_1 + n_2 \mathbf{a}_2)], \tag{15.6}$$

in which n_1, n_2 are integers, and the unit mesh, or the periodicity across the surface, is defined by a_1 and a_2; this factor represents the geometric phase difference between the incident wave ($\exp i\mathbf{k}_0\mathbf{r}_{1n_1n_2}$) and the waves [$\exp(i\mathbf{k}r_{1n_1n_2})$]/$\mathbf{r}$ diffracted by the atom in position $\mathbf{r}_{1n_1n_2}$. The summation displays strong maxima in certain directions and diffracted beams and diffractions spots are produced on the screen; from the diffraction pattern the geometry of the unit cell of the surface structure can then be deduced.

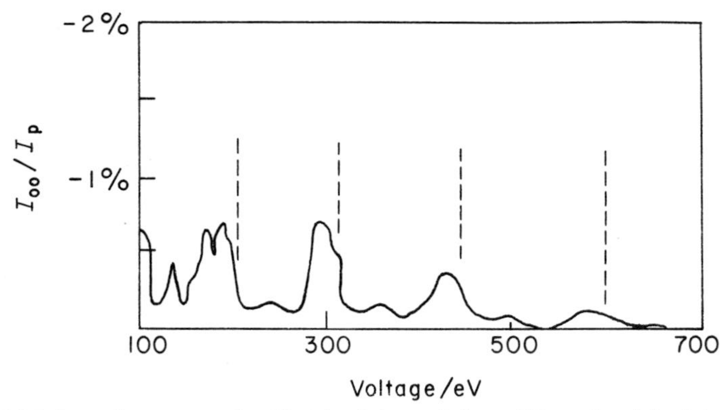

FIG. 15.6. Intensity–energy plot. Sketch of the variation of the ratio of the intensity I_{00} of the (0, 0) beam diffracted from a Ni(100) plane divided by that (I_p) of the primary beam as a function of the accelerating voltage V. The position of the primary Bragg peaks, calculated from the kinematical model, are shown by the dashed lines. The experimental maxima are at lower energies (several eV), the bands are broader and, particularly at the lower voltage, additional structural features are apparent.

The treatment can be extended to include layers within the solid by introducing an attenuation coefficient to take account of the effect of inelastic scattering. However, when the diffracted intensity divided by that of the primary beam is plotted as a function of the accelerating voltage applied to the primary beam, even some qualitative features are not predicted. Thus, with increase of voltage, i.e. deeper penetration of incident electrons, the main peaks are broadened and secondary Bragg maxima arising from multiple scattering appear at lower wavelength, thereby indicating the limitations and approximate nature of kinematic model, which can only account for the primary maximum (see Fig. 15.6). Multiple scattering must be treated by a dynamical theory.

15.5.2. The dynamical theory [11–13]

In its simplest form, a Bloch wave in the crystal is combined with the incident and reflected wave. The solution of the Schrödinger equation for the wave function of the substrate is then matched at the surface with the general solution of the incident and back-scattered plane waves in vacuum. In the more sophisticated treatment, crystal wave functions derived from the superposition of all possible Bloch functions are introduced. The result is that in the calculated intensity–energy plots the band structure contains more detailed structure characteristics over and above those of the primary Bragg peaks (see Fig. 15.6).

Moreover, since the scattering is largely confined to the first few surface layers, the electron-wave field inside the solid can be expressed by perturbation theory to include an additional set of multiple scattering events. Proper convergence is obtained for the free-electron metals of lower atomic numbers, but not with transition metals. This difficulty has been overcome by solving both intraplanar and interplanar multiple-scattering events for a pair of layers and then using the resulting transmission and reflection matrices for other layers.

Various physical parameters are involved in all dynamical calculations, viz: (i) the scattering potential, i.e. the self-consistent ion-core potential of the nucleus in an atmosphere of mobile electrons; it is obtained by a Hartree–Fock calculation that includes exchange and correlation terms [14]. (ii) The inner potential of electrons in the metal: this is less than their potential in vacuum so that the electron wave length and therefore the position of the primary Bragg peak are changed. This inner potential can be evaluated from experimental intensity–energy plots by fitting their maxima to those theoretically calculated for the same energy. (iii) The shape of the potential profile on emergence from the surface: this is not a significant effect and is usually ignored [15]. (iv) Thermal vibrations of atoms: these cause a decrease of spot intensity relative to that of the background. The effect can be estimated from the temperature variation of spot intensity to provide a Debye–Waller factor [16] from which the intensities are calculated for a rigid lattice; the results are then transformed to room temperature.

15.5.3. Data-reduction techniques [17, 18]

In these techniques, the kinematical part of the diffraction intensity is extracted from the intensity–energy data; it accounts for the Bragg-like features that originate primarily from the outermost surface layer. Determination of structure then follows the procedure adopted for X-ray crystallography. In practice, multiple scattering peaks are largely eliminated by averaging the experimental data for a series of different azimuth and scattering angles. The average intensity–energy plots so obtained give, to a close approximation, the kinematic contributions. The success of this procedure depends on the fact that the kinematic scattering factor is described in terms of the difference of the wave vector of the incident and reflected beam ($\mathbf{k}_0 - \mathbf{k}$), whereas the dynamical scattering factor is determined by the separate values of these two-wave vectors. The method, however, is only valid when there is strong inelastic scattering between the atoms and the electrons.

15.6. APPLICATION TO CHEMISORPTION [19–23]

Since the information derived from LEED diffraction patterns is predominantly concerned with the outermost layer of the surface, LEED is of considerable importance in investigations of adsorbed layers. The main requirement is lateral long-range periodicity of an ordered surface overlayer in directions parallel with the surface with unit repeating nets of linear dimensions larger than 100 Å.

Most investigations have been concerned with the low-index planes of f.c.c. and b.c.c. metals. Cleaning of the surface is usually accomplished by cycles of ion-bombardment and evacuation (10^{-8} Pa) followed by thermal annealing. Less effective is the use of high temperatures obtained by direct resistance heating or indirectly by an external oven, or by electron bombardment. Monolayer coverages of chemisorbed layers have been most frequently studied.

15.6.1. The designation of overlayer structures

In describing the surface geometry of an overlayer the unit cell of the adlayer is characterized with respect to the unit cell of the substrate. When the diffraction pattern of the overlayer is identical with that of the clean surface it is designated a (1×1) structure. A unit cell with a surface structure twice as large as that of the substrate (e.g. adatoms on alternate ion-core adsorption sites) is referred to as a primitive (2×2) or $p(2 \times 2)$ pattern (see Fig. 15.7). When this arrangement occurs along one crystallographic axis but adatoms are located at all sites along the other axis, a (2×1) structure is formed. The size of the surface net can always be determined provided that it is in registry with the substrate unit cell. In circumstances where an integral multiple relationship does not describe the structure, the nomenclature is extended. For example, $c(2 \times 2)$ denotes the presence of a rectangular centred structure, i.e. a $p(2 \times 2)$ structure in which adatoms occupy alternative adsorption sites along both axes, with an extra adatom at the centre of the rectangle. Sometimes, the surface structure must be rotated with respect to the substrate unit cell. For example, on a hexagonal lattice, every third site is uniquely situated and a surface structure rotated through 30° (i.e. having six-fold rotational symmetry) can be correlated with the substrate unit cell. The structure is designated $(\sqrt{3} \times \sqrt{3}) - R\ 30°$.

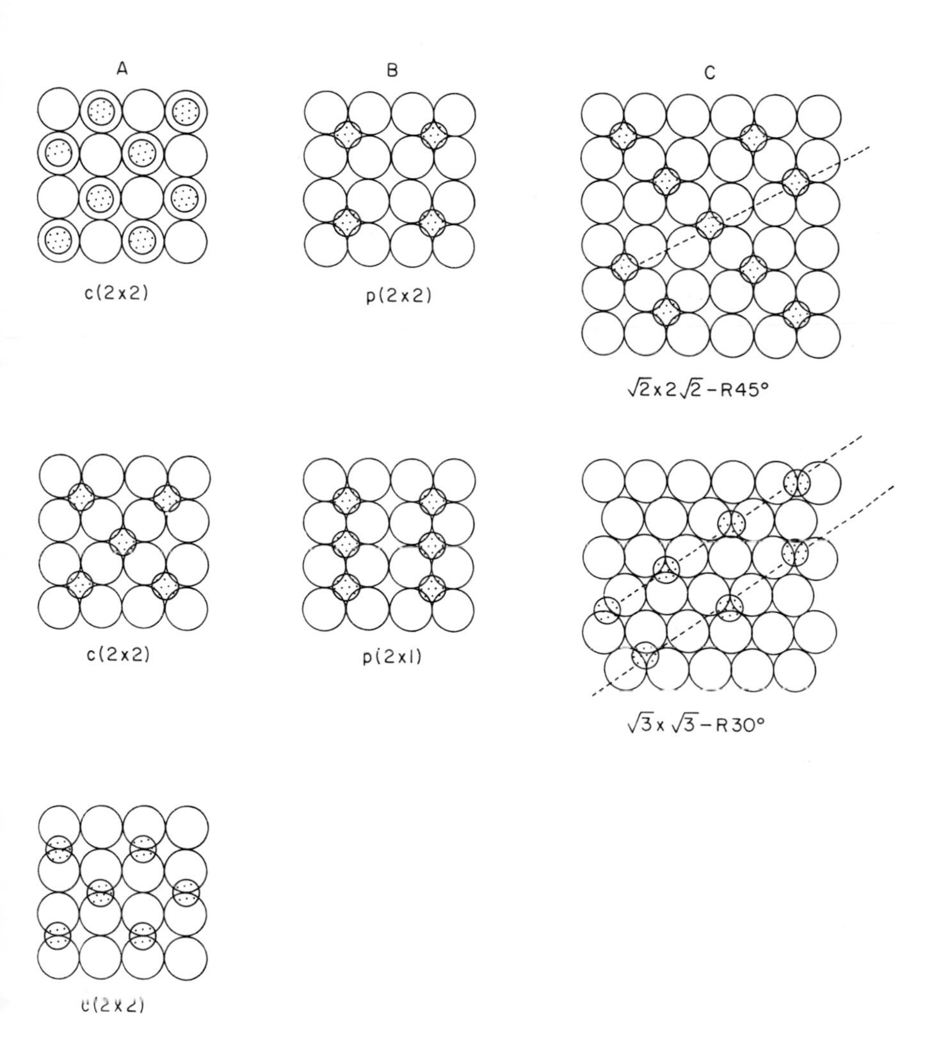

FIG. 15.7. Examples of overlayer structures. A, Three different arrangements of admolecules on a cubic substrate surface; all are c(2 × 2) structures and give identical diffraction patterns. B, A primitive rectangular structure with different spacings on the *a*-axis. C, A structure rotated with respect to unit cell of the substrate. The angle indicates the orientation of the overlayer structure. Upper structure, every other lattice site on the square face is occupied; lower structure, every third lattice site on the hexagonal face is occupied.

In general, any surface unit mesh can be translated to an equivalent position provided there is registry between surface and substrate structure. The translation usually involves distances corresponding to the lattice constants of the surface or substrate, but can also comprise contributions from each of these distances. Similarly, a unit mesh of twofold symmetry can be placed on a fourfold symmetry mesh in two equivalent ways by rotation through 90°. Such equivalent positions are equally probable and lead to registry degeneracy and formation of characteristic domains over the surface. For a domain of larger mesh and lower symmetry, fractional order LEED beams are obtained and a (2×2) structure appears to be the same as a (2×1) surface structure in three equivalent orientations of 120°, except for intensity differences.

From the extensive experimental data available for a variety of gas/metal systems some general guide lines have been formulated. The simpler gases usually have molecular dimensions close to or less than the largest inter-row distances on the substrate surface; the structure of the adlayer is then determined by maximization of adsorbate–substrate and adsorbate–adsorbate interactions. A rule of close packing generally applies, i.e. adspecies tend to form structures having the smallest unit cell with the closest packing arrangement permitted in terms of molecular dimensions and adsorbate–adsorbate interactions. Thus, the most common structures are (1×1), (2×2), $c(2 \times 2)$, $(\sqrt{3} \times \sqrt{3})$ and (2×1). In general, the unit cell vectors and the rotational symmetries of the surface and substrate structures also tend to be the same.

As has already been noted, although the structure may be identified according to the above rules, the exact location of the adatom of a monolayer with respect to different positions on the surface cannot be determined. For example, each adatom of a monolayer on the (100) plane of a f.c.c. lattice can be located above a substrate surface atom or can be recessed at the centre of the square array in a four-co-ordinated position. Both arrangements lead to a (1×1) pattern identical with that of the clean substrate although the intensities of the spots are different.

Similarly, unless the coverage has been determined, a multilayer gives the same pattern as, and may be mistaken for, a monolayer. Again, a random distribution of immobile adatoms at lower coverages than the monolayer gives the (1×1) substrate pattern and the adlayer merely contributes a different but uniformly greater background density of magnitude dependent on coverage. Should the adatoms be freely mobile, the intensity of the spots is further decreased and that of the background increased.

15.6.2. Intensity–energy structure

Since different arrangements of surface atoms can produce the same diffraction pattern, it is necessary to measure the intensities of the diffracted beam maxima in order to distinguish them. In general, all the theoretical calculations [24] on experimental intensity–energy plots using dynamical methods predict, e.g. for oxygen on Ni(100), the same co-ordination symmetry of the binding site with the adatom located above four nearest-neighbour nickel atoms. There is much less agreement on the chemisorption bond-length, largely because different values of the scattering and inner-potentials derived from various procedures, have been used in these calculations by different workers.

15.6.3. Reconstruction of the surface [25–27]

At the termination of the bulk lattice, the outermost atoms may be shifted towards the bulk owing to the absence of attraction by an overlayer of atoms above them; similarly, because the co-ordination number of surface atoms is less than that in bulk, some type of hybridization may occur in order to reduce the surface free energy. For clean metal surfaces, the diffraction patterns indicate that the surface periodicity is usually identical with that of a corresponding plane in the bulk metal. Additional diffraction spots have been observed in the cases of Pt, Ir and Au, but the abnormal surface structure that differs from that of the bulk has also been attributed to its stabilization by small amounts of impurities, such as oxygen. However, surface displacements do occur, particularly when the binding of surface to substrate atoms is directional. Any effect normal to the surface merely alters the intensity but not the pattern. Lateral effects, however, can give rise to fractional beams, although the occurrence at metal surfaces has not been convincingly proved.

Chemisorption of a gas to form predominantly covalent bonds, however, does induce atomic displacements. When the displacement is small, the LEED pattern is unchanged, but the intensity may be modified. But when the chemisorbed bond is very strong, the formation of a two-dimensional overlayer by place exchange of metal and adsorbate atoms can occur. Such surface reconstitution is frequently brought about during the chemisorption of oxygen, and metal atoms originally covered by oxygen adatoms can be thermally regenerated and form new adsorption sites for additional chemisorption. Evidence from studies of the changes of work functions accompanying reconstruction seems to be quite conclusive. In this case, e.g. when oxygen is chemisorbed on a nickel surface, fractional-order diffracted beams are observed and these have been interpreted as evidence for reconstruction. One criticism has been that the intensity of such beams are of a magnitude comparable with that of the integral-order beam. On the grounds that the atomic number of oxygen is small compared with that of nickel, the adatoms would, however, have a low scattering factor which should give rise to spots of lower intensity [28, 29]. An alternative explanation is that the fractional-order beam arises from multiple scattering of the incident electron beam.

15.6.4. Facetting of the surface

Facetting is the formation of new crystal planes inclined at an angle to the plane of the original surface. The effect can be brought about by the presence of oxygen adatoms and elevated temperatures even at low coverages [30–32]. The reason is that the surface free energy per unit area of the original oxygenated surface is decreased by formation of new planes of higher density with different crystallographic orientations and larger surface areas. Such facetting can be recognized by the change of the LEED pattern. Thus, for an incident beam normal to the original surface, all diffracted beams can be moved towards the centre of the diffraction pattern (i.e. the (0, 0) spot) by increasing the accelerating voltage of the primary beam. The pattern originating from the surface of the facet arises from an incident beam which is not normal to the surface, and its (0, 0) spot is at a different position. With increasing energy of the incident beam, facet spots move towards the specular beam position away from the centre of the screen and they can often be projected out of the field of view. Constructive interference for the facetted surface requires that [33, 34]

$$d_F(\sin\theta - \sin\alpha) = n\lambda \tag{15.7}$$

or

$$\mathrm{d}\lambda/\mathrm{d}\theta = (d_F/n)\cos\theta, \tag{15.8}$$

where α is the angle between the original and facet planes and d_F the periodicity at the surface of the facet. When a beam from the facet interferes with the (0, 0) beam from the original surface then

$$n\lambda = 2d_F \sin\alpha, \tag{15.9}$$

and, from Eq. (15.8), the relationship

$$\mathrm{d}\lambda/\mathrm{d}\theta = (d_F/n)\cos\alpha \tag{15.10}$$

is also valid.

Combining Eqs (15.9) and (15.10) gives

$$\mathrm{d}\lambda/\mathrm{d}\theta = \lambda/2)\cot\alpha.$$

Hence, from the variation of the diffraction angle θ with λ (or accelerating voltage applied to the primary beam) as the diffracted beam passes through the original (0, 0) spot, the inclination angle α can be determined, and d_F evaluated by substitution in Eq. (15.9).

15.6.5. Surface steps [35–38]

The chemisorption properties of surface steps are often substantially different from those associated with low-index planes. High-index planes can be constructed from terraces of low-index planes of equal height (and length), i.e. of facets and antiphase domains with the same periodicity (Fig. 15.8). At

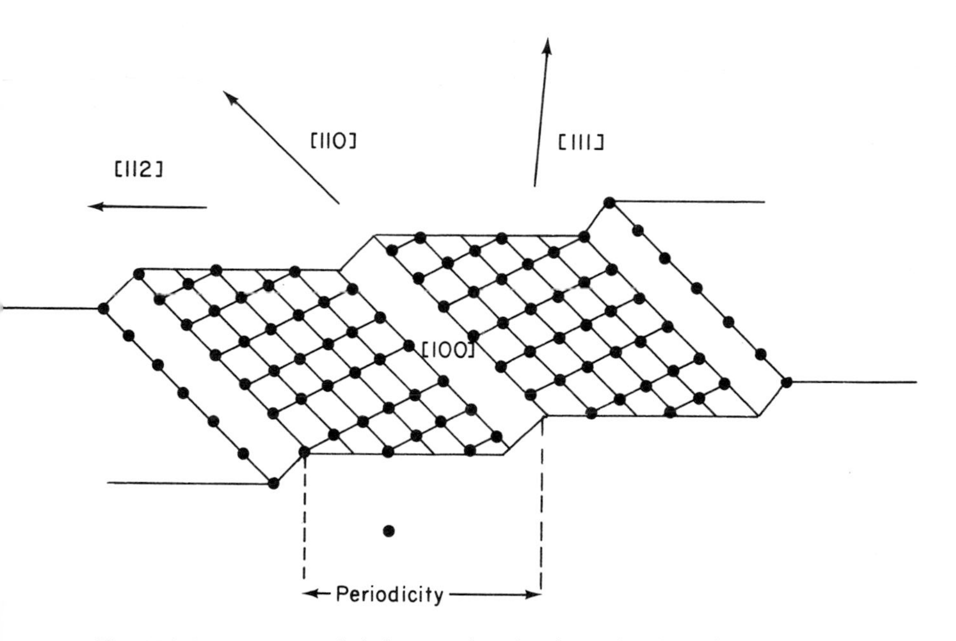

FIG. 15.8. Step structure of platinum surface showing ordered atomic steps.

certain electron energies, these regions give rise to different central spots towards which their diffraction pattern can be moved by increasing the incident electron energy; they are also responsible for the splitting of the spots. The magnitude of the splitting of the central spot is given approximately by the ratio of the electron wavelength to the width of the terrace, and the step height can be derived from the position of the diffraction pattern of the inclined surface [36, 37]. The step heights are usually equal to the lattice distance or some integer multiple of this value.

REFERENCES

1. C. Kittel, "Introduction to Solid State Physics", Wiley and Sons, Inc., New York, 1966, p. 675.
2. J. J. Lander, "Progress in Solid State Physics", Macmillan Co., New York, 1965, Vol. 2.
3. J. B. Pendry, "Low Energy Electron Diffraction", Academic Press, New York, 1974.
4. C. J. Davisson and L. H. Germer, *Phys. Rev.*, 1927, **30**, 705.
5. E. J. Scheibner, L. H. Germer and C. D. Hartman, *Rev. Scient. Instrum.*, 1960, **31**, 112.
6. J. J. Lander, F. Unterwald and J. Morrison, *Rev. Scient. Instrum.*, 1962, **33**, 784.
7. R. L. Park and H. E. Farnsworth, *Rev. Scient. Instrum.*, 1964, **35**, 1592.
8. R. L. Park, J. E. Houston and D. G. Schreiner, *Rev. Scient. Instrum.*, 1971, **42**, 60.
9. E. A. Wood, *J. Appl. Phys.*, 1964, **35**, 1306.
10. B. Lang, R. W. Joyner and G. A. Somorjai, *Surface Sci.*, 1972, **30**, 440.
11. C. G. Darwin, *Phil. Mag.*, 1914, **27**, 675.
12. H. Bethe, *Naturwiss.*, 1927, **15**, 786; 1928, **16**, 333.
13. D. S. Boudreaux and V. Heine, *Surface Sci.*, 1967, **8**, 426.
14. J. B. Pendry, *J. Phys. C*, 1971, **4**, 2501 and 3095.
15. D. W. Jepson, P. M. Marcus and F. Jona, *Phys. Rev. B*, 1972, **5**, 933.
16. C. B. Duke and G. E. Laramore, *Phys. Rev. B*, 1970, **2**, 4765, 4783.
17. C. W. Tucker and C. B. Duke, *Surface Sci.*, 1970, **23**, 411; 1972, **29**, 237.
18. T. C. Ngoc, M. G. Lagally and M. B. Webb, *Surface Sci.*, 1973, **35**, 117.
19. J. Anderson and P. J. Estrup, *J. Chem. Phys.*, 1967, **46**, 563 and 567; 1966, **45**, 2254.
20. D. L. Adams and L. H. Germer, *Surface Sci.*, 1970, **23**, 419; 1971, **26**, 109; 1971, **27**, 21; 1972, **32**, 205.
21. K. Yonehara and L. D. Schmidt, *Surface Sci.*, 1971, **25**, 238.
22. L. R. Clavenna and L. D. Schmidt, *Surface Sci.*, 1970, **22**, 365.
23. J. T. Yates Jr and T. E. Madey, *J. Chem. Phys.*, 1971, **54**, 4969.
24. J. E. Demuth, D. W. Jepsen and P. M. Marcus, *Phys. Rev. Letters*, 1973, **31**, 540; 1974, **32**, 1182; *Surface Sci.*, 1974, **45**, 733; *Phys. Rev. B*, 1975, **11**, 1460.
25. L. H. Germer and A. U. MacRae, *J. Appl. Phys.*, 1962, **33**, 2923.
26. H. E. Farnsworth, *Adv. Catalysis*, 1964, **15**, 31.
27. J. W. May, *Adv. Catalysis*, 1970, **21**, 244.
28. C. W. Tucker, *Surface Sci.*, 1971, **26**, 3111.
29. P. J. Jennings and E. G. MacRae, *Surface Sci.*, 1971, **23**, 63.
30. N. J. Taylor, *Surface Sci.*, 1964, **2**, 544.
31. L. H. Germer and J. W. May, *Surface Sci.*, 1966, **4**, 452.
32. C. C. Change and L. H. Germer, *Surface Sci.*, 1967, **8**, 115.
33. C. W. Tucker, *J. Appl. Phys.*, 1967, **38**, 1988.
34. J. C. Tracy and J. M. Blakely, *Surface Sci.*, 1969, **13**, 313.
35. B. Lang, R. W. Joyner and G. A. Somorjai, *Surface Sci.*, 1972, **30**, 440.
36. M. Henzler, *Surface Sci.*, 1970, **19**, 159.
37. J. E. Houston and R. L. Park, *Surface Sci.*, 1971, **26**, 269.
38. K. Christmann, G. Ertl and O. Schober, *Surface Sci.*, 1973, **40**, 61.

16

Infra-red Spectroscopy of Chemisorbed Molecules

Early knowledge of the interaction of an adspecies on a metal surface was largely acquired from thermodynamic interpretation of the results of adsorption equilibria and heats of adsorption, but techniques have now been developed for direct measurement of adsorbate–adsorbent interactions. One important method is the application of infra-red spectroscopy of adspecies to the investigation of their molecular vibrations, the identification of their group frequencies, and the effect of the environment on their ground-state vibrational modes.

16.1. SPECTRA OF GASEOUS MOLECULES

On transmission of infra-red radiation through a column of gas, energy transfer takes place between the electromagnetic field and the gas molecules when the incident radiation contains a frequency ν which is related to two quantized vibrational energy levels E_1 and E_2 in the molecules such that $E_1 - E_2 = h\nu$, where h is Planck's constant. Transitions between vibrational levels, accompanied by rotational transitions involving smaller energy changes, give rise to bands with fine structure in the infra-red region with a wave-number range of 2000 to 5000 cm^{-1}, provided that the vibratory movement within the molecule gives rise to a change in its dipole moment.

For a simple harmonic oscillator, the values of the vibrational energy E_v allowed by quantum theory are given by the relationship.

$$E_v = h\nu(v + \tfrac{1}{2}), \quad (16.1)$$

where v is the vibrational quantum number restricted to integer values. Allowed transitions between the various vibrational levels are those permitted by selection rules; for a simple harmonic oscillator, transitions are forbidden except between adjoining levels, v_n, $v_{n\pm 1}$.

When a molecule comprising two atoms of mass m_1 and m_2 at an equilibrium distance r_e is stretched to increase the distance between the atoms to $r + r_e$, the restoring force f is given by

$$f = -kr \tag{16.2}$$

where k is the force constant. From classical mechanics,

$$k = 4\pi^2\nu^2\mu \tag{16.3}$$

where ν is the vibration frequency (s^{-1}) and μ is the reduced mass ($1/\mu = 1/m_1 + 1/m_2$). The wave number $\tilde{\nu}$ (cm^{-1}) is ν/c, where c is the velocity of light ($\sim 3 \times 10\,cm\,s^{-1}$). For $\tilde{\nu} = 3000\,cm^{-1}$, k, the strength of the chemical bond between the masses m_1 and $m_2 = 4{\cdot}8 \times 10^5\,dyn\,cm^{-1}$.

However, the potential energy–internuclear distance relationship departs from pure harmonicity with increase of v, and

$$E_v = h\nu(v + \tfrac{1}{2}) - x(v + \tfrac{1}{2})^2 \ldots, \tag{16.4}$$

x being the anharmonicity constant. Although the correction to E_v for diatomic molecules is small, the important consequence is the relaxation of the selection rule, $v_n - v_{n\pm 1} = 1$, and the appearance of overtones and combination tones in the spectrum.

Within an n-atomic (polyatomic) molecule, the maximum number of vibrations is $3n - 6$, of which $n - 1$ are stretching or valence vibrations and $2n - 5$ are bending or deformation modes. Some of these modes may be degenerate, e.g. when vibrations of the same frequency have their axes in different directions. Examination of the infra-red spectrum of a molecule assists its identification. The so-called characteristic stretching vibration frequency of the unit A—B in the molecule X—A—B is shifted to a comparatively small extent by change in the mass of X (which may often comprise a group of atoms); but different bondings of A to X are associated with their own characteristic frequencies, e.g. for the C—H valence vibration (see Table 16.1),

$>$C—H (2960 cm^{-1}), $\quad$ =C(—)—H (3020 cm^{-1}), $\quad$ ≡C—H (3300 cm^{-1}).

TABLE 16.1

Group		Vibration frequencies/cm^{-1}
Symmetric CH_2	$>C<^{H\rightarrow}_{H\rightarrow}$	2850
Symmetric CH_3 stretch	$-C \leftarrow (H\rightarrow, H\rightarrow, H\rightarrow)$	2870
Tertiary CH stretch	$\geq C-H\rightarrow$	2890
Asymmetric CH_2 stretch	$>C<^{H\leftarrow}_{H\rightarrow}$	2925
Asymmetric CH_3 stretch	$-C \leftarrow (H\rightarrow, H\leftarrow, H\rightarrow)$	2960
Olefinic CH stretch	$=C<^{H\rightarrow}$	3020
Acetylenic CH stretch	$\equiv C-H\rightarrow$	3200

Our interest is the determination of the characteristic frequencies of a molecule chemisorbed on a metal surface and the information that these can provide regarding the structure of the adsorbed molecule and its binding energy to the surface.

16.2. TRANSMISSION SPECTROSCOPY OF CHEMISORBED MOLECULES

Most of the early investigations employed transmission spectroscopy (see Fig. 16.1). Unfortunately, metals strongly absorb infra-red radiation unless the particle size (or thickness) is less than 300 Å (and preferably within the range 50 to 100 Å); since the chemisorbed molecules are restricted to the metal surface, a method has to be devised to obtain a sufficient number of adspecies in the path of the incident beam to provide measurable intensities of the absorption bands. The intensity is determined by the Lambert–Beer law,

$$I = I_0 \exp(-\varepsilon dc), \qquad (16.5)$$

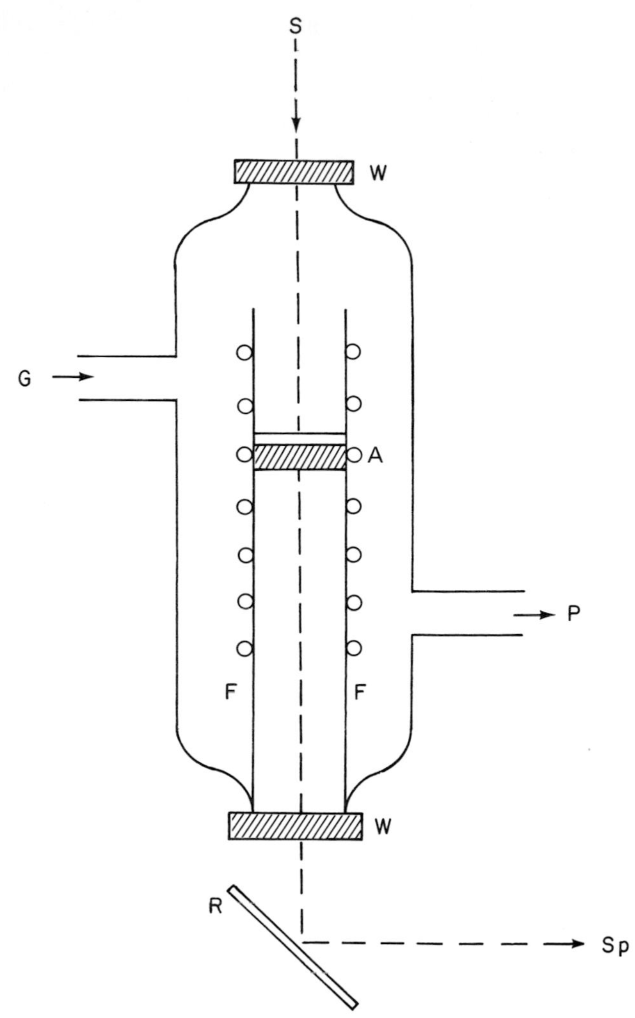

FIG. 16.1. General features of a cell for transmission-spectroscopic study of a chemisorbed gas on a supported metal adsorbent. S, infra-red source; W, CaF_2 windows; G, gas inlet; F, furnace windings; P, to pumps; A, compressed disc of SiO_2-supported metal; R, reflector plate; Sp, infra-red spectrometer.

I_0 being the incident and I the transmitted intensity. The absorbance is defined as $A = \ln(I_0/I) = \varepsilon dc$, in which d is the layer thickness (cm), c is the concentration of adsorbed molecules in the light path, conveniently expressed as the number of adspecies cm^{-3}. The extinction coefficient ε, in units of cm^2 $molecule^{-1}$, is characteristic of the adspecies but varies with the wave-number of the radiation.

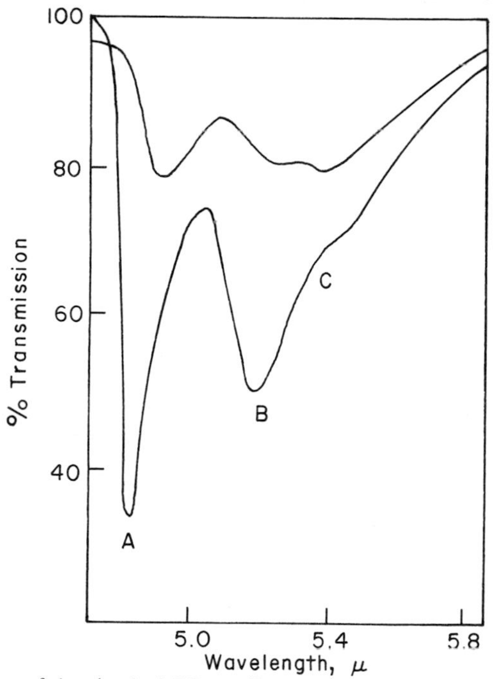

FIG. 16.2. Spectrum of chemisorbed CO on silica-supported Ni at two surface coverages. Bands: A, 2073 cm^{-1}; B, 1924 cm^{-1}; shoulder at C, 1850 cm^{-1}. Labelled curve: higher coverage.

TABLE 16.2

Metal	Support	Reference	Peak frequency/cm^{-1} Linear	Bridged(?)
Ni	SiO_2	Eischens *et al.* [1]	2050–2073	1865–1923
	Al_2O_3	Yates and Garland [2]	2030–2080	1915–1960
	Nujol	Blyholder [3]	~2080	~1940
	Film[a]	Baker *et al.* [4] and Hayward [5]	2050–2060	1900–1950
Pt	SiO_2	Eischens *et al.* [1]	2075	—
	Film[a]	Hayward [5]	2030–2050	1900–1950
Pd	SiO_2	Eischens *et al.* [1]	2060–2085	1830–1920
Ru	Film[a]	Hayward [5]	—	1980
Ir	SiO_2		~2074	—
	Al_2O_3	Lynds [6]	~2070	—
	Film[a]	Baker *et al.* [4] and Harrod *et al.* [7]	~2030	~1970
Fe	SiO_2	Blyholder and Neff [8]	~2000	—
	Film[a]	Baker *et al.* [4]	—	~1925
Cu	SiO_2	Smith and Quets [9]	~2110	—
Rh	Al_2O_3	Yates and Garland [2]	~2095	—

[a] Prepared *in vacuo* (see Table 16.3).

16.2.1. Spectrum of chemisorbed carbon monoxide

The most extensive investigations have been concerned with the adsorption of CO on transition metals. The gaseous molecule possesses a single vibrationary stretching mode giving an absorption band at 2143 cm^{-1} with some rotational structure. This wave number is shifted to a small extent when CO is physically adsorbed, and the rotational movement of the molecule is severely restricted or absent. On chemisorption, the symmetry characteristics

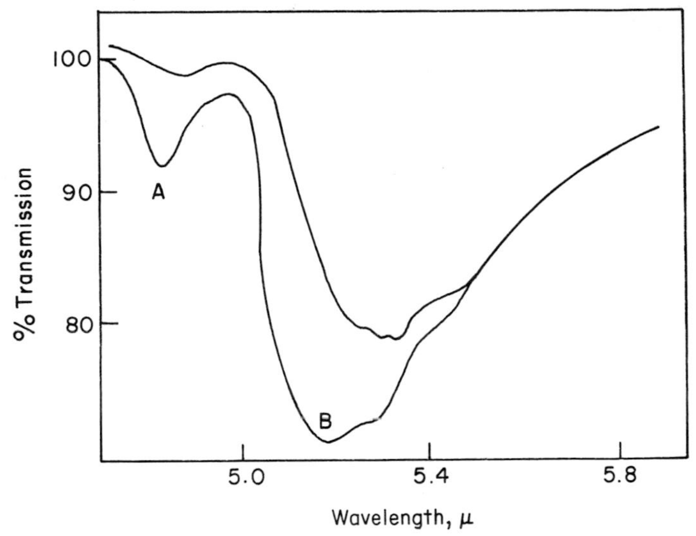

FIG. 16.3. Spectrum of chemisorbed CO on silica-supported Pd at two coverages. Bands: A, 2075 cm^{-1}; B, 1945 cm^{-1}. Labelled curve; higher coverage.

of the gaseous molecule are changed by its reaction with surface metal atoms and two main bands, between 1850 and 1940 cm^{-1} and between 1990 and 2110 cm^{-1}, are usually observed (see Table 16.2 and Figs 16.2, 16.3). The shift to lower wave numbers indicates that the C—O bond has been weakened; the number of bands suggests the formation of two chemisorbed species having different structural characteristics.

16.2.2. Dispersion of metal on inert oxide supports

The application of transmission spectroscopy to chemisorbed molecules awaited the procedure, devised by Eischens *et al.* [1], of dispersing metal as a very thin layer on an inert non-porous support, such as silica in the form of Cabosil (or Aerosil), which is commercially available in the size range 150 to

300 Å. A small particle size is essential to reduce the loss of incident radiation by scattering. The support itself strongly absorbs infra-red radiation so that some ranges of spectral frequencies cannot be examined. Fortunately, "windows" of transmission are available in those regions of the spectrum where many, but not all, absorption bands associated with the adsorbate appear. The background spectrum of support and clean metal has to be subtracted from that obtained after adsorption, e.g. by use of a double-beam spectrometer. In addition, the adsorbed layer itself causes further scattering, but this, fortunately, is not greatly dependent on the frequency of the light.

Some approximate estimations show that absorption bands should easily be observed using a good spectrometer. One gram of Cabosil has a surface area of around 100 $m^2 g^{-1}$; the maximum concentration of dispersed metal is ~10 wt %, and for a particle size of 100 Å its surface area is 1 $m^2 g^{-1}$. The surface of the metal has 10^{15} adsorption sites cm^{-2}, so that a monolayer of CO would comprise 10^{19} chemisorbed molecules g^{-1}; these have an extinction coefficient of 3×10^{-18} cm^2 $molecule^{-1}$. By compression under a pressure of $\sim 1 \times 10^4$ kg cm^{-2}, 1 g of oxide can provide a disc of 1 cm^2 cross-section and 0·1 mm thick; hence,

$$\ln(I/I_0) = 10^{19} \times 3 \times 10^{-18} \times 10^{-2} = 0{\cdot}3$$

or

$$I/I_0 = 0{\cdot}75. \tag{16.6}$$

A good spectrometer can record up to 0·99; the bands can, therefore, be clearly resolved. However, the emergent intensity is only *ca* 18 % of the incident radiation, even in the most favourable circumstances, so that an intense source of infra-red radiation is desirable. Increase of the thickness of the pellet is not helpful, since the corresponding losses by scattering, etc., are greater; likewise, increase of the metal concentration results in the production of larger particles with proportionally higher loss of intensity due to absorption and scattering by the metal itself. One advantage is that the extinction coefficient of a chemisorbed molecule is often 10 to 100 times greater than that of the free molecule.

16.2.3. Preparation and properties of supported metal adsorbent

The usual procedure for the preparation of supported metals is to form a slurry of finely divided oxide, e.g. Cabosil, in an aqueous solution of a salt of the metal, e.g. H_2PtCl_6. This compound is transformed to the corresponding platinic oxide by heating in air, and then reduced to the metal with hydrogen; finally, outgassing *in vacuo* at around 600 K is effected. Before reduction, the impregnated oxide (~ 1 g) is pelletized under a pressure of around 10^4 kg cm^{-2}; a disc of about 1 cm^2 cross-section and 0·1 mm thick is usually employed.

Other oxide supports, such as alumina and titania, have also been used, but the spectral characteristics of the two main bands depend on the chemical nature of the oxide (see Table 16.2). By comparison of these spectra with those arising from thin metal films, it would appear that silica is the most inert support. Even so, literature values for CO chemisorbed on SiO_2-supported metals from different laboratories vary over a range of ~ 30 cm^{-1}.

Variations are associated with: (i) the variable electron donor–acceptor properties of the different oxides; (ii) the different degrees of dispersion and distributions of particle size of the metal, depending not only on the chemical nature of the oxide but also on the different preparatory procedures with the same oxide; (iii) the difficulty of complete reduction to the metal since, with increasing reduction, the peaks shift to lower frequencies with simultaneous change in relative intensity of the bands; (iv) the variable extent of sintering of the metal, e.g. for the CO/Ni system the band at 2035 cm^{-1} is probably associated with amorphous metal since it shifts to 2082 cm^{-1} (with substantial reduction in intensity) as the crystallinity of the sample is improved; and (v) contamination of the surface by residual gas (at pressures of 10^{-3} to 10^{-4} Pa) during handling.

16.2.4. Evaporated metal films

Films prepared under ultra-high vacuum conditions have surfaces with minimum contamination and, provided that sintering of the metal particles is minimized, high area/weight ratios are obtainable. Various evaporation techniques have been exploited, viz.: (i) evaporation of the film on to NaCl discs, these being mounted in series in the path of the infra-red beam (the multiple transmission technique) [7]; (ii) deposition of the metal vapour into

TABLE 16.3

Method	Reference	Frequency/cm^{-1}					
In vacuo	Hayward [5]		2060		1950		
In 10^2 Pa Ar			2075			1915	
In 1 Pa CO			2058	2034		1915	1820
In 10^2 Pa CO		2084	2065	2034		1915	
Coevaporated with CaF_2			2058	2034		1915	
Evaporated into Nujol	Blyholder [3]	2080			1940		
SiO_2-supported Ni	Eischens *et al.* [1]		2073			1924	1850
$Ni(CO)_4$ (g)				2040			

a layer of unreactive Nujol oil spread over the window of the infra-red cell [3]; (iii) simultaneous evaporation of metal and an infra-red transparent salt (CaF_2) [5]; (iv) evaporation of metal, or "explosion" of metal wires [2], in the presence of a small pressure of an inert gas (argon), thereby to obtain a smaller particle size than that obtained in vacuum deposition; and (v) evaporation of the metal in presence of ~ 1 to 10^2 Pa of the gaseous adsorbate. Results for a Ni adsorbent employing such methods are summarized in Table 16.3.

16.3. RESOLUTION AND INTENSITY OF SPECTRAL BANDS

Because of the large scattering losses, resolution is not high. Losses can be minimized by reducing the particle size of the coated support to less than the wavelength of the incident light. Similarly, by substituting an atom of the adsorbed molecule with its heavier isotope (particularly a H atom by a D atom), the frequency of the absorption band is lowered and scattering is, thereby, decreased. A narrow slit-width is necessary because, unless it is less than the absorption band-width, substantial broadening of the band occurs. Although resolution is improved by decreasing the slit width, the reduction in energy received by the spectrometer is a severe limitation. Furthermore, when the slit width/band width ratio approaches a value of 0·4, the Lambert–Beer law is not obeyed.

The intensity of an absorption band of a gaseous molecule provides a valuable additional parameter in identification, since it is related to the magnitude of the change of dipole moment that occurs during vibration. However, such measurements for chemisorbed molecules provide little or no information since the absorption intensity varies in a complex manner with change of surface coverage.

Infra-red spectrometry can, nevertheless, provide estimates of surface concentrations. The maximum extinction coefficient, at the peak height of the band, is given by $\varepsilon_m = (1/c)\ln(I_0/I)$. The area enclosed by the band envelope is the integrated intensity (or integrated absorption coefficient), I_i, where

$$I_i = (1/c)\int_{\nu_1}^{\nu_2} \ln(I_0/I)\,d\nu = \int_{\nu_1}^{\nu_2} \varepsilon\,d\nu. \tag{16.7}$$

In this expression, ν_1, ν_2 are the frequencies (on either side of ν_{max}) at which ε tends to zero. The unit of concentration c is molecules cm^{-3} of the sample cross-section, and ε is expressed in $molecule^{-1}\ cm^2$. The quantity $\ln(I_0/I_i)$ is proportional to the number of adsorbate molecules in the light path, and the increase of I_i with increase of the fractional coverage of adsorbate molecules provides a measure of the relative surface concentrations. An alternative, approximate indication is the magnitude of the maximum extinction coefficient. However, this value varies with different experimental conditions, whereas the integrated absorption coefficient remains reasonably uniform.

16.4. INTERPRETATION OF SPECTRA

The most extensive results have been obtained with metals dispersed on oxide supports, although data relating to evaporated films have been reported (Table 16.3). Interpretation of the results is by direct comparison of the spectrum with that of the adsorbate in the gas, liquid or solid state. Relatively few bands are visible, because of the restricted range of transmission of the support, and the bands that are observed are broad and have no fine rotational structure. Furthermore, additional information from the amplitude of the oscillating dipoles of the adspecies cannot be obtained from intensity measurements because of the non-uniformity of the metal surface.

In general, two main bands with peak frequencies around 1925 cm^{-1} and 2050 cm^{-1} are normally observed for CO chemisorbed on transition metals. The positions and relative intensities vary with coverage and other bands often appear as shoulders on the main peaks. Carbonyl compounds involving monodentate ligand formation with transition metals usually have a C—O stretching frequency between 2000 and 2100 cm^{-1}; the chemisorbed band observed above 2000 cm^{-1} is, therefore, assigned to a linear species M—C≡O. Bidentate ligand formation in metal carbonyls give rise to bands below 2000 cm^{-1}, and the lower-frequency chemisorbed bands around 1925 cm^{-1} are generally attributed to the presence of a chemisorbed two-site bridged structure,

$$\begin{matrix} M \diagdown \\ \quad C{=}O. \\ M \diagup \end{matrix}$$

The close analogy between the peak frequencies of the carbonyls and the chemisorbed species indicates similar bonding, the dominant feature of which is back-donation of d electrons of the metal to the CO ligand in the carbonyls. Occupied d orbitals are always available in transition metal adsorbents, but the number of d electrons that partake in the formation of the surface complex varies. All of the d orbitals take part when the complex has a tetrahedral structure, whereas only the d_{xy}, d_{yz} and d_{xz} can be used in forming an octohedral complex. The two "Nujol" frequencies at 2080 and

1940 cm^{-1} may, therefore, arise, according to Blyholder [3, 10], from the different stereochemistry of a chemisorbed species on the plane surface and one that is bonded to edges, dislocations, etc., of the metal crystallites.

The linear structure should give rise to a Ni—C—O bending mode and a Ni—C stretching frequency in addition to the observed C—O stretching mode. In nickel carbonyl, the bending mode is observed at 466 cm^{-1} and the Ni—C stretch at 422 cm^{-1}. Similarly, the bridged structure should give rise to four active modes in addition to the symmetric and asymmetric Ni—C stretching frequencies. By use of the Nujol technique, spectral data down to 250 cm^{-1} have been obtained by Blyholder [11]; but he observed only one additional band at 435 cm^{-1}, which he assigned to the asymmetric Ni—C stretching frequency of the linear structure on the basis of an estimated value of 700 to 1000 cm^{-1} for this mode in the bridged structure. In confirmation, he notes that some inorganic complexes having monodentate CO ligands give rise to C—O stretching frequencies below 2000 cm^{-1}. But the force constant calculated from the Ni—C frequency of 435 cm^{-1} is much lower ($\sim 1 \times 10^5 \text{ dyn cm}^{-1}$) than that of M—C bonds in the few other chemisorbed systems that have been experimentally investigated.

Other methods of preparing the adsorbent, e.g. dispersion of the metal in KCl and KBr pellets or condensation as a highly porous metal film, also provide the necessary transparency in the range where there is no "window" with oxide-supported metals. Thus, a value of 476 to 477 cm^{-1} for the M—C stretching frequencies has been reported for the CO/platinum system; this would give a calculated force constant of the Pt—C stretch of *ca* $4{\cdot}5 \times 10^5 \text{ dyn cm}^{-1}$, i.e. considerably larger than that given above for the Ni—C vibration. It would appear, therefore, that the more acceptable conclusion is that there are two forms of chemisorbed CO, one with a linear structure and the other having a bridged two-site-adsorption configuration.

There are some other unexpected results. For example, on palladium and iron only the weakly bonded chemisorbed CO is infra-red active; the strongly-bound form, comprising some 90% of the total amount adsorbed, does not contribute to the spectrum. A possible explanation is that the adspecies lies flat on the surface. This mode should be favoured on strongly adsorbing metals, such as W, Mo, Ta, and, in fact, the absorption bands of chemisorbed CO on these metals are of extremely low intensity.

10.4.1. The CO/metal bond

In the CO molecule, three fully occupied σ orbitals give rise to a σ C—O bond with a pair of unshared electrons associated with both atoms. The four remaining electrons occupy a doubly-degenerate π orbital and σ and π antibonding orbitals, i.e. there is a triple bond comprising one σ and two π

bonds between the C and O atoms, with the one pair of electrons occupying an orbital centred on the C atom and directed away from the O atom along the internuclear axis. The lone pair takes part in the co-ordination of CO to the surface metal atom to give a weak σ M—CO bond. Back donation of electrons from the metal *d* orbitals to the CO antibonding orbitals imparts a π-character to the metal–carbon bond, which is thereby strengthened, and the C—O bond is weakened. The stretching frequency of this latter bond is, therefore, decreased from 2143 cm^{-1} in the free molecule to 2050 cm^{-1} in the adspecies with the linear structure. The force constant, likewise, decreases from $18{\cdot}6 \times 10^5$ to around $15{\cdot}0 \times 10^5$ dyn cm^{-1}, i.e. a value greater than that for the double bond in gaseous ketones (12×10^5 dyn cm^{-1}). Consequently, the triple bond still predominates in the chemisorbed CO and thus is consistent with a linear structure.

16.4.2. Isotopic substitution in adspecies

The electron distribution in deuterium is the same as that in hydrogen and, therefore, the force constant between the constituent atoms is unchanged. However, the vibrational frequencies are different because of the difference of their atomic masses. Similarly, the spectral bands arising from the chemisorption M—H vibrations are shifted to lower frequencies when the hydrogen adatom is replaced by a deuterium atom. Thus, for hydrogen chemisorbed on alumina-supported platinum, the two peak frequencies at 2105 and 2955 cm^{-1} are displaced to 1512 and 1480 cm^{-1}, respectively, by deuterium substitution; but there is little change in the band intensities. The displacement factor of 1·39 accords with that expected theoretically (1·41) from the difference in mass of the H and D adatoms.

Isotopic substitution provides additional confirmation of band assignment, and also forms the basis of an approximate method of estimating M—C force constants. Thus, on platinum the ^{12}CO frequency is 2043 cm^{-1} and that of ^{13}CO is 1991 cm^{-1}, giving an isotopic shift of 1·026. With the assumption that the system may be treated as a linear XYZ molecule with an infinitely large atomic mass for the metal atom, and considering only the stretching vibrations of the M—C and C—O bonds, the force constants are calculated to be $4{\cdot}5 \times 10^5$ dyn cm^{-1} for the M—C vibration and 16×10^5 dyn cm^{-1} for the C—O mode [1, 12].

Isotopic substitution has also been used to differentiate the effects of surface non-uniformity and lateral interactions between admolecules [1], both of which are responsible for a shift in frequency and a change of band intensity as the surface concentration of the adspecies is increased.

When each of two isotopes is separately chemisorbed, then, in the absence of lateral interactions in the adsorbed layer, the relative intensity of the two bands should be independent of the coverage, even though the surface may be non-uniform. However, coupling interaction between the vibrations of two adjacent adsorbed molecules occurs and is responsible for changes in both the frequencies and intensities of the two bands as the total surface coverage and isotopic composition are increased.

When only one isotope is chemisorbed, the in-phase vibrations (A below) are infra-red active, and the out-of-phase vibrations (B below) are inactive due to the cancellation of the changes of dipole moment of the adjacent adspecies. Thus, for the CO/Pt system,

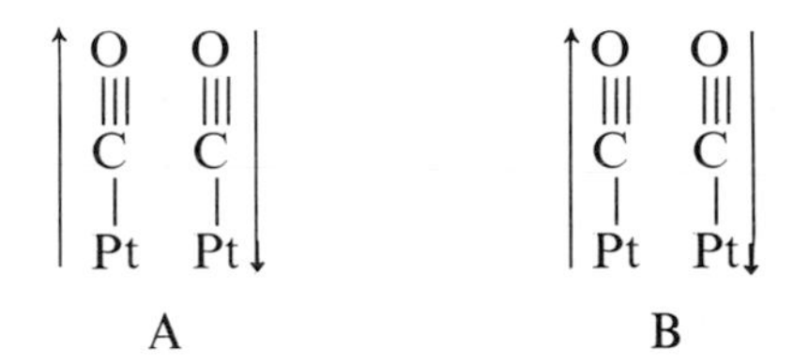

But when an isotopic mixture is chemisorbed and the ^{12}CO adspecies is adjacent to a ^{13}CO molecule, the out-of-phase vibrations then involve net changes in the dipole moment and so become infra-red active. The effect is to reduce the band intensity of one species (^{13}CO) and increase that of the other (^{12}CO). The magnitude of the change varies both with coverage and isotopic ratio since the coupling interactions increase at higher surface concentrations and also for higher isotopic ratios. A theoretical treatment of dipole–dipole coupling interaction [13] predicts that the higher the $^{12}CO/^{13}CO$ isotropic ratio, the greater is the intensity of the two bands; and at higher coverages, both the ratio and the frequency of the ^{12}CO vibration are increased. These predictions are in qualitative accord with the experimental results [13]; e.g. at full coverage, the intensity ratio was found to be 4·2 for an isotopic ratio of 1·7, and 1·1 for a ratio of 0·56; and the intensity ratios decreased at lower coverages in accord with this theory.

16.4.3. Asymmetry induced by chemisorption

Because of the symmetry along the N—N axis of the gaseous nitrogen molecule, its stretching vibration is infra-red inactive. When adsorbed on nickel, with its axis perpendicular to the surface, asymmetry is created by the surface field of the metal, and this vibration becomes infra-red active [14, 15]. Bands at 2195 cm^{-1} for $^{14}N^{14}N$, at 2160 cm^{-1} for $^{14}N^{15}N$ and at 2123 cm^{-1} for $^{15}N^{15}N$ are observed in good accord with theoretical isotopic shifts of 1·018 and 1·035, respectively; clear evidence of non-dissociative chemisorption is provided. The spectral characteristics are similar to those of chemisorbed CO, e.g. the extinction coefficient and force constant are almost the same so that the adsorbed nitrogen must be in a highly polarized state.

16.4.4. Finger-printing of adspecies

The infra-red spectra of chemisorbed species, particularly of hydrocarbon molecules, provide useful information about the nature and identity of adradicals formed by dissociative chemisorption. Identification of groups is made by comparison of the frequencies of spectral peaks with those of liquid hydrocarbons. Thus, the group frequency of MCH_3 ranges from 2870 to 2960 cm^{-1}; that of $M{-}\overset{|}{C}H_2$ from 2850 to 2925 cm^{-1} and $\begin{matrix} M \\ M \end{matrix}\!\!>\!\overset{|}{C}{-}H$ has a peak frequency at about 2890 cm^{-1} (see Table 16.1). Deformation and other modes can also provide further assistance in making assignments.

For the low-temperature associative chemisorption of ethylene on SiO_2-supported nickel, bands corresponding to the C—H stretching and the ∠HCH scissor vibrations are observed, but, at higher temperatures, the dissociative chemisorption is evidenced by loss of intensity of the C—H band due to the presence of a smaller number of C—H bonds (i.e. the atomic ratio H/C is decreased by simultaneous dehydrogenation, and a subsequent increase of intensity is effected by exposure to hydrogen). With acetylene, self-hydrogenation is confirmed by the appearance of bands characteristic of methyl and methylene groups and indicate the probable formation of surface ethyl groups [12, 16–18]. Similarly, some of the bands in the spectra of chemisorbed formaldehyde and formic acid are those of chemisorbed CO formed as a result of the dissociation of C—H bonds [7].

16.5. REFLECTANCE SPECTROMETRY

Recent advances in reflectance-absorption spectrometric techniques have greatly increased the potentialities of infra-red spectrometry as a means of investigating the properties of chemisorbed molecules. The main advantage is that this method can be applied to single crystal planes; in addition, the orientation of adspecies with respect to the plane can also be determined. Theoretical studies by Greenler [19, 20] and others [21–27] greatly stimulated interest in reflection studies. The important result that emerged is that the intensity of absorption arising from an adsorbed layer increases rapidly as the angle of the incident beam increases from normal incidence to a maximum of 88°, and then decreases on further approach to grazing incidence. At the maximum angle, the absorbance is nearly 5000 greater than at normal incidence, and 25 times that for the corresponding transmission at normal incidence. Nevertheless, at that time (1966) the absorption intensity for a single reflection was too weak for measurement, and multiple reflection techniques were developed to augment the intensity.

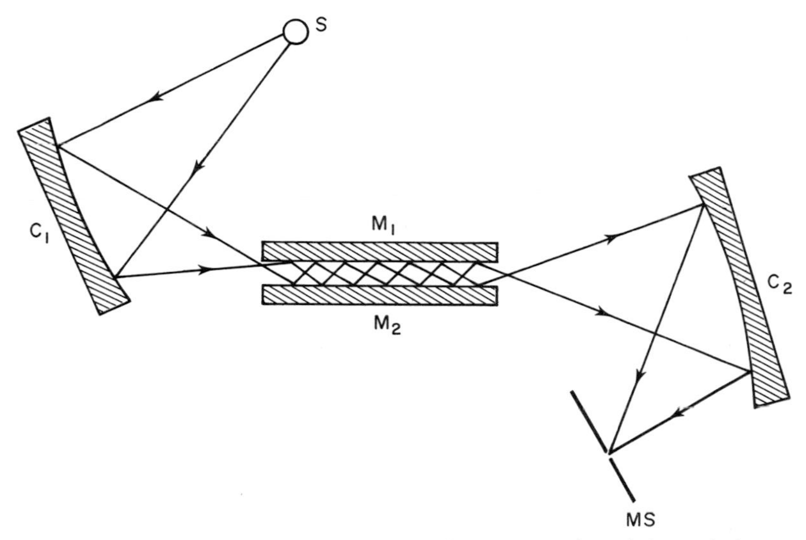

Fig. 16.4. Multiple reflection technique—schematic representation of the optical system. S, infra-red source; C_1, concave mirror for focusing source image at the entrance of two closely spaced metal adsorbent surfaces M_1, M_2; the beam undergoes multiple reflections. C_2, concave mirror for focusing the emergent rays at the monochromator entrance slit MS.

The general procedure was to focus the incident beam at the entrance of two closely-spaced metal surfaces carrying the adsorbed film (Fig. 16.4). At the exit, the reflected rays were recombined for examination by a monochromator. The number of reflections could be increased by decreasing the spacing between the metal surfaces. Optimum conditions required incident angles between 80° and 88° and 5 to 15 reflections.

However, the gain in sensitivity by multiple reflections is not large, and more recent improvements to effect an increase of the signal-to-noise ratio, the greater sensitivity of the detector, the use of signal averaging computers, and wavelength modulation to obtain the derivative spectrum, have favoured the adoption of the simpler and more convenient single-reflection technique [20]. A single-beam spectrometer may be employed but a double-beam instrument has the advantage of reducing fluctuations in the source intensity.

Chopped radiation from a monochromator that scans wavelength linearly with time at 40 cm^{-1} min^{-1} gives a resolution of 4 cm^{-1} with a slit width of 0·5 mm. The infra-red beam is arranged to have an incidence angle of between 80° and 87° with the crystal surface, and the reflected light is collected by a mirror and focused into a cooled indium photo-conductive detector. Its amplified output is recorded on punched tape via a digital voltmeter finally to provide a difference spectrum. Alternatively, spectra obtained by repeated scanning can be accumulated directly into a signal-average computer.

The recording of a wavelength modulation derivative spectra (as used in dispersive Auger spectrometry) has the advantage that the noise arising from frequencies less than the modulation frequency is reduced and the intensity of weak spectral bands are enhanced. One method of modulation is to replace the entrance slit of the monochromator by a slotted foil driven at a resonant frequency of around 75 Hz with a modulation amplitude of about 4 cm^{-1}. Alternatively, an oscillating Newton mirror behind the entrance slit has been employed to provide higher sensitivity.

16.5.1. Polycrystalline films

A number of results on the reflection-absorption spectra of CO on polycrystalline films of copper deposited on various substrates (glass, alumina, silica) have been reported [e.g. Fig. 16.5(a)] and these show good agreement

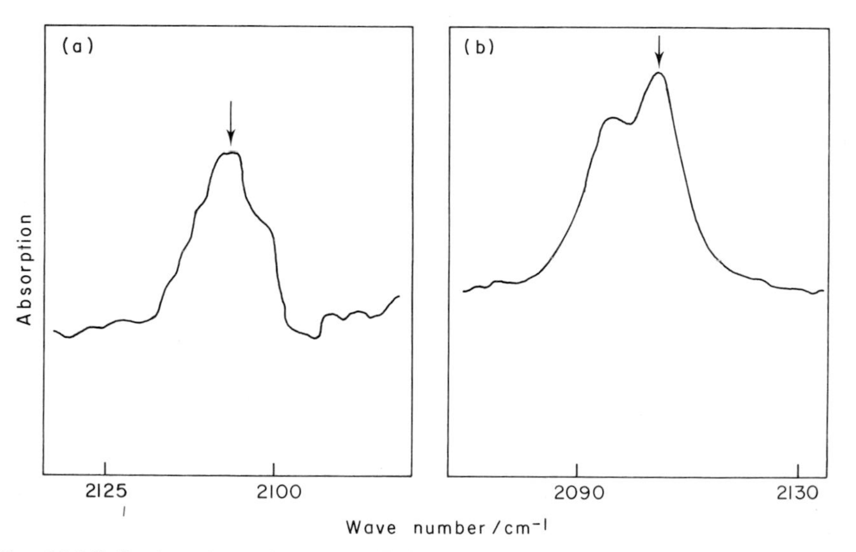

FIG. 16.5. Reflection–absorption spectra [26] of carbon monoxide adsorbed on (a) a copper film deposited on oxidized aluminium at room temperature; and (b) the Cu(311) plane at 77 K. Arrows denote a wave number of 2105 cm^{-1}; note different wave-number scales.

with transmission spectra [26]. Thus, from transmission spectroscopy at 298 K through copper dispersed on various supports, the frequency of the main CO band has been reported by various workers as 2103 to 2120 cm^{-1} (SiO_2), 2108 to 2111 cm^{-1} (Al_2O_3) and 2081 cm^{-1} (MgO). Corresponding results from reflection-absorption spectra for films deposited on various substrates are 2101 to 2105 cm^{-1} (glass), 2106 cm^{-1} (Al_2O_3) and 2182 cm^{-1} (MgO). The support effect of magnesia is clearly evident; it could indicate that the ionic nature of the support induces a high concentration of low-index planes and that, for an amorphous SiO_2 support, high-index planes are more abundant.

16.5.2. Single crystal planes

Most investigations have so far taken advantage of the high extinction coefficient of chemisorbed CO in order to obtain sufficient sensitivity. The CO/Cu system has again been the most extensively studied and results are now available for six different crystal planes [e.g. Fig. 16.5(b)]; the corresponding changes of surface potential were also simultaneously recorded. A possible correlation between the change of this parameter at high coverage and the peak frequency is apparent as is shown in Table 16.4. With one exception, the (211) plane, the change in surface potential increases regularly with increase of frequency. In addition, by assuming that the surface potential varies linearly with coverage, the heat of adsorption of hydrogen on Cu(111) has been evaluated as 40 kJ mol^{-1}.

TABLE 16.4

Crystal plane	ν/cm^{-1}	Change in surface potential/eV
(111)	2076	0·19
(100)	2085	0·12
(110)	2093	0·18
(755)	2098	0·18
(211)	2100	0·15
(311)	2102	0·19

The orientation of the adspecies can also be investigated by the reflection technique when polarized light is employed. The oscillating dipole moment of the adspecies absorbs energy within a narrow range of frequencies of the incident radiation and so gives rise to a characteristic absorption band when it possesses a dipole component parallel with the electric vector of the radiation. Consequently, the amplitude of the electric vector at the metal surface for light polarized in the plane of incidence is much greater than that for light polarized perpendicular to the plane of incidence. Indeed, the interaction intensity with the oscillating dipole of the adspecies can be as much as 10^3 to 10^5 times higher for light polarized in the incidence plane, and the relative absorption intensity is correspondingly greater. Hence, by recording the intensity of a band as a function of the angle of the plane of polarization of the incident radiation, the orientation of the dipole of the adspecies to the surface can be determined. In principle, confirmation could be obtained by transmission of the light through a very thin film deposited on a substrate that favours epitaxal growth of a single crystal plane of the metal. In this case, only dipole moment changes parallel to the surface can give rise to an absorption, i.e. the reverse situation is obtained.

Measurements on single crystal planes can now provide data on: (i) the intensity and frequency position of bands as a function of coverage (from surface potential values) with good precision and reproducibility; and (ii) information on the dipole properties of the adspecies. The contributions expected in the future will undoubtedly be of increasing importance in extending our knowledge of the structural properties of chemisorbed molecules.

REFERENCES

1. R. P. Eischens, S. A. Francis and W. A. Pliskin, *J. Phys. Chem.*, 1956, **60**, 194.
2. J. T. Yates and C. W. Garland, *J. Phys. Chem.*, 1961, **65**, 617.
3. G. Blyholder, *in* "Proceedings of the 3rd International Congress on Catalysis" (W. M. H. Sachtler, G. C. A. Schuit and P. Zweitering, eds), North Holland, Amsterdam, 1965, p. 657.
4. F. S. Baker, A. M. Bradshaw, J. Pritchard and K. W. Sykes, *Surface Sci.*, 1968, **12**, 426.
5. D. O. Hayward, *in* "Chemisorption and Reaction on Metallic Films" (J. R. Anderson, ed.), Academic Press, London and New York, 1971, Chap. 4, p. 225.
6. L. Lynds, *Spectrochim. Acta*, 1964, **20**, 1369.
7. J. F. Harrod, E. F. Rissmann and R. W. Roberts, *J. Phys. Chem.*, 1967, **71**, 343.
8. G. Blyholder and L. D. Neff, *J. Phys. Chem.*, 1962, **66**, 1464.
9. A. W. Smith and J. M. Quets, *J. Catalysis*, 1965, **4**, 163 and 172.
10. G. Blyholder, *J. Phys. Chem.*, 1964, **68**, 2772.
11. G. Blyholder, *J. Chem. Phys.*, 1962, **36**, 2036.
12. R. P. Eischens and W. A. Pliskin, *Adv. Catalysis*, 1958, **10**, 1.
13. R. P. Hammaker, S. A. Francis and R. P. Eischens, *Spectrochim. Acta*, 1965, **21**, 1295.
14. R. P. Eischens and J. Jacknow, *in* "Proceedings of the 3rd International Congress on Catalysis" (W. M. H. Sachtler, G. C. A. Schmidt and P. Zweitering, eds), North Holland, Amsterdam, 1965, p. 627.

15. R. van Hardewald and A. van Montfoort, *Surface Sci.*, 1966, **4**, 396.
16. B. A. Morrow and N. Sheppard, *J. Phys. Chem.*, 1966, **70**, 2406.
17. J. B. Peri, *Disc. Faraday Soc.*, 1966, **41**, 121.
18. J. Erkelens and Th. J. Liefkens, *J. Catalysis*, 1967, **8**, 36.
19. R. G. Greenler, *J. Chem. Phys.*, 1962, **37**, 2094; 1969, **50**, 1963.
20. R. G. Greenler, R. R. Rahn and P. J. Schwartz, *J. Catalysis*, 1971, **23**, 42.
21. A. M. Bradshaw and J. Pritchard, *Surface Sci.*, 1969, **17**, 372.
22. A. M. Bradshaw and J. Pritchard, *Proc. Roy. Soc. A*, 1970, **316**, 169.
23. M. A. Chester and J. Pritchard, *Surface Sci.*, 1971, **28**, 460.
24. C. S. Alexander and J. Pritchard, *J. Chem. Soc., Faraday Trans.*, 1974, **68**, 202.
25. J. Pritchard and M. L. Sims, *Trans. Faraday Soc.*, 1970, **66**, 427.
26. J. Pritchard, T. Catterick and R. K. Gupta, *Surface Sci.*, 1975, **53**, 1.
27. H. Papp and J. Pritchard, *Surface Sci.*, 1975, **53**, 371.

17
Field-emission Microscopy

According to the free-electron theory of metals, the lowest electron energy levels are fully occupied by pairs of electrons of opposite spin, the highest filled level at 0 K being the Fermi level of energy E_F. In classical theory, the minimum energy required to eject an electron from the metal is the difference in energy between the electron at the Fermi level and at rest at an infinite distance from the metal surface in vacuum, and, by definition, equals the work function Φ. In quantum theory, there is a finite probability of an electron in the metal tunnelling through this potential barrier, but, because the barrier width is semi-infinite, this process does not occur (see Fig. 17.1). However, the electrostatic potential of an electron at the surface approximates to the work function (4 to 5 eV) and decays exponentially with distance into vacuum to virtually zero at 4 or 5 Å; it is, therefore, subjected to a strong electric field of about 1 V $Å^{-1}$ or 10^8 V cm^{-1} that prevents its escape from the metal. However, by application of an external attractive field of this magnitude, cold emission of electrons can be effected.

17.1. THEORY OF METHOD [1–3]

The basic feature of field-emission microscopy is the application of a high external electric field to a metal tip, thereby modifying the shape of the potential barrier and reducing its width (see Fig. 17.1). Electron tunnelling through the barrier and, consequently, electron emission from the metal surface can then take place at room temperature. The high field ($F \approx V/r \sim 10^8$ V cm^{-1}) is obtained by using a tip of very small radius of curvature

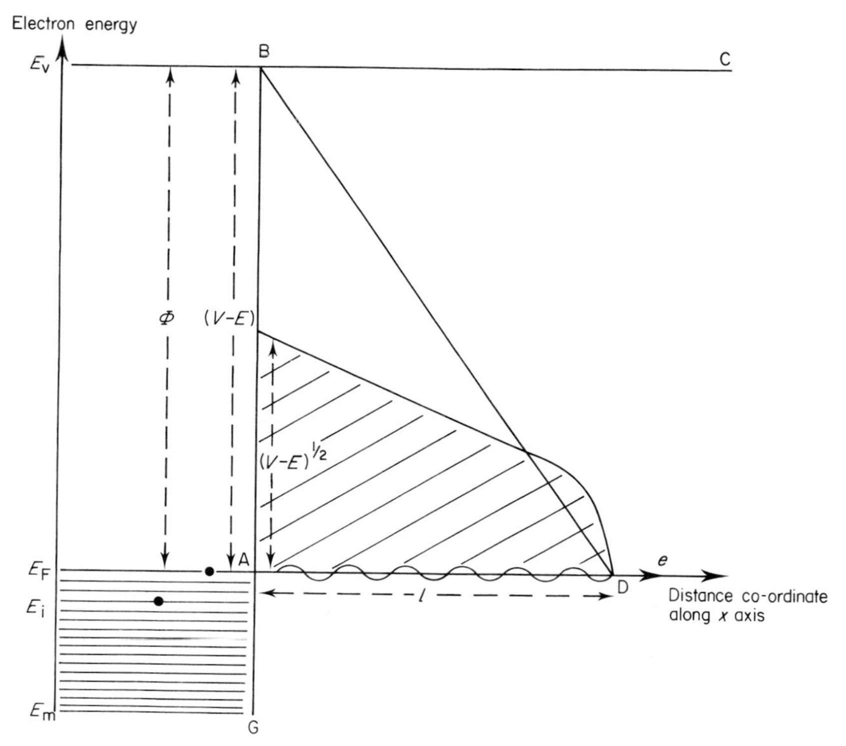

FIG. 17.1. Potential energy diagram for electrons in a metal with and without an applied field. BG, metal surface; AB, potential barrier of height equal to the work function of the metal $\Phi = E_v - E_F = V - E$, where E_v and E_F are the vacuum level and Fermi level of the metal. V and E are, respectively, the potential and kinetic energies of the tunnelling electron. In the absence of an applied field, the barrier is represented as ABC, where BC (the barrier width) is semi-infinite. On application of a field of strength $F = \Phi/l$, the barrier has the triangular profile ABD of finite width l at the Fermi level through which the electron may tunnel. The shaded area, roughtly triangular in shape, is approximately given by $\int_0^l (V - E)^{\frac{1}{2}}\,dx \approx \frac{1}{2}\Phi^{\frac{3}{2}}/F$. The tunnelling probability of an electron in the energy level E_i below E_F is proportional to $\exp[-(\Phi + E_F - E_i)^{\frac{3}{2}}/F]$.

($r = 10^{-4}$ to 10^{-5} cm) and a potential of 3 to 15 kV between it and a suitably placed anode.

The tunnelling probability $P(E)$ through the barrier depends on the energy state E of the electron in the metal, so that the number of electrons with energy between E and $E + dE$ that are emitted from 1 cm^2 surface area per second is given by $N(E)\,P(E)\,dE$. For a one-dimensional, free-electron model, $P(E)$ can be approximated [3] to

$$P(E) = a \exp\left[-b\int_0^l (V - E)^{\frac{1}{2}}\right]dx, \qquad (17.1)$$

in which the kinetic and potential energies of the tunnelling electron are denoted by E and V, respectively, l is the width of the barrier, and a and b are constants. In the simplest model, the barrier shape is triangular with co-ordinates $(V - E) = \Phi$ at the metal surface ($x = 0$), and $(V - E) = 0$ at a distance outwards from the surface at $x = l$. The area evaluated from the above integral (see Fig. 17.1) is also roughly equal to that of a triangle of height $(V - E)^{\frac{1}{2}} = \Phi^{\frac{1}{2}}$ and breadth ($l = \Phi/F$), where F is the electric field. The integral can, therefore, be approximated to $\Phi^{\frac{3}{2}}/2F$.

The flux of the emitted electrons is the product of $P(E)$ and the number $N(E)$ of the electrons in the energy state E that arrive at unit area of the metal surface per second; hence, the total emission current j is given by

$$j \propto \int_{E_M}^{E_F} N(E) \exp(-b\Phi^{\frac{3}{2}}/2F)\, \mathrm{d}E, \qquad (17.2)$$

where E_M is the lowest energy level in the metal. Since the probability of emission of an electron from an energy state E_1 below the Fermi level is proportional to $\exp[-b(\Phi + E_F - E_1)^{\frac{3}{2}}/2F]$, the contribution of electrons from low-lying levels to the total emission current j decreases exponentially with increase of $(E_F - E_1)$ and becomes negligible for a difference of ~ 2 eV. The greatest part of the emission, therefore, comprises electrons originating from energy states close to the Fermi level.

A more rigorous derivation [4–6] of Eq. (17.2), including the effect of the image potential and adatom polarizability at the surface, leads to the Fowler–Nordheim equation, viz.

$$j = 6{\cdot}2 \times 10^6 (\Phi/E_F)^{\frac{1}{2}} (\Phi + E_F)^{-1} F^2 \exp(-6{\cdot}8 \times 10^{17} \Phi^{\frac{3}{2}}/F), \qquad (17.3)$$

in which $F = KV/r$ V cm^{-1} and K is a constant ($\sim 0{\cdot}2$) depending on the exact geometry of the tip.

17.1.1 Effect of surface geometry [7]

The image potential arises from a redistribution of the electron density at the surface. The electron wave function is oscillatory within the metal but decays exponentially outside the surface to give a net negative charge [see Fig. 17.2(b)]. An equal but opposite charge is induced within the metal; the electrostatic interaction between the electron and its positive image charge is called the image potential. The more closely-packed low-index planes have larger image potentials and higher work functions. The effective work function, which includes the image potential, therefore varies with the geometry of the different crystal planes present at the surface of the tip, e.g.

Φ W(100) is nearly 0·3 eV higher than Φ W(111). Since the emission current depends exponentially on $\Phi^{\frac{3}{2}}$, even such small differences as a few hundredths of an electron volt effect marked changes in the emission current.

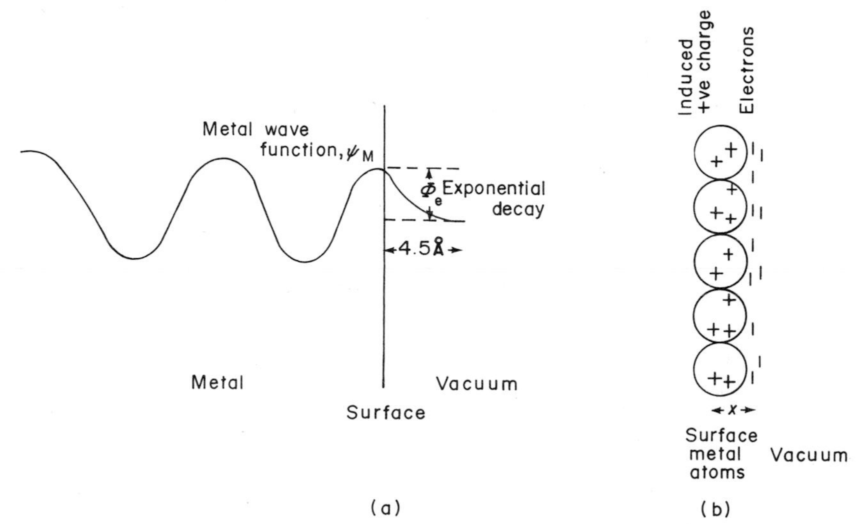

FIG. 17.2. Image potential. (a) At the metal surface, the oscillatory wave function of an electron in the metal is transformed go an exponential decay function, such that the potential energy decreases from Φ_e (the effective work function) to zero within 4·5 Å of the surface.
(b) The electron charge outside the surface induces an equal but opposite charge within the metal; the net charge is greater the higher the surface atomic density. The image potential is $e^2/4x$, where e is electronic charge and x the distance across the double layer of charge.

17.2. THE MICROSCOPE [1–3]

The essential feature is a glass envelope, part of which has been suitably coated internally to form a fluorescent screen, surrounding with axial symmetry a finely-pointed metal cathode tip (see Fig. 17.3). The anode at ground potential is often an aquadag-coating incorporated in the screen. On application of the electric field, electrons are emitted from the tip with small kinetic energies, but they are rapidly accelerated aong the lines of force of the applied field, perpendicular to the surface of the tip, towards the screen. The intensity of fluorescence is proportional to the electron flux that impacts with the screen.

The hemispherical tip comprises a single crystal displaying a large number of well-defined crystallographic planes having different work functions. The image on the fluorescent screen is brighter for the lower-density planes, from

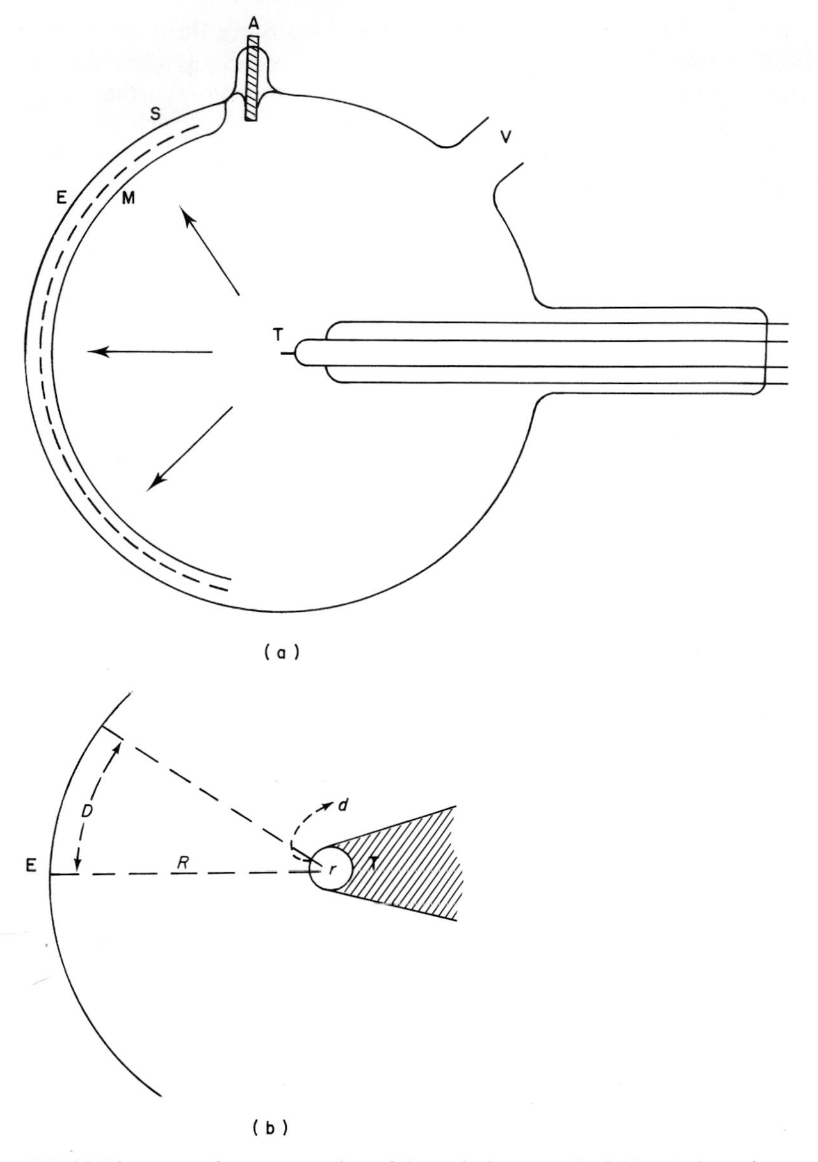

FIG. 17.3. (a) Diagrammatic representation of the main features of a field-emission microscope. E, glass envelope; S, fluorescent screen with conductive layer M; T, the cathode emitter tip sealed to a loop which can be electrically heated through the supporting leads; its temperature is derived from measurement of the resistance; A, anode connected to conductive backing; V, outlet to pumps.

(b) Magnification of microscope. Radius of curvature of hemispherical tip $= r$; radius of spherical glass envelope $= R$. Magnification is the ratio of the length of the arc D on the screen to that of d on the tip, where $D/d = R/r \approx 10^2\ \text{cm}/10^{-4}\ \text{cm}$ in a typical instrument.

which the electron emission is larger; consequently, the assembly of different planes gives rise to a highly magnified pattern of varying intensities on the screen. The magnification is given by the ratio of the tip-to-screen distance to the radius of curvature of the tip, and is of order 10^5 to 10^6. The crystallographic indices of the various planes can be unequivocally assigned from the symmetry characteristics of the pattern and the angular separation of the planes (see Fig. 17.4).

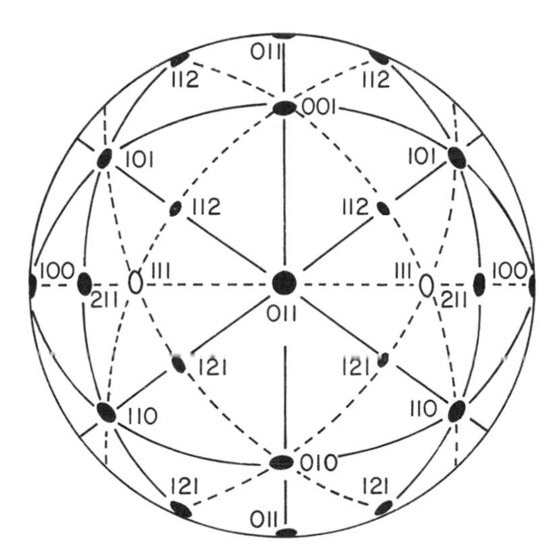

FIG. 17.4. Orthographic projection of a body-centred cubic crystal.

A probe-hole technique can be employed for separate examination of electrons originating from a single plane of the tip. The electrons pass through the hole, which selects the specific plane, and are focused by an electrostatic lens and collected in a hemispherical Faraday cage. Energy analysis of their distribution can be simultaneously effected using the normal retarding field method.

The resolution of the microscope [8, 9] is about 20 Å, the limit being imposed by the statistical distribution of the tangential velocities of the electrons of the free-electron gas. The resolution is largely independent of the applied voltage. Higher fields extract an increasing number of electrons from energy states below the Fermi level and widen the distribution, but the increase of the average energy parallel to the surface is balanced, almost exactly, by the decrease of the time of flight of the electron from the tip to the screen.

17.3. APPLICATIONS TO CHEMISORPTION

17.3.1. Change of work function

The electron density distribution at the surface of a metal is modified by the chemisorption of a gas and the chemisorbed molecules are polarized due to charge transfer between the metal and the adspecies. The complete monolayer, comprising n_m surface dipoles per cm^2 surface, each having an effective dipole moment μ debye, can be considered as a parallel-plate condenser. The change of work function (in eV) of the metal is then related to the dipole moment of the adspecies through the classical electrostatic equation

$$-\Delta\Phi = 1{\cdot}2 \times 10^{18}\pi\, n_m \mu_m, \tag{17.4}$$

the convention being that μ_m is positive for excess positive charge outwards from the surface and the change in surface potential is $\Delta\chi = -\Delta\Phi$. For an adspecies of small dipole moment, $\mu_m = 0{\cdot}1$ debye, with $n_m \approx 10^{15}\ cm^{-2}$, $\Delta\Phi = 0{\cdot}4$ eV. Experimental values of the change of work function (and, hence, of μ_m) can, therefore, be measured with considerable accuracy.

In practice, however, there are various complications.

(i) The adspecies itself is polarized by the applied field and the induced dipole moment μ_i contributes to the measured change of work function. As a crude approximation, $\mu_i = \alpha F^2/2$, where F is the field strength ($\sim 10^8\ cm^{-1}$) and α the polarizibility of the adspecies in cm^3. For the dissociative chemisorption of nitrogen, hydrogen and oxygen on many transition metals, the polarizability of the adatom is probably similar to that of the free atom (1·1 to 0·7 $Å^3$). The induced work functions of these adatoms may, therefore, approach 0·2 eV, and so represent a substantial contribution to the change of work function.

(ii) Values of $\Delta\Phi$ are evaluated by application of the Fowler–Nordheim equation to the results of measurements of the current density j as a function of the applied voltage. In the application of this equation to adsorbed layers a continuous double-layer of charge is assumed, but at low coverages the adsorbed phase forms a set of discrete localized potentials; since the current density is extremely sensitive to local variation in charge distribution, the $\Delta\Phi$ values are subject to some uncertainty.

(iii) The double-layer potential is at a constant distance from the surface for a full monolayer; but at low coverages this is not so and somewhat lower $\Delta\Phi$ values are obtained.

(iv) Up to moderate coverages, $\Delta\Phi$ is usually a linear function of the surface concentration; at higher coverages, departures from linearity are caused by mutual depolarization of the aligned dipoles, the magnitude of which depends not only on the polarizability of the molecule but also on the geometric arrangement of the adspecies on the surface.

17.3.1.1. *Experimental procedure*

Measurements of the effective work function of the clean tip and that obtained after chemisorption are required. Two methods have been employed. A constant applied voltage is maintained and the emission current is recorded before and after adsorption. The method is rarely preferred, since the electron emission is highly sensitive to departures from the Fowler–Nordheim equation, which was derived for an idealized one-dimensional model.

In the second procedure, the current j is recorded as a function of the applied voltage V. At low fields, the plot of $\ln(j/V^2)$ against $1/V$ is linear [see Eq. (17.3)] to a very good approximation with a slope proportional to $\Phi^{\frac{3}{2}}$. The value Φ_m of the clean tip is first obtained and then gaseous adsorbate is introduced via an appropriate inlet source, e.g. by electrical heating of a platinum sleeve on which adsorbate had been condensed, or by thermal decomposition of a suitable compound such as cupric oxide (for oxygen), zirconium hydride (for hydrogen), etc. The new value $\Phi_A^{\frac{3}{2}}$ is then evaluated from the change of slope, and $\Delta\Phi = \Phi_m - \Phi_A$. Low fields are employed to minimize the variable contribution of the field-induced dipoles of the adspecies as the applied voltage is increased.

Measurement of the changes of work function due to adsorption on individual crystal planes [10] is accomplished using the probe-hole technique. The change $\Delta\Phi$ brought about by a monolayer of adspecies on the surface of the tip varies for the different crystal planes; moreover, at lower pressures, the amount adsorbed is not the same on each plane. Since only the total adsorption on all planes of the tip can be assessed (with limited accuracy), separation of the two contributions to $\Delta\Phi$ is not possible. Moreover, the image pattern from planes of high work function may not be visible and the dark region is unchanged when the chemisorbed layer brings about an increase in Φ_M or a decrease of insufficient magnitude. Consequently, the information available from single-plane investigations is limited by these two factors.

17.3.2. Mobility of adsorbed species [11–15]

Gaseous adsorbate is introduced via an inlet source so positioned that the incoming molecules are directed to one side of the tip (see Fig. 17.5). The microscope is maintained at liquid-hydrogen, or liquid-helium, temperature so that: (i) the vapour pressure of residual gases is less than 10^{-13} or 10^{-28} Pa,

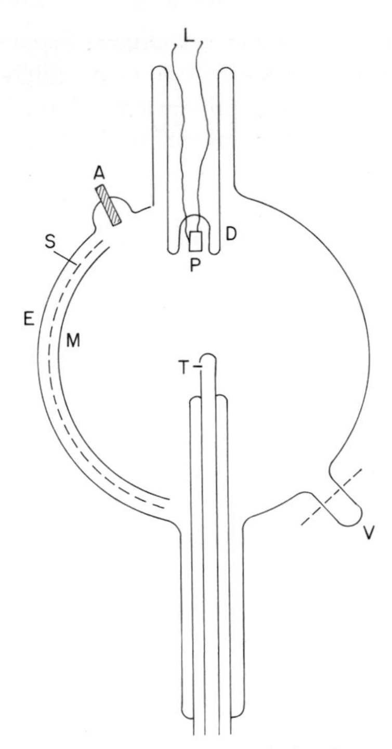

FIG. 17.5. Schematic representation of microscope design for surface diffusion investigation. D, Dewar flask for low-temperature condensation of gaseous adsorbate on the platinum sleeve P; L, leads for resistance heating of P. Other symbols are as for Fig. 17.3, but note the new position of the tip T.

respectively; and (ii) the sticking probability of the gaseous molecule on impact with both the tip and the walls of the glass envelope is unity. Adsorption is, therefore, confined to those crystal planes of the tip that are in direct line with the trajectory of the incoming molecules; all other molecules are immediately condensed on the walls. The ultra-high vacuum is, therefore, automatically maintained and the other planes remain completely free of adsorbed molecules.

The tip is then heated by passing an electric current through the supporting loop; simultaneous measurements of the resistance are made to monitor the temperature. The adspecies acquire mobility and spread over the clean crystal planes of the tip. From the visual change of the pattern, the diffusion constant and the activation energy of migration have been determined, and different types of diffusion at different temperature and/or coverages have been distinguished (see Chap. 19). In addition, desorption energies have been estimated from the temperature coefficient of the desorption rates. Moreover, reversible chemisorption can be distinguished from an irreversible incorporation process. Thus, the pattern obtained after chemisorption of oxygen can be transformed to that of the clean metal by raising the temperature of the tip and, thereby, effect desorption of the oxygen adatoms; the original pattern can then be regenerated by re-exposure to gaseous oxygen. When an irreversible oxidation takes place, however, the pattern from the oxidized surface remains unchanged after heating and introduction of more oxygen.

17.4. ELECTRON RESONANCE TUNNELLING

A basic assumption in the derivation of Eq. (17.3) is that the shape of the potential barrier to electron emission from an adsorbed layer is not changed when the high fields are applied; indeed, for many systems, within the limit of accuracy of the measurements, the emission may be assumed to be independent of the shape of the barrier. However, for a narrow potential barrier and high applied voltages, the shape can be modified and resonance tunnelling of electrons through the chemisorbed layer can take place. Nevertheless, the Fowler–Nordheim equation is still valid for electron emission from a clean metal surface. A change in slope of the plot of $\ln(j/V^2)$ against $1/V$ due to chemisorption confirms that the barrier height has been altered, and a decrease of work function should always be accompanied by an increase in total emission current, as indeed has been found for most gas/metal systems. However, when nitrogen is chemisorbed on the W(100) plane of tungsten [16], both the effective work function and the total emission current are simultaneously reduced. The explanation for this (and some other anomalous results) is that the adspecies modifies the shape as well as the height of the barrier and so provides a momentum source and sink for the tunnelling electrons. The only effective parameter in the Fowler–Nordheim equation is the height of the barrier, and Eq. (17.3) cannot, therefore, accurately describe the total energy distribution of the emitted electrons. The problem must be considered in more detail.

In a system comprising a foreign metal adatom the highest energy-bound state of which overlaps the conduction band of the metal, interaction between the adatom and metal atoms raises the electronic energy levels of the adatom. The ionization energy of the adatom is decreased because the conduction electrons of the metal are attracted to the ion cores in an attempt to screen the field of the positively charged ions within the metal; at the same time, these free electrons partially screen the field arising from the electrons of the adatom within the metal. In other words, an adatom electron is simultaneously subjected to an attractive potential from the ion-core and a repulsive potential arising from its outer ring of electrons. The net effect is the formation of an exchange and correlation hole near the metal surface, thereby creating an attractive force. As a result, there is a shift and broadening of the originally unperturbed electron ground-state level of the free foreign metal atom to form an atomic energy band or virtual state. The field-emitted electrons from within the metal that have the same energy as that of the virtual state in the adatom then undergo interference with the de Broglie waves; an enhanced or resonance transmission of tunnelling electrons through the potential barrier results. The electrons tunnelling through the adatom do not suffer any decrease of probability amplitude. However, since the second barrier to their emission is now much lower than the original single barrier, there is a large local increase in the total energy distribution of the field-emitted electrons undergoing resonance transmission. The enhancement factor, or the ratio of the local energy distribution in the presence and absence of the adatom, can range from unity to 10^4. Moreover, the energy distribution of emitted electrons is considerably more affected by the potential at the surface than by the total electron density obtained by integration over the complete energy range; consequently, resonance tunnelling is highly surface sensitive and, therefore, of great value in the examination of surface phenomena.

In order to obtain a quantitative assessment of the effect of a foreign adatom, Duke and Alferieff [17]† set up a model in which the adatom was replaced by a one-dimensional square well at a small distance outside the metal surface; an additional potential was, thus, formed, comprising an attractive interaction together with a repulsive (delta-function) pseudopotential to ensure orthogonization of the tightly bound occupied electron orbitals (see Fig. 17.6). The eigenstates and eigenvalues of the conduction electrons of the metal and of the highest-energy bound electrons in the adsorbate were then explicitly calculated. By the usual procedure of matching the logarithmic derivatives of the wave functions at each end of the barrier, the tunnelling probability was evaluated. This value was then inserted into the Sommerfield equation for a free-electron model to obtain the electron flux through the adatom. Some important general conclusion emerged from this analysis.

When the adatom–metal interaction was varied by changing the depth and/or width of the potential well so as to provide different energy levels for the highest-energy occupied state of the adatom, the energy of the enhanced band in the total energy distribution followed the change of energy of the occupied states. An increase of the effective work function of the metal, or a decrease in the distance of the adatom from the surface, both gave rise to larger enhancement factors. Similarly, when the height of the tunnelling barrier was reduced by increasing the adatom–metal interaction, there was a greater broadening of the virtual atomic states.

† See also Clark and Young [18].

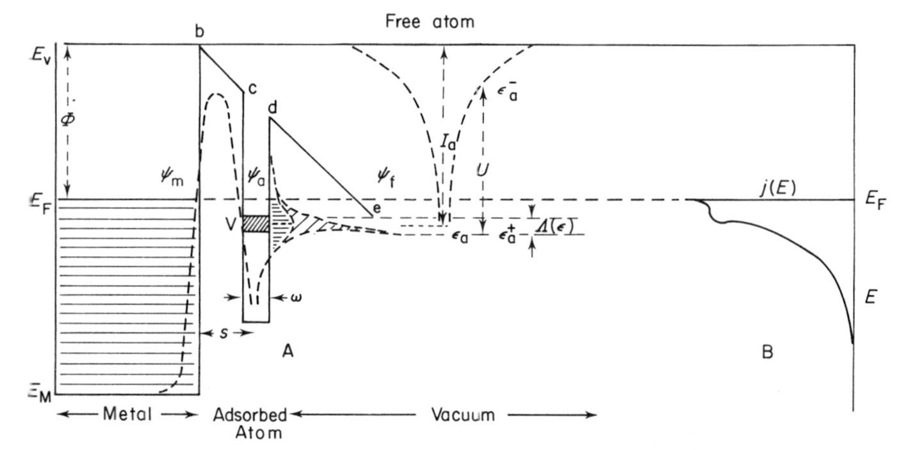

FIG. 17.6. Electron energy diagram for an adsorbed atom at a metal surface. E_v is the vacuum energy level, E_F the Fermi level and E_M the lowest energy state in the metal; Φ is the work function.

In section A, the full lines bcde represent the model of Duke and Alferieff [17] with the adatom replaced by a square well of width ω at a distance s from the surface. The electron wave function in the metal is ψ_m, that of the adsorbed atom ψ_a, and that of the emitted electron ψ_f. The narrow energy state of the free adsorbate ε_a with (ionization energy I_a) corresponds to an energy state E_d in the d band of the metal. Interaction produces broadening and a downward shifting to give a virtual state V; $\Lambda(\varepsilon)$, the level shift function, is the net energy level shift of half-width $\Gamma/2\,(\approx \hbar/\tau$, where τ is the tunnelling time of an electron on the adatom into the metal) and includes upward shifts due to the intra-atomic Coulombic repulsion U and the attractive image potential. The dashed curve is an approximate representation of the combined atomic and metal potentials for an adatom at the metal surface.

In section B, the total energy distribution of emitted electrons is indicated; superimposed on the normal exponential decay is a small peak due to enhancement of the probability of emission by resonance tunnelling at energies corresponding to those of the virtual state.

Some typical calculations were given. For example, for an atomic bound state at 4 eV within the conduction band of the metal, at an adatom–surface distance of 2 Å, the enhancement factor was of order 10^3. Nevertheless, because the resonance broadening was so large, the slope of the Fowler–Nordheim plot and, hence, the effective work function, was unchanged. When this bound state was moved through the Fermi energy surface by increasing the applied field, the resonances sharpened considerably and the slope of the Fowler–Nordheim plot was reduced, i.e. a decrease of the work function was accompanied by the normal increase of the total emission current.

Moreover, since the energy shift depended on the magnitude of the adatom–metal interaction, the more tightly-bound *d* electrons suffered smaller shifts than did the *s* electrons. Furthermore, the change in total energy per electron was the same for a two-electron adatom as for a one-electron atom, and this was also true for a singlet and triplet state, i.e. the triplet–singlet splitting observed in resonance tunnelling was the same as that in the free atom.

The theoretical treatment was later developed and extended by Gadzuk [19]† using a different approach, and his calculations have been applied to the experimental results of Plummer and Young [21]‡ for the adsorption of alkaline-earth metal atoms on the (110), (112), (111) and (310) planes of tungsten. Results obtained for the barium adatom–tungsten system were given (see Fig. 17.7).

The free barium atom has a $6s^2$ ground state at 5·2 eV, with first and second excited states at 4·09 eV ($^3D6s5d$) and 3·8 eV ($^1D6s6d$), all referenced to vacuum level. On adsorption on the W(111) plane of lowest work function (4·39 eV), the $6s^2$ state was shifted by almost 1 eV and broadened to 0·75 eV due to overlap with the triplet and singlet excited states; the positions of these two latter free-adatom states were unchanged and two sharp peaks of width (0·1 eV) appeared at 4·1 and 3·8 eV in the total energy distribution plot. On the (112) face of higher work function, the ^3D peak was absent, but the ^1D peak was still visible. On the (110) face of highest work function (3·68 eV), only the broad peak at 4·5 eV originating from the $6s^2$ state remained. The adatom ground state was, therefore, regarded as a mixture of the free-atom ground and first two excited states. The binding energies and adsorption heats would then depend on the number and the energy of states below the Fermi level and would be different for crystal planes the work functions of which were sufficiently different.

This interpretation of these results may, however, require modification, since for Ba adatoms on Mo(110) [having virtually the same work function

† See also Penn *et al.* [20].
‡ Cf. also Clark and Young [22].

as W(110)] the broad band is absent and the two sharp peaks that do appear have a smaller separation (0·1 eV) than that (0·3 eV) between the first two excited states of the free barium atom [21, 22].

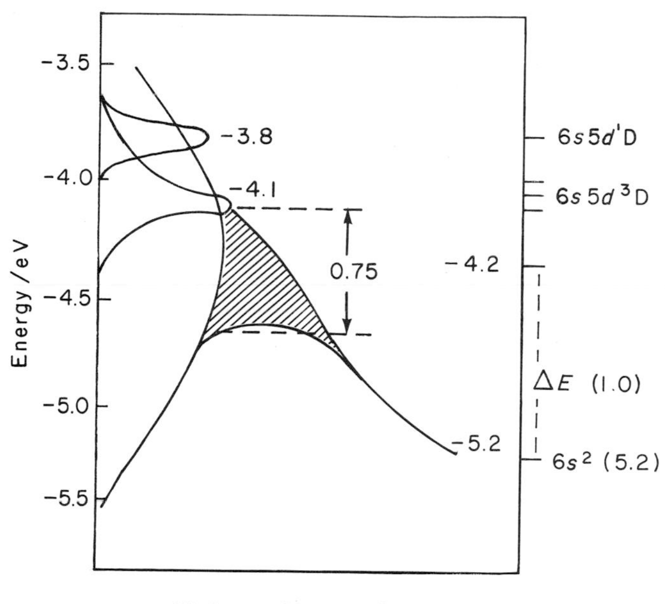

FIG. 17.7. Representation of the broadening and shifting of energy levels by interaction of a Ba adatom with the W(111) plane.

In another resonance tunnelling investigation [23] concerned with the chemisorption of hydrogen on W(100), a peak at 0·9 eV and a shoulder at 1·1 eV below the Fermi level were visible at lower coverages; these might have originated from two β_2 sub-states of different energies; but with increasing extent of adsorption, the peak and shoulder disappeared and two new broad bands, one centred at 2·6 eV below the Fermi level and the other extending downwards from this level, developed. Apparently, therefore, the two β_2 sub-states are converted to the β_1 state, i.e. the latter is not formed by adsorption over the β_2 state. There was no evidence for the existence of other sub-states that appear as multiple-peak structures in flash desorption spectra; such peaks, therefore, probably arise from nearest-neighbour and next-nearest-neighbour interactions between the adatoms.

17.5. INELASTIC TUNNELLING

A tunnelling electron may undergo an inelastic collision with an adsorbed molecule; and, because of dipole interaction between the electron and a localized vibrational mode of energy $\hbar\omega$ in the molecule [24], excitation of its ground-state vibration brings about a decrease in energy $\hbar\omega$ of the inelastically scattered electron. The total energy distribution is the sum of the elastic and inelastic contributions; the elastic component is, therefore, displaced downwards to give a step at an energy $\hbar\omega$ below the Fermi energy (see Fig. 17.8). The strength of the inelastic component is directly proportional to the

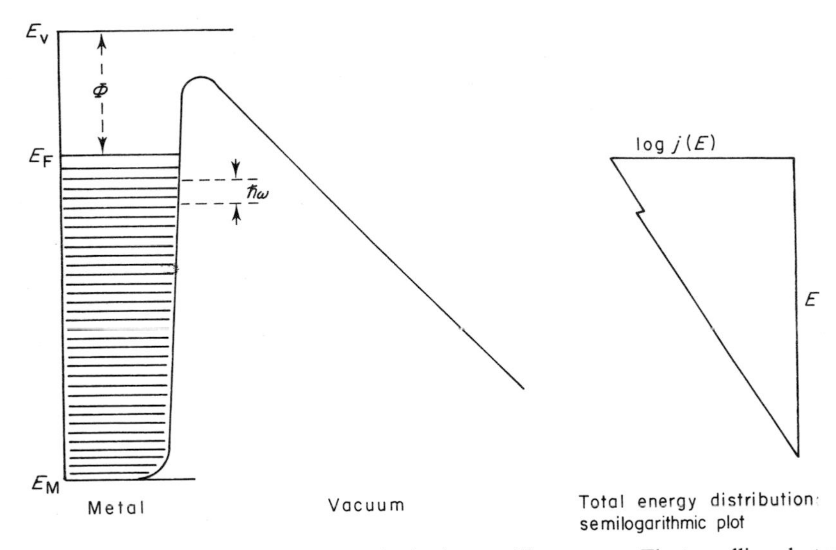

FIG. 17.8. Pictorial representation of the inelastic tunnelling process. The tunnelling electron is inelastically scattered by an adsorbed molecule with excitation of a localized vibrational mode. The electron loses energy $\hbar\omega$, which appears as a step in the semi-logarithmic plot of the total energy distribution [or of the enhancement factor $R(E)$].

square of the dipole matrix element of the vibrator, and for small adsorbed molecules represents only about 1 % of the total electron distribution (which is best plotted in terms of the enhancement factor $R(E)$). The signal-to-noise ratio of the inelastic signal is, therefore, very small, but has recently been greatly enhanced by improvements in the deflection analyser [25]; use of second derivatives and computer processing of data greatly improves the situation.

An example of the potentialities of this technique is given [23]. The stretching vibrations of the hydrogen and deuterium molecules have energies of

0·55 and 0·4 eV, respectively. For the γ-state of hydrogen and deuterium on W(111) at 77 K, inelastic energy losses at 0·55 and 0·4 eV, respectively, were observed; furthermore, as expected, these disappear on heating to 200 K. The energies of these vibrational modes, therefore, confirm the molecular character of the adspecies. In contrast, for hydrogen on the W(100) plane at 77 K in the β_2 region ($\theta < 0{\cdot}2$ monolayer), two different vibrational modes with energies 0·14 and 0·07 eV were detected. For deuterium, the peaks shifted to 0·1 and ~0·05 eV in accord with the mass ratio; they were attributed, therefore, to H—W and D—W modes. At full coverage, these loss peaks disappear probably due to changes in vibrational structure.

Experimental peaks are much easier to detect when the adsorbate is a large organic molecule since the relative strength of their inelastic contributions approaches unity [26]. These peaks are believed to arise from inelastic tunnelling and are attributed to vibrational modes, because they are only slightly shifted by changing the applied field, whereas electron levels would have been displaced according to $E' = E + eFz$, with $dE'/d(eF) = z$, where z is the approximate molecule–surface separation distance. However, experimental reproducibility is lacking and theoretical interpretation of the results is at present obscure.

Nevertheless, inelastic tunnelling could become a useful means of investigating vibrational modes of adspecies, particularly since the wavelength range may, in principle, be extended from the far infra-red to the visible region of the spectrum.

REFERENCES

1. E. W. Müller, *Z. Phys.*, 1937, **106**, 541; 1951, **131**, 136; *Ergeb. exakt. Naturwiss.*, 1953, **27**, 290.
2. R. Gomer, "Field Emission and Field Ionization", Harvard University Press, Cambridge, Mass., 1961.
3. R. Gomer, *Adv. Catalysis*, 1955, **7**, 93.
4. R. H. Fowler and L. W. Nordheim, *Proc. Roy. Soc. A*, 1928, **119**, 173.
5. L. W. Nordheim, *Proc. Roy. Soc. A*, 1928, **211**, 626.
6. R. D. Young, *Phys. Rev.*, 1959, **113**, 110.
7. R. Smoluchowski, *Phys. Rev.*, 1941, **60**, 661.
8. E. W. Müller, *Z. Phys.*, 1943, **120**, 270.
9. R. Gomer, *J. Chem. Phys.*, 1952, **20**, 1772.
10. Z. Sidorski, I. Pelly and R. Gomer, *J. Chem. Phys.*, 1969, **50**, 2382.
11. R. Gomer, R. Wortman and R. Lundy, *J. Chem. Phys.*, 1957, **26**, 1099, 1147.
12. R. Gomer and J. K. Hulm, *J. Chem. Phys.*, 1957, **27**, 1363.
13. R. Gomer, *Disc. Faraday Soc.*, 1959, **28**, 23.
14. R. J. Klein, *J. Chem. Phys.*, 1959, **31**, 1306.
15. G. Ehrlich and F. G. Hudda, *J. Chem. Phys.*, 1961, **35**, 1421.
16. T. A. Delchar and G. Ehrlich, *J. Chem. Phys.*, 1965, **42**, 2686.

17. C. B. Duke and M. E. Alferieff, *J. Chem. Phys.*, 1967, **46**, 923.
18. H. E. Clark and R. D. Young, *Surface Sci.*, 1968, **17**, 95.
19. J. W. Gadzuk, *Phys. Rev. B*, 1970, **1**, 2110; 1971, **2**, 1772; 1974, **10**, 5030; *Surface Sci.*, 1974, **43**, 44.
20. D. Penn, R. Gomer and M. H. Cohen, *Phys. Rev. B*, 1972, **5**, 768.
21. E. W. Plummer and R. D. Young, *Phys. Rev. B*, 1970, **1**, 2088.
22. H. E. Clark and R. D. Young, *Surface Sci.*, 1968, **17**, 95.
23. E. W. Plummer and A. E. Bell, *J. Vac. Sci. Tech.*, 1972, **9**, 583.
24. D. J. Flood, *J. Chem. Phys.*, 1970, **52**, 1355.
25. C. E. Kuyatt and F. W. Plummer, *Rev. Scient. Instrum.*, 1972, **43**, 108.
26. L. W. Swanson and R. Gomer, *Surface Sci.*, 1970, **23**, 1.

18

Field-ion Microscopy

18.1. GENERAL PRINCIPLES [1–5]

The field-ion and field-emission microscopes are essentially the same in construction, but: (i) the tip of the ion microscope is made the anode and the fluorescent screen is maintained at ground potential; and (ii) the apparatus is filled with an inert gas, usually helium, at a pressure of 10^{-1} Pa. On application of an external electric field (*ca* 5×10^{8} V cm^{-1}), the helium atoms are ionized at the surface of the tip and then accelerated towards the screen by the field to produce an image comprising a set of bright spots. For a tip of radius of curvature of 5×10^{-4} cm and a tip-to-screen distance of 100 cm, the magnification is 10^{6} to 10^{7}. The spots define specific locations at the surface of the emitter, since ionization is preferentially effected at protruding atomic positions at which both the field intensity and the gas supply are highest. The image contrast, therefore, arises from the variation of the local ionization rate over the tip surface, the rate being dependent on the intensity of the local field, the gas concentration in the ionization zone, and the surface electronic properties of the emitter.

The accommodation coefficient of the helium atoms is quite small ($\sim 0{\cdot}02$) and the bright spots arise from the presence of helium atoms which are normally weakly adsorbed on metals. However, as the helium atoms approach the tip, they are polarized in the applied field and, since they are then strongly accelerated towards the tip, their kinetic energy is increased to about twice that in absence of the field. In consequence, not only is the trapping cross-section of the atoms increased, but also their binding energy to the surface. In addition, the high field penetrates the metal surface and the surface metal atoms acquire an induced positive charge. Dipole–dipole interaction between these atoms and the polarized helium atoms then further contributes to the stabilization of the helium atoms at the surface. Their bonding is a maximum when they are directly above the metal atoms; consequently, the image pattern is not changed by the presence of the adsorbed atom except in those rare cases of imperfect registry of the adsorbate with the substrate. Finally, the presence of helium adatoms is responsible for greater penetration of the ionization zone and the ionization rate is increased. As a result, both the intensity and the contrast of the spots are enhanced.

Ions are formed only from helium atoms that are within a zone of width 3 to 5 Å outside the metal surface. On initial trapping, the adatoms retain sufficient kinetic energy to migrate over the surface in a series of hops, the heights of which decrease with loss of kinetic energy until the atoms are finally equilibrated with the surface. The ionization zone is, therefore, restricted to helium adatoms traversing the surface by this hop-site mechanism.

It is essential that the rate of ionization of the helium atoms in the gas phase should be negligibly small. Gas-phase ionization requires an energy equal to the ionization potential I of the free atom; in the presence of the applied field F, this is decreased by the polarization of the atom by an amount $\frac{1}{2}\alpha F^2$, where α is the polarizability of the atom. The ionization energy of the free atom is, therefore, $I - \frac{1}{2}\alpha F^2$. At the metal surface, the high field reduces the width of the potential barrier and the electrons freed by surface ionization of the helium atoms tunnel through the barrier to energy states predominantly at, or slightly above, the Fermi level. The ionization energy is, therefore, $(I - \Phi) - e^2/4x^2$, where Φ is the effective work function of the metal, and the final term is the interaction energy of the helium ion with its image within the metal. The energy of the ion at a distance x from the surface is further decreased by an amount $Fex + \frac{1}{2}\alpha_+ F^2$, where α_+ is the polarizability of the ion. Hence [2], electron tunnelling occurs at a distance x_c without change of energy when

$$Fex_c = (I - \Phi) - \frac{e^2}{4x_c^2} + \tfrac{1}{2}F^2(\alpha - \alpha_+). \tag{18.1}$$

The width of the ionization zone is given by x_c and has a value of 3 to 5 Å. The tunnelling frequency is not high; but, since gas-phase ionization requires an energy of $I - \frac{1}{2}\alpha F^2$, which is much larger than that for surface ionization given in Eq. (18.1), the rate of ionization at the emitter surface is rapid while that in the gas phase is negligibly small.

The main advantage of field-ion microscopy is the great improvement in resolution, which is 2 to 3 Å compared with 20 Å in the field-emission microscope; in principle, this atomic resolution means that single atoms at the surface can be located. In electron emission, the resolution is limited by the high tangential velocity of the emitted electrons, whereas the helium ions are emitted with thermal velocities; lateral spreading is, therefore, much smaller and can be further reduced by maintaining the emitter at liquid-hydrogen temperatures. The pressure of the gaseous helium is usually 10^{-1} Pa, so that the mean free path of the atoms is approximately the same as the distance ($\sim$100 cm) traversed by the helium ions from the tip to the screen. Since the ion-current is largely controlled by the rate of arrival of atoms at the tip, higher gas concentrations increase the image intensity, but there is a resultant loss of definition brought about by the scattering of the ions by the gas atoms; lower pressures have no advantage, since no further reduction of scattering is effected and the image intensity is lower.

18.2. APPLICATIONS TO CHEMISORPTION [2]

The interpretation of, and conclusions derived from, much of the earlier work are less certain than was originally thought. For example, when nitrogen is chemisorbed on a tungsten tip, additional bright spots are observed on the fluorescent screen using helium as the imaging gas and high fields. Since the presence of nitrogen atoms should enhance the rate of surface ionization, the additional features were attributed to the presence of these adatoms. However, later work, again employing high fields and helium as the imaging gas, indicated that the spots were produced by tungsten atoms displaced from their original sites by a field-induced process in presence of the adsorbate. Similarly, it had been previously concluded that the chemisorption of nitrogen and of carbon monoxide caused surface reconstruction; this conclusion was drawn because the pattern did not return to that of the clean tip after field-desorption of the adsorbate using high fields and helium as the imaging gas. However, for the lower fields that can be used for argon imaging, little surface damage takes place. Indeed, there is now no evidence that adatoms of oxygen and nitrogen and admolecules of CO produce detectable changes in the image; this non-visibility of many adsorbates severely limits the use of the field-ion microscope. Reconstruction was, therefore, the result of using helium and high fields. Nevertheless, this latter procedure can be employed to investigate gross topographical changes, such as the facetting and creation of terraces brought about by heating a chemisorbed oxygen layer on a tungsten tip, either in the absence or in the presence of an external field.

Since many of the results of the earlier investigations on chemisorbed atoms were obtained using helium imaging at quite high fields, surface damage had probably taken place and the conclusions reported are therefore of doubtful validity. In the more recent work, however, this problem has been recognized and largely eliminated, particularly in investigations in which the adspecies is a (visible) metal adatom.

The most definitive field-ion microscopic work is now largely concerned with the measurement of surface diffusion coefficients and of the kinetics and energetics of cluster formation of metal adatoms [5]. This subject is discussed in some detail in Chapter 19.

REFERENCES

1. R. Gomer, "Field Emission and Field Ionization", Harvard University Press, Cambridge, Mass., 1960.
2. G. Ehrlich, *Adv. Catalysis*, 1963, **14**, 256.
3. E. W. Müller and T. T. Tsong, "Field-ion Microscopy, Principles and Applications", Elsevier Publishing Co., New York, London and Amsterdam, 1969.
4. D. W. Basset, *in* "Surface and Defect Properties of Solids" (M. W. Roberts and J. M. Thomas, eds), The Chemical Society, London, 1973, Vol. 2, p. 34.
5. E. W. Müller and T. T. Tsong, *Prog. Surface Sci.*, 1974, **4**, 1.

19

Mobility of Adspecies

19.1. THE MIGRATION PROCESS

When an adspecies acquires thermal energy from the lattice phonons of the adsorbent in an amount sufficient to surmount the potential barrier between adjacent adsorption sites (the locations of minimum potential energy), it can migrate from its initial site to an adjacent unoccupied site by a hop-site mechanism. The hop distance equals the wavelength of the oscillating energy function, or the interatomic distance between adjacent ion-cores; the amplitude of the oscillations is small because of the marked screening effect of the conduction electrons of the metal. During migration, the adsorbate–metal bond at the initial site is weakened and an incipient bond to be formed at an adjacent unoccupied site is strengthened, thereby producing a transition-state complex which is subsequently transformed to the final state with the adspecies strongly bonded to the adjacent site. The activation energy to form the complex, or the migration energy, is considerably less than the energy to rupture the adsorbate–adsorbent bond with desorption of the adspecies. Mobility, therefore, ensues at a temperature well below that required for measurable rates of desorption; the smaller the heat of chemisorption the lower is the energy necessary for migration. Similarly, since chemisorption heats normally decrease with coverage, mobility is freer at higher coverages. For a uniform surface, every adspecies is equally mobile at each coverage, but on a non-uniform surface mobility can be restricted over the high potential sites while free mobility ensues over sites of low adsorption potential.

19.2. DIFFUSION COEFFICIENTS AND MIGRATION ENERGIES

19.2.1. Polycrystalline surfaces

Earlier investigations were confined to the migration of foreign metal atoms over polycrystalline tungsten filaments; ill-defined average values of the diffusion coefficient and migration energy were obtained.

19.2.1.1. *Early macroscopic measurements*

An initial concentration of adatoms was deposited, at room temperature or below, on part of the surface so as to produce a sharp boundary between the adsorbed layer and the remaining clean metal surface. The temperature of the filament was then raised to a constant value at which migration proceeded at a convenient rate; the decrease of concentration of the adsorbed layer and increase on the clean part of the filament were monitored as a function of time. The total loss of adatoms q from the adsorbed layer (and gain by the originally bare region) at a time t up to a distance x along the filament from the initial boundary ($x = 0$) of width l is given by the usual diffusion equation

$$q = lc_0(Dt/\pi)^{\frac{1}{2}}, \tag{19.1}$$

in which D is the diffusion coefficient and c_0 the initial concentration (adatoms cm^{-2}) of the uniform adsorbed layer. When c_0 is of known magnitude, D can be evaluated explicitly. The activation energy for diffusion is evaluated from measurements of D at a series of constant temperatures of the filament and insertion of these values in the equation

$$D = D_0 \exp(-E_m/RT), \tag{19.2}$$

where E_m is the energy per g atom of the adsorbate. Temperatures are measured by an optical pyrometer, or evaluated from the resistance of the filament and its temperature coefficient of resistivity.

Various techniques were employed: (i) to obtain sharp boundaries and (ii) to measure surface concentrations. For example, for thorium adatoms on tungsten, an atomic beam was directed to the front face of a flat tungsten filament, which was maintained at a low temperature, so that the back face remained bare. Thorium adatoms [1] decrease the tungsten work function by an amount linearly related to their concentration. At the temperature of migration, the thermal electron emission current from the filament, as it decreased from the front face and increased from the back face, was measured as a function of time until a uniform concentration over both faces was obtained (see Fig. 19.1).

For caesium adatoms [2], advantage was taken of the fact that at low coverages ($\theta < 0{\cdot}03$) and high temperatures, positive caesium ions are thermally desorbed. The tungsten filament with adsorbed layer of caesium was surrounded by three concentric, but electrically separated, metal cylinders.

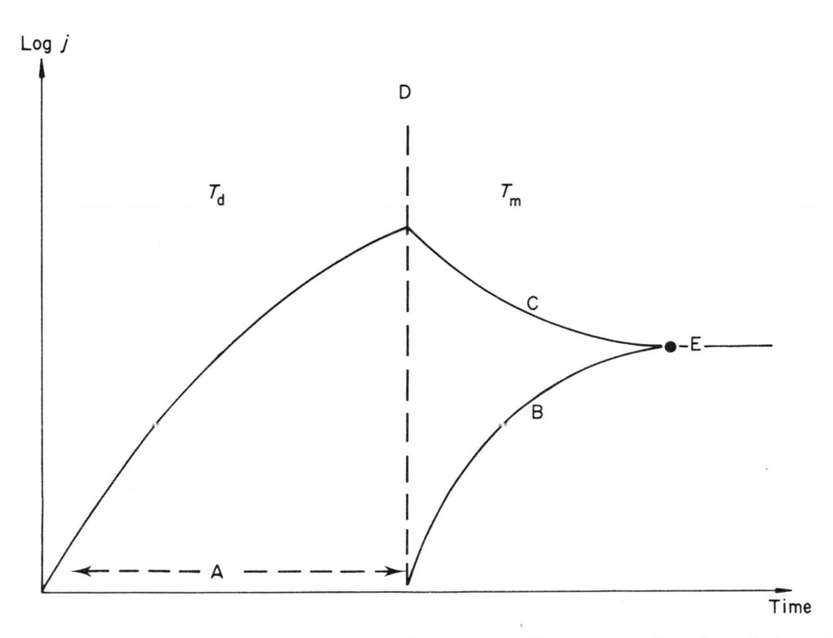

FIG. 19.1. Thermal emission current j from a flat tungsten filament as a function of time. A, increase of j with increasing concentration of thorium atoms deposited on the front face of filament at the lower temperature T_d. The temperature is increased at point D on cessation of deposition from T_d to the temperature of migration T_m. B, increasing emission current from the back face as adatoms migrate from the front to the back face. C, decreasing current due to loss of adatoms. At E there is a uniform concentration of adatoms on the back and front faces.

The caesium atoms were first confined to the middle section of the filament by increasing its temperature while maintaining a positive potential on the middle cylinder and negative potentials on the outer ones. Then, at the selected diffusion temperature, a positive potential was maintained on all three cylinders to prevent loss by desorption during migration. After a suitable time interval, the filament was flashed at high temperature with negative potentials applied to the cylinders. The positive currents received separately by each of the cylinders gave direct measures of the adatom concentrations on the middle and on the two outer sections of the filament.

19.2.2. Single crystal planes

19.2.2.1. *Field-emission microscopy*†

The migration energy of an adatom varies for different crystallographic planes of the same metal adsorbent. The specific diffusion parameters for a particular plane can be determined in the field-emission microscope. As has been noted, the tip displays a large number of well-defined single crystal planes that can be identified crystallographically from the pattern produced on the fluorescent screen by the emitted electrons. Using a directed beam of gaseous adsorbate, with the tip and glass envelope at liquid-helium temperatures, adsorption can be restricted to planes in the trajectory of the beam, leaving all other planes bare. The migration characteristics are then derived from the changes of the pattern caused by movement of the adatoms from the adsorbed layer to the originally bare planes.

TABLE 19.1. Migration and desorption energies of chemisorbed species over the (100) and (110) planes of a tungsten tip.

Type	E_m/(kJ mol^{-1})	E_d^c/(kJ mol^{-1})	E_m/E_d^c
H, boundary-free	42–68	273–336	0·20
(110) boundary[a]	25	252	0·1
O, boundary-free	126	546	0·24
(110) boundary	105	525	0·2
(100) boundary[b]	97	525	0·18
CO, boundary-free	252	378	0·66
(110) boundary[c]	151	218	0·70
N, boundary-free	147	672	0·22
(100) boundary[d]	54	672	0·12

[a] Gomer *et al.* [3]. [b] Gomer and Hulm [4]. [c] Gomer [5] and Klein [6]. [d] Ehrlich and Hudda [7].

The observations in the field-emission microscope are limited to the movement of a diffusion boundary over a crystal plane, although preferred directions of migration can also be distinguished. Three different types of surface diffusion have been identified and their diffusion coefficients and migration energies measured.

Boundary diffusion of physisorbed molecules [5, 6]. The adsorbed layer, deposited at liquid-helium temperature on one side of the tip, comprises a primary immobile chemisorbed layer and a molecular physisorbed layer on top of it. At higher temperatures (around 50 K for nitrogen, ~30 K for oxygen,

† See Chapter 7.

and <20 K for hydrogen), physisorbed molecules diffuse to the boundary edge of the primary layer and are chemisorbed on the bare sites on the clean part of the tip. The chemisorbed layer is extended and the distance travelled by the boundary can be observed by the change of emission current.

The diffusion coefficient D_m of the mobile molecules is evaluated from measurements of the mean distance $\bar{x}$ travelled by the boundary during a period at time τ, since the Einstein equation

$$.\bar{x}^2 = D_m \tau \tag{19.3}$$

was found to be valid. Migration takes place by a hop-site mechanism with a jump frequency ν_m (of *ca* $10^{12}\ s^{-1}$), a jump distance a (of about 3 Å), and an activation energy of E_m mol^{-1}. Equation (19.3) can, therefore, be transformed to

$$D_m = a^2 \nu_m \exp(-E_m/RT). \tag{19.4}$$

The average distance $\bar{x}$ traversed by an adspecies depends on its average lifetime τ_d on the primary layer given by

$$\tau_d = [\nu_d - \exp(E_d/RT)]^{-1}, \tag{19.5}$$

where E_d is the energy of desorption of the physisorbed molecule. Above a critical temperature T_c, the boundary become stationary, since the molecule is desorbed just before it reaches the boundary having travelled an average distance $\bar{x}_c$. Hence, by combination of Eqs (19.4) and (19.5),

$$\bar{x}_c^2 = a^2 \nu_m \exp(-E_m/RT_c) \nu_d^{-1} \exp(E_d/RT_c). \tag{19.6}$$

With the reasonable assumption that $\nu_m \approx \nu_d$, then

$$E_d = E_m + 38{\cdot}6\, T_c \log_{10}(\bar{x}_c/a), \tag{19.7}$$

where E_d, E_m are, respectively, the energy of desorption and of diffusion in J mol^{-1}. Values of E_m are evaluated from measurements of $\bar{x}$ at a time τ at various temperatures below T_c, and E_d can then be calculated from Eq. (19 7) by inserting a value of a derived from the surface geometry of the crystal plane.

Boundary diffusion of chemisorbed atoms. At higher temperatures, desorption of the physisorbed layer is complete, and migration takes place in the primary chemisorbed layer. Mobility is most rapid over the closest-packed plane [the (001) plane of a b.c.c. metal], since the bonding energy of the adsorbate is least. The migrating adatoms are finally trapped on the surrounding more open planes at which the bonding is greater. The values of D_m and E_m vary for different adsorbates. For hydrogen, a sharp boundary is observed below 180 K and the adatoms spread radially outwards from the central (001) plane, the rate being most rapid along the [110]–[211] direction. It must be noted that a sharp boundary is only observed when the distance between the trapping sites is less than about 20 Å, the resolving power of the microscope.

Boundary-free diffusion of chemisorbed atoms. At still higher temperatures, and at very low coverages, the number of adatoms is insufficient to saturate the trapping sites; a site-to-site migration with a higher E_m value then occurs. Since no sharp moving boundary is visible, this migration is described as a boundary-free diffusion.

In Table 19.1 are given the migration energies E_m of chemisorbed adatoms of hydrogen, oxygen, nitrogen and molecular carbon monoxide over a tungsten tip involving both modes of diffusion, together with their desorption energies E_d^c.

A few general conclusions are indicated. (i) For hydrogen, oxygen and nitrogen adatoms, the E_m/E_d^c ratio for boundary-free diffusion is nearly the same (~0·2); it is smaller for boundary diffusion, being roughly half this value for hydrogen and nitrogen. The ratio is much higher for the carbon monoxide molecule (~0·7) possibly because its larger size prevents it from squeezing its way between two adjacent tungsten atoms as is possible with the other adatoms. (i) The ratio E_m/E_d^c is smaller for the close-packed than for the more open planes. (iii) Some correlation may exist between the entropy of activation $\Delta S^{\ddagger}$ for diffusion and the surface geometry of the planes, the entropy being derived from the pre-exponential factor $D_0 = a^2 \nu \exp(\Delta S^{\ddagger}/R)$.

19.2.2.2. *Field-ion microscopy*†

The field-ion microscope, with its much higher resolution, permits the observation of the motion of a single adatom; moreover, the effect of interaction between adatoms and ço-operative phenomena within the adsorbed layer can be investigated. Migration of adatoms is brought about by raising the temperature of the tip in the absence of the applied field; subsequent imaging, after a suitable time interval, is effected by cooling the tip to liquid-hydrogen temperature before applying the field. In this way, perturbation of the surface by the field during migration is avoided.

A series of n atomic displacements of the adatom during the time interval τ is recorded and from these measurements the value of $\bar{x}^2$ is evaluated. Since

† See Chapter 18.

$\bar{x}^2 = na^2$ and $D_m = na^2/\tau$, then, as before,

$$D_m = \bar{x}^2/\tau; \tag{19.3}$$

the number of jumps executed by the adatom during the time τ is n, and a is the jump distance between adjacent adsorption sites. Some adatoms are reflected at the boundary edge and n must be sufficiently small so as to make the effect of such reflections negligible. Very small values must also be avoided, otherwise $\bar{x}^2$ becomes dependent on the initial position of the adatom.

TABLE 19.2. Activation energies for surface diffusion of metal adatoms [9].

Substrate	Adatom	Activation energies/kJ mol^{-1} (110) plane	(211) plane	(321) plane
W	W	84, 92	54	80, 84
Ir	W	—	96	100
	Re	50	113	84
	Ir	—	89	92
	Pt	< 40	67	92
Rh	Rh	23	—	—

Migration energies of some metal adatoms on some specific crystal planes of a few metals have been reported (Table 19.2) [8, 9]. An explanation of the results obtained was first formulated in terms of pair-wise bonding. Thus, as as an adatom moves from one site to an adjacent one, the change of co-ordination with the surface atoms of the metal adsorbent is smaller on the closely-packed W(110) plane than on the more open W(111) plane, and the activation energy of migration should be lower on the smoother (110) surface. Although some parallelisms between the energy and surface geometry were found, it became evident that a simple model based solely on pair-wise bonding (and the similar broken-bond theory) was inadequate [10–13].

For example, the motion of a tungsten adatom over the W(211) and W(321) planes is one-dimensional along the [111] direction, which comprises a row of close-packed protruding atoms. There is a small change in co-ordination for movement along this direction compared with that across an atomic row. The pair-bonding theory predicts this particular result, but, however, does not account for the exceptionally high value of E_m over the smooth W(110) plane. An additional, highly anomalous feature is that the pre-exponential factor D_0 for diffusion over both the W(110) and W(321) plane has the normal value of 10^{-3} to 10^{-4} cm^2 s^{-1}, whereas over the (211) plane an extremely low magnitude of 10^{-7} cm^2 s^{-1} is obtained, thereby suggesting that some co-operative process is involved (see Table 19.3).

TABLE 19.3. Self-diffusion of tungsten atoms [13].

Plane	E_m/kJ mol^{-1}	D_0/(10^{-3} cm^2 s^{-1})
(110)	84	2·1
(211)	55	3·8 × 10^{-4}
(321)	82	1·2

Similar anomalies in other self-diffusion studies have also been noted. Thus, the motion of rhodium adatoms over the close-packed Rh(110) plane of the f.c.c. metal [14] requires a small activation energy and mobility is observed at temperatures below 53 K, whereas high E_m values are found for self-diffusion over the close-packed W(110) plane of this b.c.c. metal. Similarly, the migration energies of W adatoms over the W(211) and W(321) planes are widely different, being 54 and 82 kJ (g atom)$^{-1}$, respectively, whereas for Rh adatoms over the corresponding (110) and (331) planes they are practically the same (see Table 19.4).

The general observation [15, 16], that formation of dimers and clusters of adatoms commonly occurs, indicates that there are strong attractive interactions between adatoms; the diffusion parameters of single adatoms and dimers are found to be markedly different. Thus [17], for single tungsten adatoms over the W(211) plane, $D_0 \approx 10^{\ 3}$ cm^2 s$^{\ 1}$ and $E_m \approx 76$ kJ (g adatom)$^{-1}$ whereas for adatoms pairs $D_0 \approx 10^{-12}$ cm^2 s^{-1} and $E_m \approx$ 30 kJ (g adatom)$^{-1}$. Apparently, the attractive interaction [15,16] between the adatoms of the dimer largely compensates for the decrease of interaction of each adatom with the lattice as the dimer moves towards the saddle-point of the energy barrier to migration. The range of interaction between adatoms extends over a distance of 5 to 7 Å, so that movement of an adatom along a one-dimensional channel may be modified by the presence of other adatoms within this distance of the channel.

The complex nature of the interaction between adatoms and that between the adatoms of the lattice is emphasized by the marked chemical specifity

TABLE 19.4. Self-diffusion of rhodium atoms [14].

Plane	E_m/kJ mol^{-1}	D_0/(10^{-3} cm^2 s^{-1})
(111)	15	0·2
(311)	52	2
(110)	58	300
(331)	62	10
(100)	85	1

of adatom polymers [9]. Two-dimensional clusters are formed from tungsten and from rhodium adatoms, but irridium and platinum adatoms produce one-dimensional chains; although rhenium adatom-pairs have never been observed, both triangular and linear trimers have frequently been seen.

It is, therefore, essential that measurements of the surface mobility of a single adatom should be made under conditions where no other adatoms are in close proximity, in order that the specific effect of the surface geometry of a crystal plane can be separately investigated.

19.3. ENTROPY OF MOBILE AND IMMOBILE CHEMISORBED LAYERS

In principle, comparison of the magnitude of the integral molar entropy of a chemisorbed layer, derived from experimental measurements of the integral molar free energy and the integral molar enthalpy, with the corresponding entropy theoretically derived from statistical-mechanical calculations, provides a criterion for mobility (or immobility) of the chemisorbed adspecies. The procedure has been discussed in some detail in Chapter 8 and it is concluded that little confidence can be placed on the validity of the conclusions drawn.

REFERENCES

1. W. H. Brattain and J. A. Becker, *Phys. Rev.*, 1933, **43**, 428.
2. J. B. Taylor and I. Langmuir, *Phys. Rev.*, 1933, **44**, 423.
3. R. Gomer, R. Wortman and R. Lundy, *J. Chem. Phys.*, 1957, **26**, 1147.
4. R. Gomer and J. K. Hulm, *J. Chem. Phys.*, 1957, **27**, 1363.
5. R. Gomer, *Disc. Faraday Soc.*, 1959, **28**, 23.
6. R. Klein, *J. Chem. Phys.*, 1959, **31**, 1306.
7. G. Ehrlich and F. G. Hudda, *J. Chem. Phys.*, 1961, **35**, 1421.
8. G. Ehrlich, *Adv. Catalysis*, 1963, **14**, 378.
9. D. W. Bassett, *in* "Surface and Defect Properties of Solids" (M. W. Roberts and J. M. Thomas, eds), The Chemical Society, London, 1973, Vol. 2, p. 34.
10. G. Ehrlich and F. G. Hudda, *J. Chem. Phys.*, 1966, **44**, 1039.
11. G. Ehrlich, *J. Chem. Phys.*, 1966, **44**, 1050.
12. G. Ehrlich and C. F. Kirk, *J. Chem. Phys.*, 1968, **48**, 1465.
13. D. W. Bassett and M. J. Parsley, *J. Phys. D*, 1970, **3**, 707.
14. G. Ayrault and G. Ehrlich, *J. Chem. Phys.*, 1972, **57**, 1788; 1974, **60**, 281.
15. D. W. Bassett, *Surface Sci.*, 1970, **23**, 240.
16. T. T. Tsong, *Phys. Rev. B*, 1972, **6**, 417.
17. W. R. Graham and G. Ehrlich, *Phys. Rev. Letters*, 1973, **31**, 1407.

20

Electron-impact Auger Spectroscopy

Ionization of an inner-shell electron state of a metal atom can be accomplished by bombarding a metal surface with a beam of electrons of energy of about 10^3 eV. The electron is ejected from the solid and an inner-shell vacancy is produced. One mode of de-excitation of this excited state is the Auger process [1, 2], which leads to further electron emission.

20.1. PRODUCTION OF AN AUGER ELECTRON

The process involves: (i) the ionization of one of the inner-shell electron states by a high-energy primary beam ($\sim$3 eV); (ii) the occupation of the electron vacancy by an electron from a higher-energy inner-shell level with consequent release of energy; (iii) transference of this energy by a radiationless process to a second inner-shell electron at a lower-energy level; and (iv) ejection of this electron out of the solid with excess kinetic energy. Alternatively, de-excitation may take place by emission of a quantum of X-ray radiation to give a spectral peak in an X-ray fluorescence spectrum (Fig. 20.1). This emission is governed by the usual X-ray selection rules, i.e. the quantum number of the orbital angular momentum changes by ± 1. The Auger process, however, is controlled by the electrostatic forces between the hole and its surrounding electron cloud and is, therefore, less severely restricted by selection rules.

A specific sequence of events leading to the production of an Auger electron is: (i) ejection of an electron from the K shell of the atom; (ii) the filling of the vacancy or hole by an electron from the L_I shell; and (iii) transference of the energy released to an electron in the L_{III} shell and its consequent emission. Such a sequence is denoted as a KL_IL_{III} process, or the Auger electron is referred to as a KL_IL_{III} electron. The number of possible transitions depends largely on the atomic number Z of the atom. For a primary beam energy of around 1·5 keV, and for $Z < 20$, the Coulomb interactions between

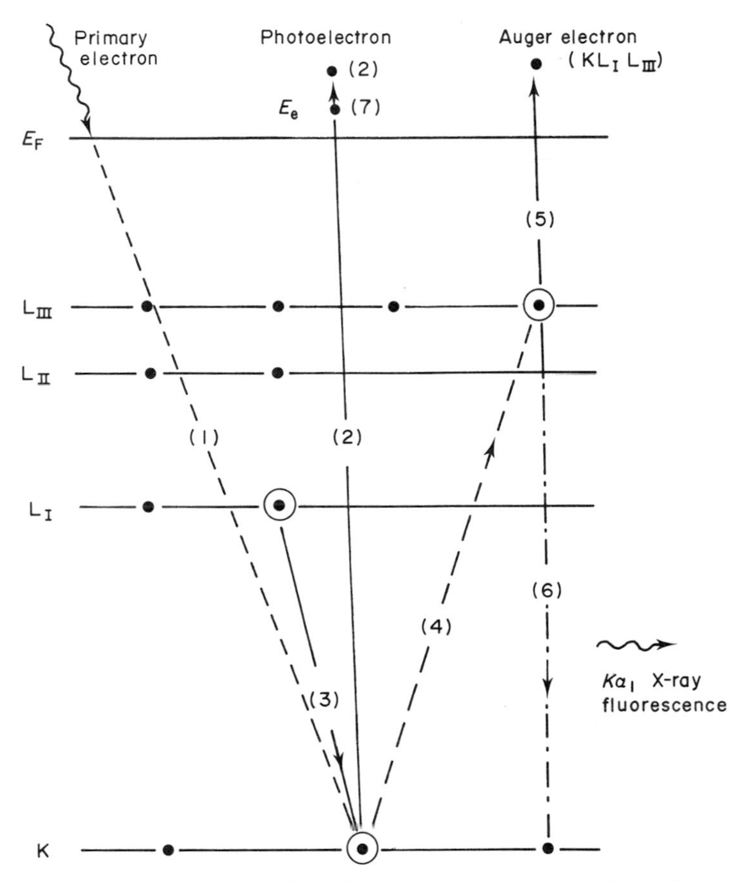

FIG. 20.1. Auger electron emission and X-ray fluorescence. 1. Ionization of a K electron by primary beam. 2. Photoemission of a K electron leaving core vacancy (see 7, below). 3. Core vacancy filled by transition of an L_I electron with liberation of energy. 4. Energy transferred to an L_{III} electron by radiationless transition. 5. Emission of an Auger electron (KL_IL_{III}) with kinetic energy E_k.

Competing de-excitation processes are: 6. De-excitation by an L_{III} electron transition to the core vacancy and emission of X-ray fluorescence, $K\alpha_1$. 7. K electron transition to excited level E_e above E_F (for low primary electron energy); energy is adsorbed by the metal atoms and an X-ray absorption spectral peak is produced.

electrons predominate, thereby leading to Russel–Saunders or LS coupling to give six possible transitions, two of which are degenerate; hence, five spectral peaks appear. For high values of $Z(>80)$, the spin–orbit interactions are much the stronger and the X-ray *jj* coupling restricts the number of allowed transitions to six. For atomic numbers between these two broad limits, both types of transitions are possible and nine peaks may appear.

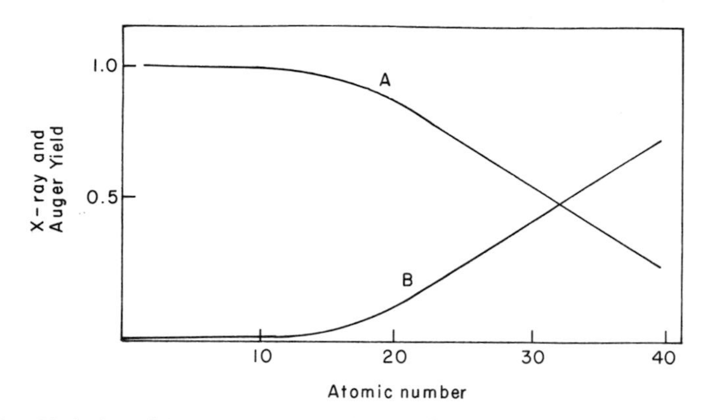

FIG. 20.2. Variation of Auger (curve A) and X-ray fluorescent (curve B) yields per K electron vacancy with increase of atomic number Z of the chemical element.

The energy of a specific Auger transition increases markedly with increase of Z; consequently, transitions that occur for lower values of energy of the primary beam may change from KLL to LMM, and so on, as Z becomes larger [2]. At higher incident beam energies, peaks due to initial double ionization also become more evident. Furthermore, the Auger electron emission and radiative de-excitation to give X-ray fluorescence are in competition; the Auger process is highly efficient for small Z but increasingly less so for larger values until at $Z = 33$ the two yields are the same. With further increase of Z, the fluorescence yield becomes increasingly dominant at the expense of the Auger emission [2] (Fig. 20.2).

20.2. THE AUGER SPECTROMETER

The essential features comprise an electron gun generating a primary beam of electrons of energy ~2 to 3 keV and flux $\sim 10^5$ electrons s^{-1} incident on the sample, together with an energy analyser to determine the distribution of kinetic energy E_k of the emitted Auger electrons [3–10] (Fig. 20.3).

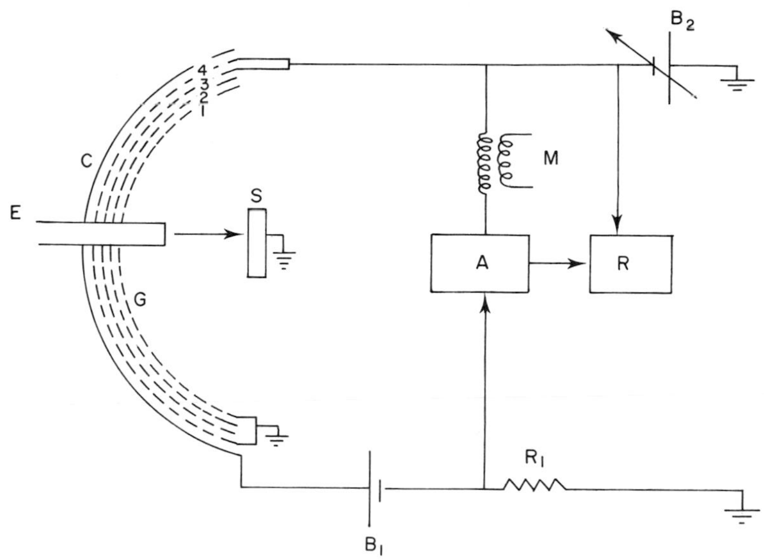

FIG. 20.3. Diagrammatic representation of an Auger spectrometer with a four-grid retarding-field analyser. E, Electron gun to provide primary electron beam of energy ~2 to 3 keV; S, metal sample; G, concentric four-grid system (a retarding potential is applied to G_2G_3 by a variable d.c. supply B_2); M, modulator superimposing an a.c. signal (1 to 10 V amplitude) at a frequency ω to the retarding grids; C, collector of elastically scattered electrons from sample; A, lock-in amplifier providing reference signal; R, X–Y recorder; B_1, battery connected to collector (~300 V); R_1, sensing resistor. The apparatus is essentially a post-acceleration LEED apparatus.

In normal operation, the electron beam is focused to give an irradiation area of 0·2 mm diameter. The power generated at the surface is around 30 W cm^{-2}; consequently, local heating up to 100 °C is possible, and marked radiation damage may ensue. Surface charging also occurs but is usually neutralized by secondary electrons. Since the Auger process is not dependent on the energy distribution of the primary beam, filtering is not necessary.

X-ray irradiation in the required energy range may also be used, but the photon flux (~10^6 photons s^{-1}) is considerably smaller than the electron flux above; this is only partly compensated by the X-ray ionization probability being several orders of magnitude higher. The scan time is, consequently, much longer, but the background spectrum is flatter and the surface damage is less. The region under observation may also be considerably reduced (~25 μm in diameter).

In the retarding field analyser [6], the current flux received by the collector is controlled by a retarding potential V_r, which only allows the entry of electrons with energy E greater than eV_r. These include not only the unscattered Auger electrons, but also Auger and other electrons which have

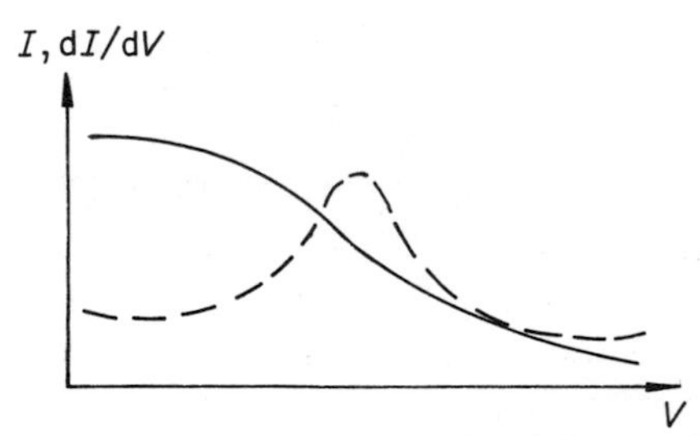

FIG. 20.4. Current I as a function of retarding potential V given as full line. The derivative dI/dV plotted against V is shown as the dotted line.

suffered multiple scattering before emission; consequently, the energy spectrum has a fairly intense broad background on which the much smaller peaks of the unscattered Auger electrons are superimposed. Separation of the Auger peaks is normally effected by imposing an a.c. signal of 2 to 10 V on the retarding d.c. potential at a frequency ν and applying a phase-sensitive detection to the collected current [4, 5]. The derivative dI/dV_r is then displayed at the second harmonic frequency 2ν (Fig. 20.4). The Auger peaks are, thereby, easily distinguished from the background, as illustrated in Fig. 20.5 which shows the LMM spectra of copper, where $dN(E)/dE$ ($\triangleq dI/dV_r$) is plotted as a function of E; $N(E)$ is the number of electrons of energy E_k ($=eV_k$) received by the collector.

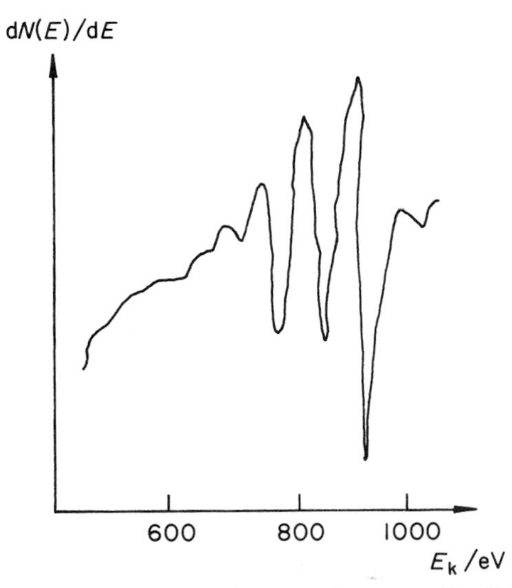

FIG. 20.5. LMM Auger spectra of copper using a retarding field analyser.

Coaxial cylindrical analysers [9] have a signal-to-noise ratio about 100 times larger, and high-speed recording is possible (scan times of >1000 eV s^{-1} compared with 2 eV s^{-1} for retarding fields). Alternatively, the input flux may be reduced to 0·1 μA, to avoid local heating and radiation damage, at the expense of an increasing time period to scan the spectrum. The transmission intensity is a function of $EN(E)$ and the spectra obtained differ from those obtained by the retarding field analyser which provides the energy distribution $N(E)$. The resolving power of both analysers is usually around 100 to 200, although the retarding field method is capable of much higher resolution (>10000). Concentric hemispherical analysers [10] are also employed; these have a resolution of ~ 1000 but a low scan speed (~ 1 eV s^{-1}).

20.3. BAND SHAPE AND WIDTH OF SPECTRAL BANDS

The Auger spectra of free molecules comprise fairly sharp peaks; their half-widths are virtually independent of the energy distribution of the primary beam because ionization by electron impact takes place in less than 10^{-16} s, whereas the lifetime of the hole is an order of magnitude longer. The energy spread, or lifetime broadening, is due to the operation of the uncertainty principle. It is greatest for a Coster–Kronig transition, in which the energy is rapidly transferred between two sub-shells having the same principle quantum number; the half-width may then be as large as 5 to 10 eV, compared with an instrumental resolution of around 0·1 eV. However, for shells of low ionization energy that require lower incident energies, the half-width may be only a fraction of an electron-volt.

The corresponding spectrum of a solid displays much less structure, since the discrete energy levels of the free atom are broadened into bands of considerable width due to delocalization of the energy levels. Within each energy band there is a marked variation of the electronic density of states; the line shape of the band reflects this initial electron energy distribution. The contour of a peak is further modified by inelastic scattering losses arising from inter- and intra-electron transitions, excitation of bulk and surface plasma, and phonon excitation of the lattice with a resultant tailing of the energy distribution at the lower energies.

20.4. AUGER YIELD

The rate of emission of Auger electrons, or the Auger yield, is directly related to the probability of ionization of inner-shell electrons by electron impact.

Ionization cross-sections have been estimated by a Born approximation method similar to that used in X-ray excitation [11]. For an incident energy E_p less than the ionization potential E_I of the inner-shell electron, the cross-section is zero, but it increases rapidly for further increase of E_p to a maximum value at a ratio $E_p/E_I = 2$ to 3 and then decreases slowly [12] (Fig. 20.6).

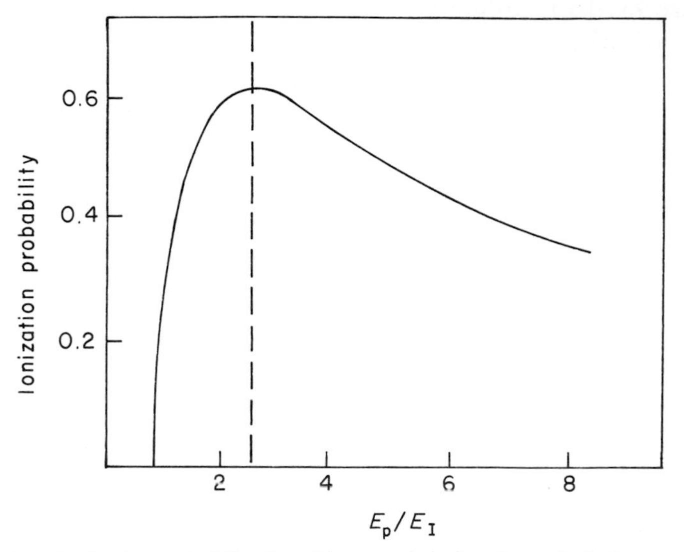

FIG. 20.6. Ionization probability (in arbitrary units) of an inner-shell electron as a function of the ratio E_p/E_I, where E_p is the energy of the primary electrons and E_I is the ionization energy of the inner-shell electron. The maximum value occurs at a ratio of *ca* 2·5.

The maximum cross-section for K-shell ionization of an adatom on a metal is around 10^{-19} cm^2 and, hence, for a monolayer of 10^{15} atoms cm^{-2} the ionization probability per incident electron is about 10^{-4}. The other factor affecting the Auger yield is the probability of filling the hole by an Auger transition; this term, however, approximates to unity, i.e. nearly all the holes produced by the incident radiation are filled by Auger electrons.

20.5. ESCAPE DEPTH

The peaks in the Auger energy spectrum are produced by Auger electrons that have been elastically scattered. They appear over an energy range of 10 to 1000 eV, but those with energies of several 100 eV are of primary interest, i.e. the energy of an Auger electron is usually one third to one quarter of that of the incident electrons. The attenuation of Auger electrons that are created only at a short distance from the surface determines the intensity of the peaks. Energy losses suffered by these electrons before emission are due to inelastic scattering processes. These include: (i) interband electron transitions from valence and inner energy levels to unoccupied higher states; (ii) plasmon losses, or collective excitation of free electrons in the conduction band due to their own mutual interaction and also with the incident electrons; and (iii) quasi-electric scattering involving phonon or lattice vibration excitation, which, however, merely effects slight heating of the solid.

The mean free path of an Auger electron is defined as the average distance travelled before it is inelastically scattered. The attenuation undoubtedly varies in the few outermost layers, but, for homogeneous layers, it can be approximated to

$$i_A = i_0(1 - \exp(-x/\lambda)), \tag{20.1}$$

where i_A is the Auger emission current, λ the mean free path of the Auger electron, or its escape depth for normal incidence of the primary beam, and x is the distance of a particular solid layer from the surface. The escape depth is, therefore, the thickness of a layer that reduces the number of electrons emitted by a factor of 1/e, and is small compared with the penetration depth of the primary beam.

The main factor that determines the escape depth is the energy of the Auger electrons. Cross-sections for losses by inelastic scattering in the Auger energy range are not generally available, but experimental values have been obtained by monitoring the intensity of the substrate Auger electron yield as a function of the thickness of a uniform overlayer of a foreign metal on the substrate. Low-temperature condensation involving sticking probabilities of around unity and electron deposition have been most commonly used. For example, for silver deposited on a gold substrate [6], the escape depth is between 4 and 8 Å for Auger electrons of energy 71 and 362 eV, respectively. In general, an appropriate value is 7 ± 2 Å for 500-eV electrons; the thickness then rapidly increases with increasing energy [13, 14]. Fortunately, many Auger peaks have energies of around 500 eV, where the variation of escape depth with energy is comparatively small.

For a clean surface, the Auger current decreases with distance x inwards along the normal to the surface according to $i_A - i_0 \exp(-x/x_0)$, where x_0 is the escape depth. The emission current, measured at an angle θ normal to the surface, is given by

$$i_A(\theta) = \text{const.} \int_0^x \exp(-x/\lambda \cos\theta)\, dx = \text{const.}\ \lambda \cos\theta,$$

where λ is the mean free path of the electrons. For chemisorption studies, the contribution of the metal substrate to the Auger current can, therefore, be minimized by using a glancing angle of incidence (Fig. 20.7). The roughness of the surface decreases the effectiveness of this procedure for $\theta > 80°$.

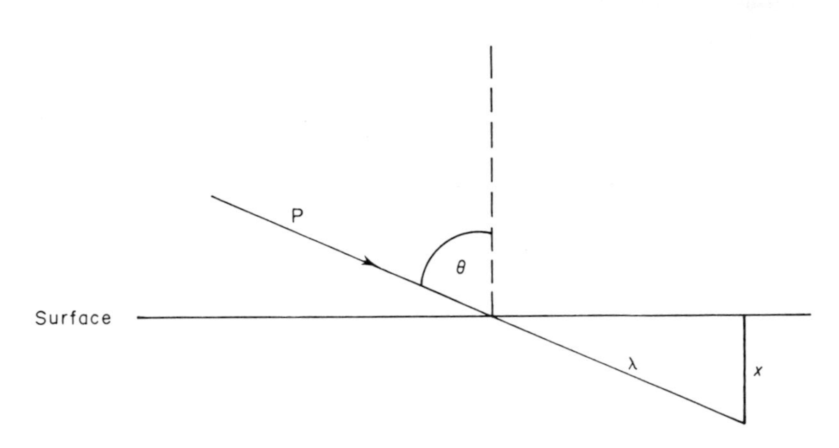

FIG. 20.7. Variation of the escape depth x of Auger electrons with an incidence angle θ of the primary electron beam P and a mean free path λ of the electrons in the sample.

20.6. BINDING ENERGY AND ENERGY OF AUGER SPECTRAL PEAK

For a KL_IL_{III} process, the energy released is $E_K - E_{LI}$, and the energy required to eject the KL_IL_{III} Auger electron is $E_{LIII} + \Phi_m$, where Φ_m is the work function of the metal. In a practice, see Fig. 20.8, the difference between the work function of the analyser grid Φ_A and that of the metal, i.e. $-(\Phi_A - \Phi_m)$, has to be included.

The quantity measured experimentally is the kinetic energy E_k of the emitted electron, corresponding to the Auger spectral peak energy; hence, to a first approximation.

$$E_k = E_K - E_{LI} - (E_{LIII} + \Phi_A). \tag{20.2}$$

The energy of the Auger peak gives the difference in binding energy of an electron in the inner-shell states involved in the Auger process (referenced to the Fermi level of the metal grid of the analyser). However, additional energy is expended to overcome the attractive force between the positively

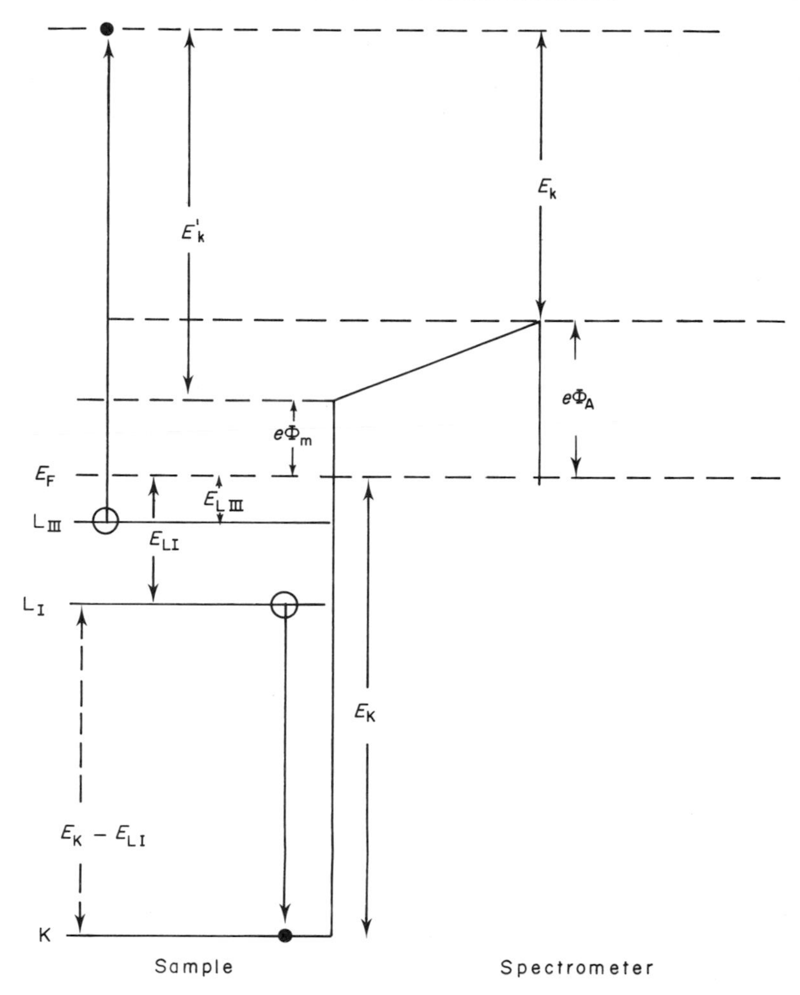

FIG. 20.8. Energy level diagram of Auger-electron emission. The energy released by filling a K core vacancy with an L_I electron is $E_K - E_{LI}$. The energy required to eject an L_{III} electron is $E_{LIII} + e\Phi_m$, where Φ_m is the work function of the metal sample. The energy of the Auger spectral peak E_k is the excess kinetic energy of the emitted L_{III} electron determined by the spectrometer with a grid analyser. Hence,

$$E_k = E'_k + e\Phi_m - e\Phi_A,$$

where Φ_A is the work function of the analyser and E'_k is the energy transferred to the L_{III} electron minus the energy to eject it from the metal, i.e.

$$E'_k = E_K - E_{LI} - (L_{III} + e\Phi_m).$$

From the diagram,

$$E_k = E'_k + e\Phi_m - e\Phi_A,$$

or

$$E_k = E_K - E_{LI} - E_{LIII} - e\Phi_A,$$

where E_k is of the order 10^2 eV and $\Phi_A \approx 5$ eV.

charged ion and the ejected electron, i.e. the electron binding energy is larger than that when it is bound to the parent neutral atom and approximates to that of a neutral atom of higher atomic number $Z + \Delta Z$, the correct E_{LIII} term being $E_{\mathrm{LIII}}(Z + \Delta Z)$, where ΔZ is probably closer to 0·5 than to unity. There are also relaxation energy contributions arising from rearrangement of the atom accompanying the initial ionization and the transition of the L_I electron to the K-shell vacancy.

The Auger spectrum gives only a fair approximation to the initial density of electron states; the results do not accord with those calculated from the delocalized three-dimensional band-structure of the solid. The reason is that the Auger process causes a strong localized disturbance in the emitter so that the **k** momentum vector is not a good quantum number. The experimental density of states derived from the Auger spectrum therefore refers to a local density of states modified by some type of selection rules.

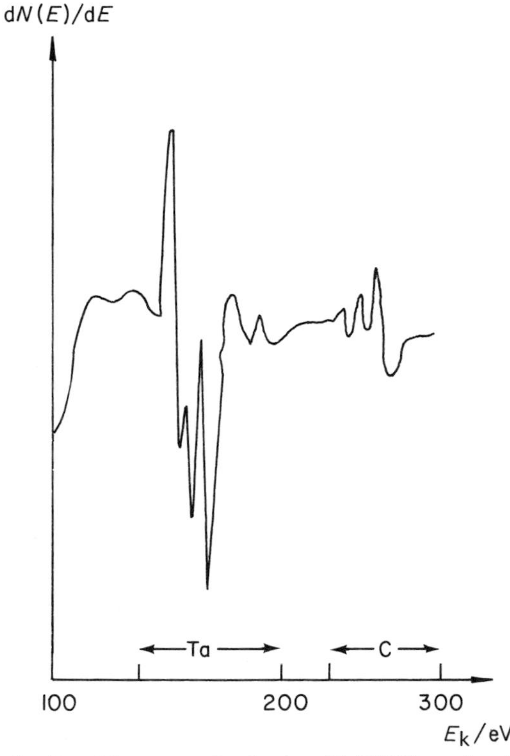

FIG. 20.9. First derivative of the energy distribution $dN/(E)/dE$ as a function of E_k for a monolayer of chemisorbed carbon monoxide on the (100) plane of tantalum at room temperature. The Auger peaks of carbon are close to those of tungsten carbide; this similarity probably indicates dissociative chemisorption of CO to C and O adatoms. The oxygen peaks do not appear below $E_k \sim 500$ eV.

20.7. APPLICATIONS TO CHEMISORPTION

20.7.1. Detection and identification of contaminants

From the viewpoint of chemisorption, the most important contribution of Auger spectroscopy (AES) is that it provides an excellent and rapidly executed criterion of the cleanliness of the surface of the metal absorbent. The presence of a contaminant comprising only 1% of a monolayer can be detected; AES is, therefore, generally more sensitive than LEED. Moreover, the chemical identification of foreign atoms on a surface is specific and definite. As already noted, an energy analysis of the emitted Auger electrons gives the difference in the electron binding energies in inner-shell states. These atomic states have well-established atomic-like orbitals and are highly characteristic of the element. Tabulated charts of Auger electron energies as a function of atomic number are available [15] and immediate identification of any impurity, except hydrogen and helium, can, therefore, be made. As an example, Fig. 20.9 is a spectrum of a chemisorbed overlayer of CO on tantalum; the Ta and C peaks are well separated and easily characterized.

Although some bulk metals are now available in a high state of purity (approaching 1 p.p.m. impurity), the segregation of trace impurities at the surface during annealing can be detected [16]. Carbon is particularly troublesome [17]; its removal by heating in oxygen at *ca* 1500 K is an effective procedure, but resegregation on cooling the metal from these high temperatures sometimes occur. Sulphur and oxygen are also common surface impurities in highly purified metals. One result of the use of the Auger technique is the discovery that certain LEED structures, previously assigned to the clean surface of the solid, are probably stabilized by small amounts of a surface impurity.

20.7.2. Determination of surface concentration

A quantitative assessment of the concentration of foreign surface species can also be made. Relative surface concentrations are related to the Auger current i_A by an equation of the form $i_A = Kn_i(1 + s)$, where n_i is the concentration of the ith adspecies, s is the back-scattering factor, usually constant but of unknown magnitude, and K is related to the ionization cross-sections, the values of which are again not available except semi-quantitatively. The Auger current is, therefore, proportional to the number of adspecies present on the surface, provided that complicating features, such as their partial incorporation in bulk or aggregation, are absent and that the operation conditions are maintained strictly constant. Peak heights are often used as a measure of the current but it should be confirmed that the band shape does not vary with coverage. Alternatively, the peak-to-peak amplitude in the $dN(E)/dE$ plot can be taken to be proportional to the surface concentration with the assumption that the band shape is Gaussian [19].† But an external calibration by an independent method is necessary to obtain an absolute value of concentration.

An example of the use of Auger spectra is the determination of the surface composition of binary alloys. This composition may deviate considerably from that in bulk; it is determined from the spectra by assuming that the peak intensities of the bands of the pure components can be linearly extrapolated to obtain their concentrations at the surface of the alloy [22, 23]. Unfortunately, electron impact may change the surface composition of some alloys (e.g. Cu/Ni), but with Pd/Ag alloys, the results are internally consistent and the composition of the surface is within 2% of that of the bulk.

There are also published reports on the use of Auger spectroscopy to follow the temporal changes of concentration of adspecies. Thus, the desorption kinetics and the evaluation of the isosteric heat of nitrogen on polycrystalline tungsten at 1150 to 1400 K have been studied by recording the nitrogen peak heights at 1-s intervals [24]. Other kinetic processes [25–27] that have been investigated are: the measurement of sticking probabilities, the rates of diffusion of adatoms into the bulk substrate, and the efficiency of electron-stimulated desorption. Catalytic studies using peak heights as a direct measure of surface concentrations during the reaction have also provided additional information on the kinetic mechanism of surface processes [24].

The main disadvantage of Auger spectroscopy is the probability of causing surface damage or change of surface concentration by electron impact and, hence, to obscure the interpretation of the results. When other means are available for the determination of surface concentrations with acceptable accuracy, these are preferred, particularly since external calibration of the spectra normally involves such procedures.

Nevertheless, the detection, identification and assessment of surface concentrations are major advantages. The sensitivity and speed of analysis, particularly in scanning the surface, are high. The technique poses few experimental difficulties, and both LEED and AES studies can be carried out on the same crystal surface in the same apparatus.

20.7.3. Chemical shifts

As a result of chemisorption, partial charge transfer occurs, e.g. from the surface metal atoms to an electro-negative adspecies. The binding energies of the electrons of the metal atom are, thereby, increased and the core-level

† See also a series of papers by Seah [19, 20] and Seah and Hondios [21].

energies are shifted; the energies of the emitted Auger electrons are, therefore, decreased and the spectral peaks are displaced (cf. Fig. 20.10).

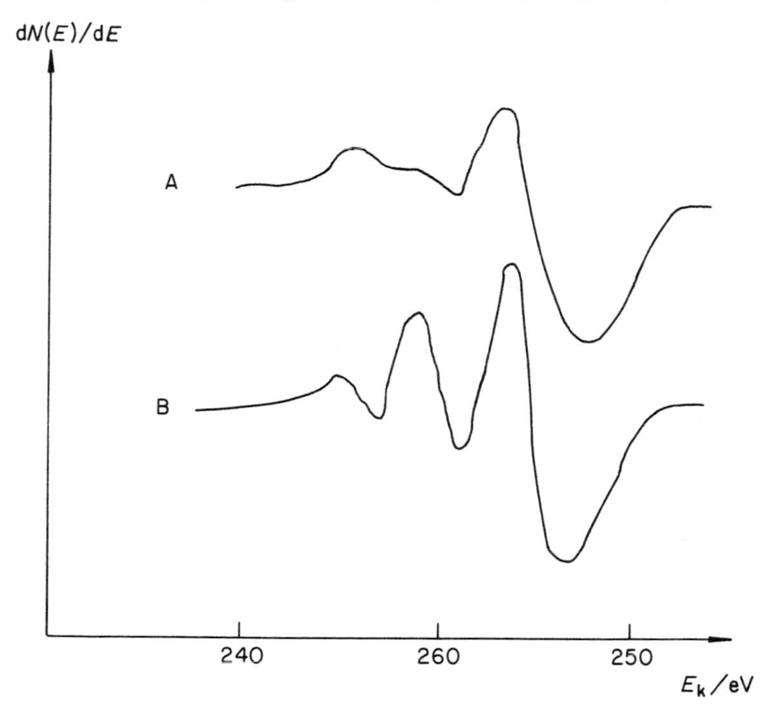

FIG. 20.10. The carbon KLL spectra for chemisorbed carbon monoxide on the (112) plane of tungsten (curve A) and tungsten carbide W_2C (curve B), shown as a first derivative plot of $dN(E)/dE$ against E_k. The positions of the main peaks are displaced and their shapes are modified to varying extents.

The shift of the Auger peak is larger the greater the transference of charge, but the relationship is complicated by crystal field effects. Moreover, since the wave functions of the valence electrons are modified by the change in chemical bonding, the transition probability of a specific Auger process is affected and causes a change of shape of the Auger band, particularly evident on the low-energy tail.

A simple interpretation [28] is provided by assuming that the inner-shell electron is within the charged shell of the valence electrons. The transfer of a metal valence electron to the chemisorbed species effects an increase of the inner-shell electron binding energy by an amount ΔE, given by

$$\Delta E \approx e^2\left(\frac{1}{r} - \frac{1}{R}\right) \tag{20.3}$$

in which e is the electronic charge, r is the radius of the valence shell and R is the internuclear distance between the atoms concerned. An Auger shift corresponding to this energy change should result. However, the calculated ΔE value is, as expected, much too high, since chemisorption normally involves only partial electron transfer. Quantum mechanical procedures using Pauling's electro-negativities to estimate the degree of ionicity (or charge transfer) give greatly improved values and a rough parallelism between chemical shift and extent of transfer is then evident.

Since Auger emission involves three electronic levels, the interpretation of chemical shifts is difficult. However, some qualitative information has been reported. For the chemisorption of oxygen on some transition metals, the shift of the Auger electrons of the metal is to lower voltages, reaching a maximum of 4 to 6 eV at monolayer coverage and indicating an increasing positive charge on the metal atoms [29, 30]. Surprisingly, no shift to higher voltages of the oxygen Auger spectrum with increasing coverage takes place. For carbon monoxide as absorbate, a much smaller transference of negative charge from the metal is observed, being 0·5 to 1·0 eV at monolayer coverage on tantalum and even less on tungsten. As yet, the main value of Auger shifts has been to follow the various stages from chemisorption to surface and bulk reactions as in the oxidation of metals [31–33], and in the formation of carbide following adsorption of carbon monoxide (Fig. 20.10). Photon excitation and high-resolution analysers will undoubtedly be increasingly employed in future investigations, thereby providing more detailed results that may be susceptible to more quantitative interpretation.

REFERENCES

1. P. Auger, *J. Phys. Radium*, 1925, **6**, 205.
2. J. J. Lander, *Phys. Rev.*, 1953, **91**.
3. K. Siegbahn, "ESCA Applied to Free Molecules", North Holland, Amsterdam, 1969.
3. L. N. Tharp and E. J. Scheibner, *J. Appl. Phys.*, 1967, **38**, 4355.
4. L. A. Harris, *J. Appl. Phys.*, 1968, **39**, 1419 and 1428.
5. R. E. Weber and W. T. Peria, *J. Appl. Phys.*, 1967, **38**, 4355.
6. P. W. Palmberg and T. N. Rhodin, *J. Appl. Phys.*, 1968, **39**, 2425.
7. P. W. Palmberg, *Appl. Phys. Letters*, 1968, **13**, 183.
8. C. C. Chang, *Surface Sci.*, 1971, **25**, 53.
9. P. W. Palmberg, G. K. Bohn and J. C. Tracy, *Appl. Phys. Letters*, 1969, **15**, 254.
10. P. J. Bassett, *J. Phys. E*, 1974, **7**, 461.
11. C. R. Worthington and S. G. Tomlin, *Proc. Phys. Soc. A*, 1956, **69**, 401.
12. H. E. Bishop and J. C. Rivière, *J. Appl. Phys.*, 1969, **40**, 1740.
13. C. J. Powell, *Surface Sci.*, 1974, **44**, 29.
14. I. Lindau and W. E. Spicer, *J. Electron Spectr.*, 1974, **3**, 409.

15. Y. Strausser and J. J. Uebbing, "Varian Chart of Auger Electron Energies", Varian Corp., Palo Alto, Calif., 1970.
16. T. W. Haas, J. T. Grant and G. J. Dooley, *J. Vac. Sci. Tech.*, 1970, **7**, 43.
17. R. W. Joyner, J. Rickman and M. W. Roberts, *Surface Sci.*, 1973, **39**, 445.
18. P. B. Needham, T. J. Driscoll and N. G. Rao, *Appl. Phys. Letters*, 1972, **21**, 502.
19. M. P. Seah, *Surface Sci.*, 1972, **32**, 703; 1973, **40**, 595.
20. M. P. Seah, *J. Phys. F*, 1973, **3**, 158.
21. M. P. Seah and E. D. Hondios, *Proc. Roy. Soc. A*, 1973, **335**, 191.
22. G. Ertl and J. Küppers, *Surface Sci.*, 1971, **24**, 104.
23. G. A. Somorjai and S. H. Overburu, *Faraday Disc. Chem. Soc.*, 1975, **60** (for Au/Ag and Pb/In alloys).
24. R. W. Joyner, J. Rickman and M. W. Roberts, *J. Chem. Soc., Faraday Trans. I*, 1974, **70**, 1825.
25. H. P. Bonzel, *Surface Sci.*, 1971, **27**, 387.
26. H. P. Bonzel and R. Ku, *J. Chem. Phys.*, 1973, **58**, 4617; **59**, 1641.
27. D. A. Shirley, *Phys. Rev. A*, 1973, **7**, 1520.
28. C. R. Brundle, *in* "Surface and Defect Properties of Solids" (M. W. Roberts and J. M. Thomas, eds), The Chemical Society, London, 1972, Vol. 1, p. 194.
29. R. J. Fortner and R. G. Musket, *Surface Sci.*, 1971, **28**, 504.
30. F. J. Szalkowski and G. A. Somorjai, *J. Chem. Phys.*, 1972, **23**, 271.
31. G. Schön, *J. Electron Spectr.*, 1973, **2**, 75.
32. A. P. Janssen, R. C. Schoomaker, A. Chambers and M. Prulton, *Surface Sci.*, 1974, **45**, 45.
33. J. E. Castle and D. Epler, *Proc. Roy. Soc. A*, 1974, **339**, 49.

21

X-Ray Photoelectron Spectroscopy

21.1. GENERAL PRINCIPLES [1]

The metal surface is irradiated by an X-ray beam comprising the K_α radiation of Mg(1254 eV) or of Al(1487 eV); if the incident energy $h\nu$ exceeds the binding energy of an inner-shell electron, the electron is excited and is subsequently photoemitted from the surface with a range of kinetic energies, the distribution of which is determined by using the type of analysers and collectors employed in Auger spectroscopy. The photoemission involves one electronic level only, and the kinetic energy E_k of the ejected core electron is related to its binding energy E_B according to Eq. (21.1):

$$E_k = h\nu - E_B - \Phi_A, \tag{21.1}$$

where Φ_A is the work function of the analyser and E_B is referenced to the Fermi energy E_F of the metal (Fig. 21.1). The surface sensitivity of this technique is again due to the strong inelastic scattering undergone by the ejected electron. The escape depth of 20 to 30 Å may be greater [2] than that usually encountered in Auger spectroscopy, i.e. the measured signal may include a greater contribution from layers below the outermost layer.

For binding energy determinations, the primary beam is generally not monochromatized and the resolution is limited by the natural width (~ 1 eV) of the K_α radiation. However, for measurements of chemical shifts, a monochromatic X-ray source giving a peak resolution of <0.3 eV is necessary [3].

By comparison with electron impact Auger spectroscopy, ionization by

X-ray excitation is more efficient than that by electron bombardment, since the capture cross-section for X-rays is substantially larger; the signal-to-noise ratio is also better because of the less copious emission of secondary electrons that contribute to the large background in Auger spectra. Moreover, surface damage by X-ray irradiation is smaller.

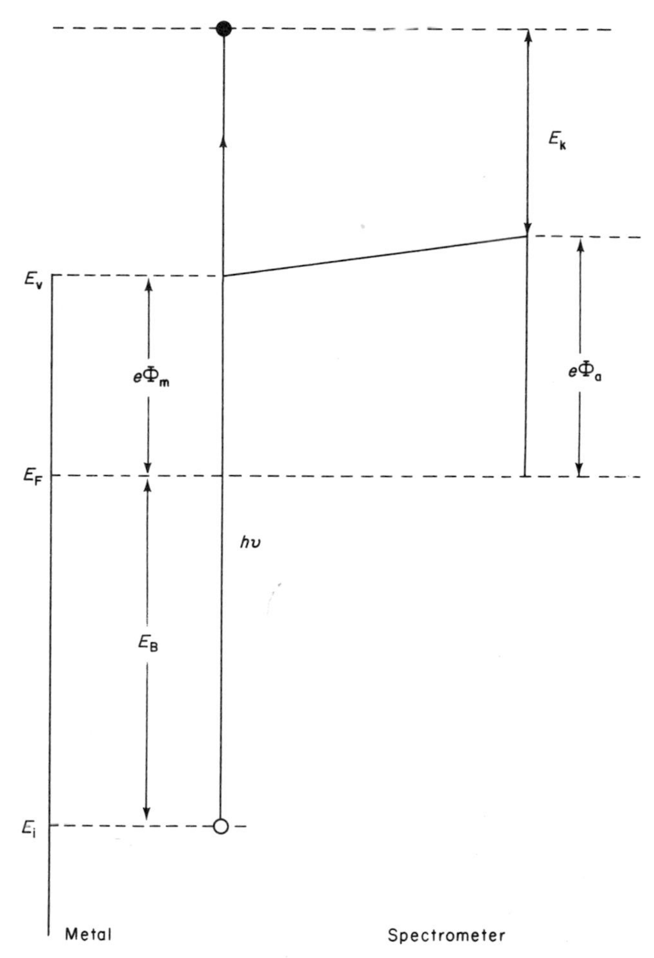

FIG. 21.1. Energy level diagram of an X-ray photoelectron spectrometer.
The photon energy of the X-rays, $h\nu$, is absorbed by an inner-shell electron. E_F is the Fermi energy level of the metal and $e\Phi_m$ is its work function; E_v is the vacuum level; E_k is the kinetic energy of an ejected inner shell electron; and $e\Phi_a$ is the work function of the analyser. Hence,

$$E_k = h\nu - e\Phi_a - E_B,$$

where E_B is the binding energy of the ejected electron, referenced to E_F.

The technique as applied to free molecules was developed by Siegbahn [1] following improvements of the resolution and sensitivity of β spectrometers and given the name, electron spectroscopy for chemical analysis (ESCA). However, its application to surface analysis has not been extensive because it is much less sensitive than Auger spectroscopy and requires greater experimental refinement. Moreover, LEED/Auger systems in which the LEED grid system can be adapted to detect and examine electron emission are readily available; a further advantage is the ease of generation and focusing of high-intensity electron beams; consequently, the strong signals obtained allow rapid scanning and analysis of surface composition.

XPS can be used to investigate the density of occupied states in the valence bands of metals [4]. But, since the valence electrons are ejected with high kinetic energies (>1 keV), their escape depth is >10 Å^2, and the technique is no longer surface sensitive.

21.2. APPLICATIONS TO CHEMISORPTION

Information about chemisorbed species has been obtained from measurements of the chemical shifts brought about by the change of surface chemical environment. The displacements are more readily observed and their interpretation is more susceptible to theoretical analysis than the corresponding shifts in Auger spectra, because the photoemission involves one electronic energy level instead of transitions between three Auger levels. Quite small shifts are measurable since both $h\nu$ and Φ_A can be maintained constant during the measurements. As an example, three different states have been distinguished for carbon monoxide chemisorbed on molybdenum and tungsten [15].

TABLE 21.1

	Mo			W		
	Mo(3*d*)	O(1*s*)	C(1*s*)	W(4*f*)	O(1*s*)	C(1*s*)
β	277·7	530·0	282·7	30·7	529·9	282·6
Virgin	277·7	531·2	284·6	30·7	530·9	284·8
γ	277·7	534	288	30·7	533	287

In Table 21.1, the various binding energies (eV) referenced to the Fermi level of the metal were measured with an accuracy of ± 0·2 eV. It is evident that the lower the heat of adsorption (γ < virgin < β), the higher is the binding energy of the adsorbate core-levels. The resolution for distinguishing states of similar energies is, however, insufficient and more effective X-ray monochromatization would be necessary to detect the substates visible in thermal desorption spectra. The unexpected constancy of the Mo(3*d*) and W(4*f*) binding energies in Table 21.1 suggests that the electron transfer accompanying the formation of the chemisorbed bond may extend to some depth in the metal substrate; and the comparable variation of both the O(1*s*) and the C(1*s*) binding energies may be a consequence of strong interaction of the adjacent C and O atoms at the surface, i.e. the β state is present as a molecular adspecies [5].

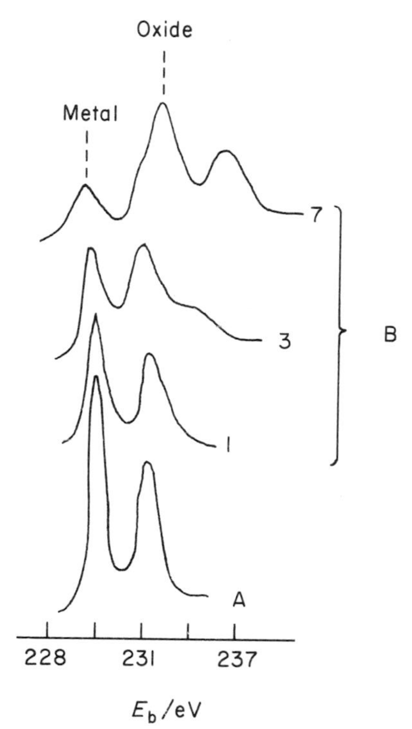

FIG. 21.2. XPS spectra of clean molybdenum (3*d* orbital) (curve A) and after increasing uptakes of oxygen, (curves B). The number beside each curve indicates the number of "monolayers" of oxygen that has reacted with the metal.

XPS has also been frequently employed to investigate the formation of oxides when metal surfaces react with oxygen, since much larger shifts (up to 10 eV) are observed [5, 6] (Fig. 21.2). Nevertheless ultra-violet spectroscopy has proved to be more informative for studying chemisorbed layers, but XPS may well play a most effective role in the future.

REFERENCES

1. K. Siegban, C. Nordling, A. Fahlman, R. Norderg, K. Hamrin, J. Hedman, G. Johansson, T. Bergmark, E. Karlsson, J. Lindgren and B. Lindberg, "ESCA, Atomic, Molecular and Solid State Structure by Means of Electron Spectroscopy", Almquist and Wilksells, Uppsala, 1967.
2. M. Klasson, J. Hedman, A. Berndtson, R. Nilsson and C. Nordling, *Physica Scripta*, 1972, **5**, 93.
3. K. Siegbahn, D. Hammond, H. Felner-Feldegg and E. F. Barnett, *Science*, 1972, **176**, 245.
4. D. A. Shirley (ed.), "Electron Spectroscopy (Proceedings of an International Conference at Asilomar, 1971)", North Holland, Amsterdam, 1972.
5. C. R. Brundle, *Surface Sci.*, 1975, **48**, 99.
6. K. S. Kim and N. Winograd, *Surface Sci.*, 1974, **43**, 625.

22

Ultra-violet Photoelectron Spectroscopy

22.1. GENERAL PRINCIPLES

The first successful application of ultra-violet photoelectron spectroscopy to a chemisorbed overlayer on a metal surface was reported by Eastman and Cashion [1, 2] in 1971, although the technique had been employed for many years to determine molecular ionization potentials of gaseous molecules [3]. It is now extensively used in detailed studies of the properties of adspecies and the nature of the chemisorption bond; the method represents a major experimental advance in surface chemistry investigations.

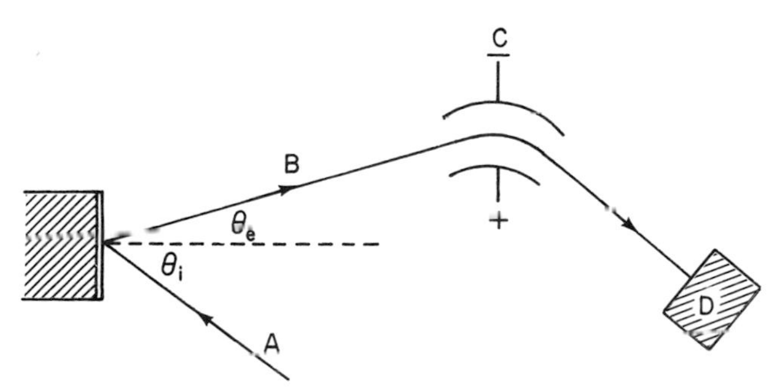

FIG. 22.1. Experimental arrangement for ultra-violet photoelectron spectroscopic investigation of a chemisorbed layer on a metal surface. A, Photon source, He(I) (21·2 eV) at incident angle θ_i to normal of surface with chemisorbed layer; B, emission of electrons with kinetic energy E_k at emission angle θ_e; C, electrostatic analyser; D, photomultiplier and detector of number of electrons s^{-1} $N(E_k, \theta_i, \theta_e, h\nu)$, or $N(E_k)$ averaged over all angles for a single photon source.

The chemisorbed layer on a metal adsorbent is irradiated with monochromatic ultra-violet light of photon energy sufficiently large to bring about excitation of some of the valence electrons of the surface complex and their subsequent photoemission from the surface region with a distribution of kinetic energies (Fig. 22.1). The form of the distribution curve is determined with the aid of a cylindrical electrostatic, or retarding field,

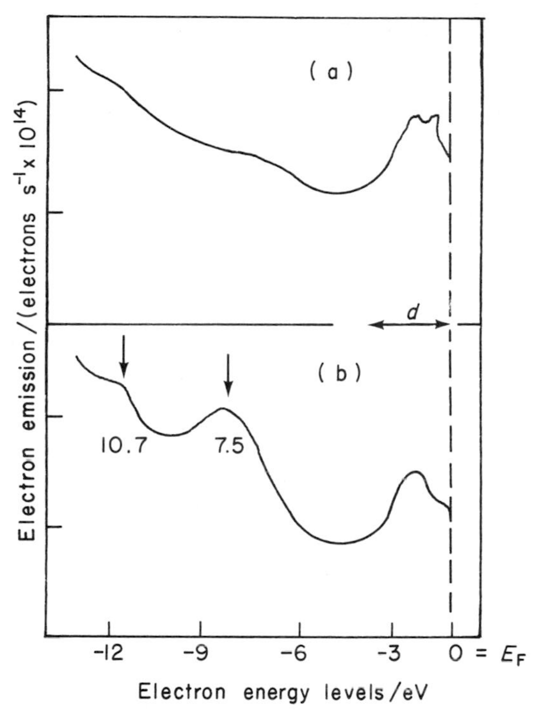

FIG. 22.2. He(I) UPS spectrum of (a) clean polycrystalline Ni; and (b) after chemisorption of 7 Langmuirs of CO. Energies of peak positions for adsorbed CO are 7·5 eV and 10·7 eV below $E_F = 0$. The position of the Ni *d* bands is indicated at *d*.

analyser and a detector of the type that had been largely developed in Auger spectroscopy. The UPS spectrum is a plot of the photoemission intensity (usually as number of electrons per second in a kinetic energy range E_k – ($E_k + dE_k$) as a function of the kinetic energy E_k of the electrons (Fig. 22.2).

In the early work, the primary photon beam was generated in a gas discharge tube by applying a voltage of 3 to 15 kV across two metal electrodes in the presence of a low pressure of hydrogen. The required wavelength was selected by an ultra-violet monochromator. The spectrometer was fitted with lithium fluoride windows which restricted transmission to photons with energies less than 11·6 eV. In current investigations, windowless spectrometers [4]† are invariably employed. They are connected via a capillary to an inert-gas discharge tube and to differential pumping systems which provide an ultra-high vacuum in the spectrometer and, simultaneously, a constant gas pressure of *ca* 5 Pa in the discharge tube. The resonance lines, He(I) (21·2 eV) and He(II) (40·8 eV), and less frequently Ne(I) (16.7 eV) and Ne(II) (26·9 eV), provide the primary irradiation beam.

22.2 ESCAPE DEPTH OF PHOTO-EMITTED ELECTRONS

In a metal, the attenuation length of the incident ultra-violet light is of order 10^2 Å, but the excited electrons undergo strong inelastic scattering and only those generated within a depth of *ca* 5 Å in the metal are photo-emitted [6].

It is not possible to extract the contribution of the outermost layer to the total emission, because the variation of electron density with distance inwards from the surface and the effect of interference between volume and surface emission are uncertain. Some theoretical estimates for an ideal free-electron metal and an incident beam at glancing angle ($\sim 10°$) with the surface suggest that virtually all of the emitted electrons originate from the surface layer. In practice, the substrate is usually a transition metal, and probably only 40 to 50% of the measured signal is generated by the outermost layer.

22.3. INTENSITY OF SPECTRAL PEAKS

The probability of effecting ionization by the primary beam increases roughly as the third power of the photon energy. However, the cross-section for photo-ionization varies with the size of the orbital such that its maximum value moves to higher energies with decrease of orbital size. Thus, for low incident energy, an orbital of small ionization cross-section may not give rise to a spectral peak that is quite strongly developed at higher photon energies. Similarly, the escape depth of emitted electrons varies with the energy of the incident beam, and variation in peak intensities result. Figure 22.3 shows an example of the different spectra obtained using He(I) and He(II) radiation.

An additional factor is that chemisorbed molecules are invariably preferentially oriented at the adsorbent surface (in contrast to the random orientation of free gas molecules), thereby causing the photo-emission

† See also Eastman *et al.* [5].

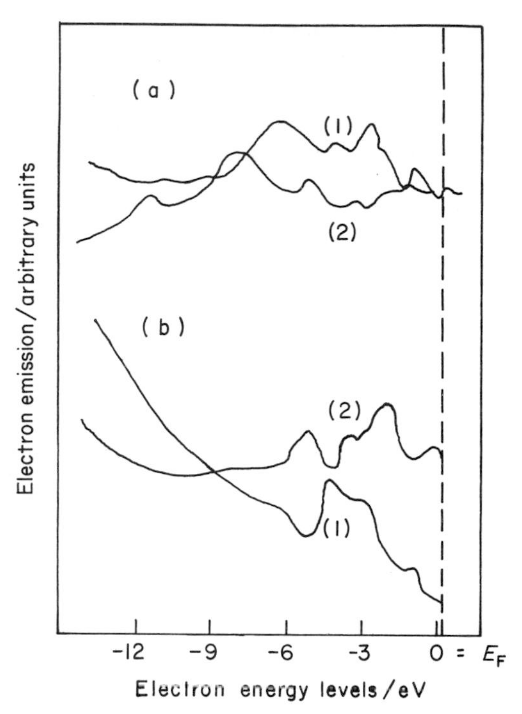

FIG. 22.3. Change of spectral features with primary energies. Top curves (a), $h\nu = 21{\cdot}2$ eV; lower curves (b) $h\nu = 40{\cdot}8$ eV. Curves (1) refer to a clean W(110), and curves (2) are difference spectra obtained after adsorption of 10 Langmuir of CO.

intensity to bc angular-dependent. For localized adsorbate levels, angular-resolved surface emission can, therefore, provide information about the symmetry of both the adsorption site and the bonding orbitals [7, 8]. The theoretical aspects of this process are being developed and experimental data are now being acquired, but the interpretation of the results is complicated by other factors that affect the energy distribution of the emitted electrons [7–9].

22.4. REFERENCE ENERGY LEVEL

The photoelectron binding energy in a free molecule is derived from the energy of the spectral peak measured with respect to zero energy of the emitted electron at rest at infinity (the vacuum level). For a molecule chemisorbed on a metal, the reference state is the Fermi energy of the metal ($E_F = 0$), since the photoemitted electron is raised from the Fermi level and then has to surmount the electrostatic barrier associated with the double layer of charge at the metal surface (the surface potential). The barrier height varies with the atomic density of the ion cores at the surfaces of the different crystal planes of the metal and is also modified by the nature and

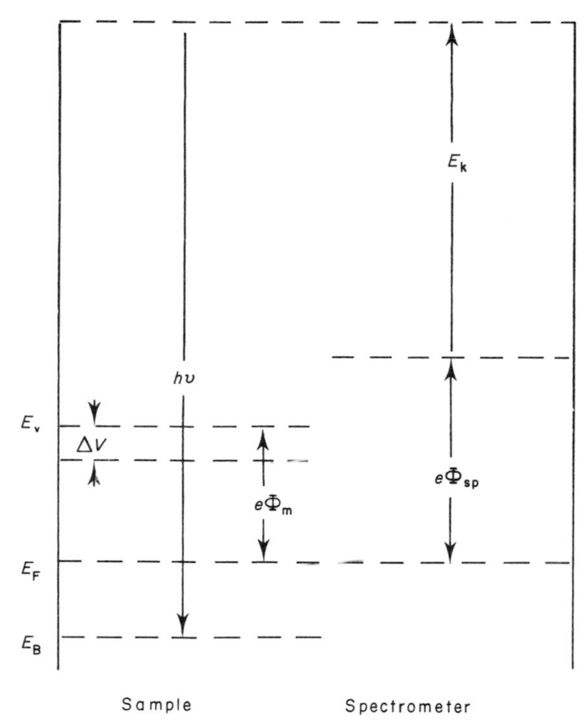

FIG. 22.4. A valence electron of the surface molecule in an energy level E_B below the Fermi level E_F absorbs the incident photon energy $h\nu$. The work function of the metal is $e\Phi_m$ and that of the spectrometer is $e\Phi_{sp}$; ΔV is the surface electrostatic barrier or surface potential. The kinetic energy of the ejected electron is E_k and its binding energy in the surface molecule is E_B referenced to the Fermi level. Hence,

$$E_B = h\nu - e\Phi_{sp} - E_k = E_B^g - E_F,$$

where E_B^g is the binding energy referenced to the vacuum level E_v.

concentration of the adsorbate. There is still controversy about the correct procedure for transformation of peak energies of a chemisorbed molecule to those of the corresponding peaks of the free molecule; usually, the effective work function of the metal/adsorbate system is added to the chemisorption peak energy, since an experimentally accessible reference level is not available (Fig. 22.4).

22.5. RELAXATION ENERGY

The experimentally-determined quantity is the kinetic energy E_k of the emitted electron from a particular orbital. For the jth orbital

$$E_k = h\nu + E_0^N - E_{je}^{N-1}, \tag{22.1}$$

where $h\nu$ is the known incident photon energy, E_0^N is the energy of the electron in the initial ground state of the neutral chemisorbed adspecies with its full complement of N electrons, whereas E_{je}^{N-1} is the energy of the final excited state comprising an electron vacancy in the jth orbital together with the remaining $N - 1$ electrons in orbitals which are not identical with the corresponding orbitals in the initial groundstate. According to Koopmans' theorem of frozen orbitals [10], the $N - 1$ orbitals are exactly the same in the final excited state as in the initial ground state; in a Hartree–Fock calculation (for a one-electron state) of the orbital energy E_j' this theorem is assumed to be valid.

The experimental binding energy [see Eq. (22.1)]

$$E_{Bj} = h\nu - E_k = E_{je}^{N-1} - E_0^N, \tag{22.2}$$

however, cannot be equated to the orbital energy of the initial state, E_j', because the $N - 1$ orbitals of the final excited state relax towards the positive electron vacancy in the jth orbital. The energy change accompanying the relaxation of the excited state is termed the relaxation energy E_R. The orbital energy of the jth electron in the initial neutral atom is, thererore, $E_j' = E_{Bj} + E_R$, i.e. the experimental binding energy $E_{Bj} = h\nu - E_k$ is greater than the orbital energy by an amount equal to the relaxation energy. For a free molecule, the relaxation energy can be evaluated theoretically with fair accuracy, but for a chemisorbed molcule there is an additional surface-induced relaxation term arising from the relaxation of the free electrons of the metal towards the positive hole on the adsorbate.

An approximate treatment of this problem using a Born–Haber cycle has been proposed [11]. The difference in binding energy of the jth electron in the free molecule (E_{Bj}) and in the surface complex (E_{Bj}^s) is put equal to the difference in binding energy of this electron to the surface ion formed by removal of the jth electron (E_j^{ion}) and its binding energy (E_j^{mol}) to the surface neutral molecule in its ground state, i.e.

$$E_{Bj} - E_{Bj}^s = E_{js}^{ion} - E_j^{mol}. \tag{22.3}$$

The total energy shift, given by $E_{Bj} - E^{s}_{Bj}$, includes a chemical shift (since the energy levels of the adsorbate are shifted by the act of chemisorption) and the difference of the relaxation energy terms for the free and chemisorbed molecule. The following assumptions are then made: (i) the relaxation of electrons towards the positive hole of the adsorbate does not alter the bonding energy of the neutral adsorbate; (ii) the relaxation energy of the outer electrons of the adsorbate is the same as that for the free molecule; and (iii) the orbital under consideration is not involved in bonding, nor is its energy shifted by bonding of other orbitals. In these circumstances, the energy difference $E_{Bj} - E^{s}_{Bj}$ arises from the screening of the positive hole by the free electron of the metal; the bonding energy of the ion, therefore, equals the sum of the bonding energy of the molecule and the energy arising from the screening process.

A calculation of the energy change when a proton is brought up to a metal surface has been carried out using the jellium model for a metal with a high free-electron density [12]. The maximum change is 5·7 eV at 1·5 Å increasing to 8·7 eV at 0·5 Å. The energy change decreases as the size of the positive hole is made larger and becomes zero for complete delocalization of the hole.

22.6. OPTICAL DENSITY OF STATES

The UPS spectrum which reflects the distribution of kinetic energies of photo-emitted electrons is not an exact representation of the density of states. This (optical) density of states (ODS) is, however, directly related to the energy distribution curve (EDC) when the excitation process involves an indirect electronic transition for which the Bloch wavevector **k** is not conserved. In practice, direct transitions with **k** conservation usually take place and the EDC representation should be transformed to an ODS one.

The density of states N(E) is obtained by integrating over all states having different **k** vectors within an energy band and summing these over all bands. The number of electrons excited from an energy level E_i to a higher state E_j is proportional to the product of the density of the filled states N_i, the density N_j of the unoccupied states at E_j, and the transition probability from E_i to E_j levels. The total optical density of states is the summation over all excitations from E_i to E_j for which **k**-momentum is conserved. The final states having energies E are then selected to provide the internal distribution. To refine the correlation of the ODS with the measured EDC, an escape probability can be introduced to take account of the fact that the electron has to overcome the surface electrostatic barrier (i.e. its momentum normal to the surface must exceed a critical value). This refinement, however, is only significant for low photon energies. Finally, the transition probability of excitation should be included; but again this term is generally neglected so that the calculated ODS represents a joint density of both occupied and unoccupied states. Changes of special features brought about by variation of the incident photon energy are then attributed to changes in the unoccupied band density. Fortunately, for excitation energies above 20 eV, the variations are sufficiently small and the EDC plot consequently provides an acceptably good representation of the density of occupied states.

22.7. APPLICATIONS TO CHEMISORPTION

22.7.1. Evaluation of bonding energies

The energy difference between the spectral peaks arising from the ejection of an electron from the jth orbital of the free molecule (E_{Bj}) and that of the chemisorbed molecule (E^s_{Bj}) is given by

$$E_{Bj} - E^s_{Bj} = \Delta E_B + \Delta E_R + \Phi_m. \tag{22.4}$$

The effective work function Φ_m of the metal adsorbent is included in order to transform E^s_{Bj} referenced to the Fermi level to the vacuum level. The term ΔE_B is the difference between the binding energy of the jth electron in the ground state of the free and chemisorbed molecule, and ΔE_R is the difference of the relaxation energies of the excited states of the free and chemisorbed molecule, the contribution being larger for the latter.

The following procedure provides an experimental estimate of ΔE_R [13, 14]. For various small molecules that undergo physisorption (but not chemisorption) on metals at 77 K, the binding energy difference ΔE_B is assumed to be zero. The energy separation of a spectral peak in the UPS spectrum of the molecule in the free physisorbed state is, therefore, equal to ΔE_R. It is roughly constant with a value between 1 and 2 eV irrespective of the chemical nature of the adsorbate molecules, and is the same for each orbital of the molecule. For chemisorption, ΔE_B is assumed to be zero for all orbitals that do not take part in the bonding of the adsorbate to to the surface, but their corresponding relaxation energy $\Delta E'_R$ is somewhat larger, i.e. their energy levels undergo an additional energy shift. But for an orbital that does take part in surface bonding there is a larger difference in the magnitude of the energy displacement, given by $\Delta E_B + \Delta E'_R$; consequently, ΔE_B can be separately evaluated.

22.7.2. Multiplet states

The earlier UPS investigations were largely concerned with the chemisorption of CO on various metals. Most of the data has been summarized in Table 22.1 [1, 15–22].† Although the peak maxima obtained using different crystal planes and polycrystalline specimens of the same metal have slightly different energies, the variation is about the same as that recorded by various workers for the same adsorption system; consequently, only averaged values have been listed. For comparison with the energies of the spectral peaks of gaseous adsorbate, an average value of Φ_m for each metal and a rough estimate of ΔE_R of 1·4 eV have been included.

TABLE 22.1. Spectral peak energies[a] of two main bands for CO chemisorbed on various metals

Metal	First peak/eV	Second peak/eV	Energy difference/eV
Ni	7·8 (14·1)	11·0 (17·3)	3·2
Ru	7·6 (13·6)	10·7 (16·6)	3·1
Pt	9·0 (15·8)	11·6 (18·4)	2·7
Pd	7·9 (14·3)	10·8 (17·2)	2·9
Cu	7·6 (13·5)	10·7 (16·6)	3·1
Mo	7·3 (12·9)	10·7 (16·3)	3·4
W	8·4 (13·2)	11·5 (17·4)	3·1
Gas	13·9	16·8	2·9

[a] Values are referenced to the respective Fermi levels of the metals; those in parentheses are with respect to vacuum level by inclusion of the metal work function and a constant relaxation energy (~1·4 eV).

The spectrum of gaseous CO comprises three main bands, each with fine structure, with peak energies around 13·9, 16·8 and 19·6 eV; that of chemisorbed CO displays two prominent bands corresponding to the displacement of the two lower-energy peaks of the free molecule (Fig. 22.2). The peak energies of the adspecies are closely similar in magnitude and the energy differences between the maxima of the chemisorption peaks and those of the two lower gaseous peaks are almost the same. It may be concluded that: (i) the type of bonding of the CO molecule; and (ii) the two bonding orbitals involved, are the same irrespective of the different electron configurations of the metal adsorbents.

Thermal desorption spectra of CO chemisorbed on most transitions provide proof of the presence of two different states,‡ the α-state being less

† For the spectrum of gaseous CO, see Price *et al.* [23].
‡ See Chapter 5.

strongly bonded to the surface than the β state. The existence of these two states is evident in the UPS spectrum under favourable conditions (Fig. 22.5).

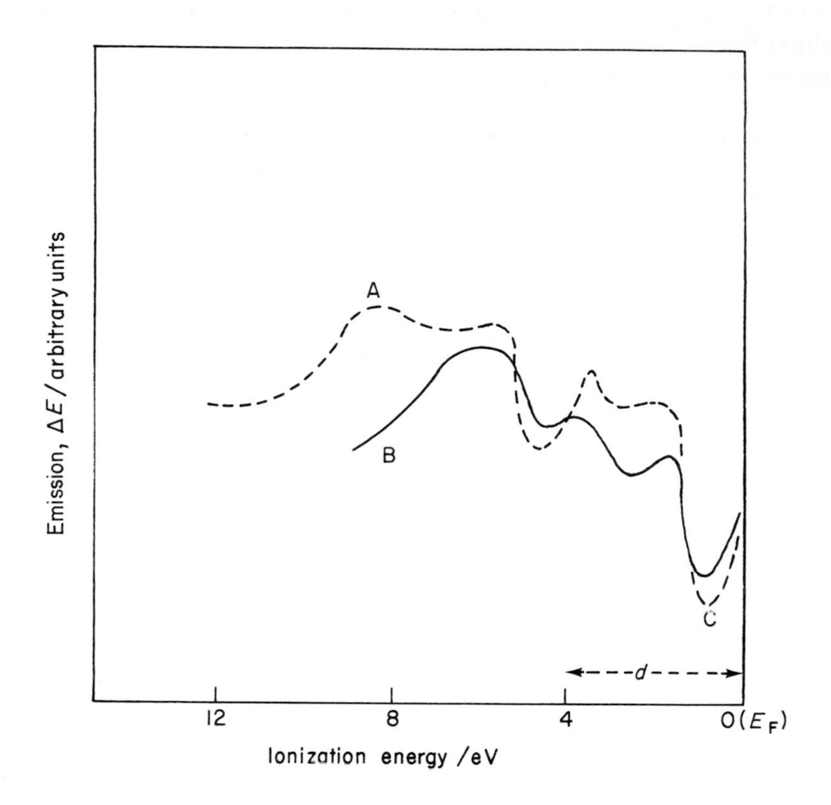

FIG. 22.5. Photoelectron spectrum of CO on W(100) as a difference plot [i.e. spectrum of α-CO − spectrum clean W(100)]. Incident radiation $h\nu = 21.2$ eV; reference level $E_F = 0$. A, α-CO spectrum from CO monolayer at room temperature with peak at 8·3 eV; B, β-CO spectrum following thermal desorption of α-CO state; new peak at 6·2 eV; C, surface state on W(100) at 0·4 eV; *d*, width of *d* band.

The spectrum of a weakly bound α state formed at low temperature on the (100) plane of tungsten displays a peak at 8·3 eV, which corresponds to the gas-phase 1π peak and indicates that the adspecies retains its molecular state. At 300 K, some desorption takes place and the 1π peak decreases in intensity (see Fig. 22.6). At 700 K, the 1π level is absent, and a new peak appears at 11·4 eV that indicates 4σ or 5σ bonding of CO to the metal. The β state at 1100 K undergoes dissociation [20, 24, 25], since its spectrum may be simulated by a linear sum of the photoelectron spectrum of C and O adatoms both of which have a c(2 × 2) structure. Moreover, the peaks obtained at 1·4 and 6·4 eV correspond to W—O levels, and those at 3·1 and 5·1 eV to C—W levels. Similar results have been obtained for the CO/molybdenum system.

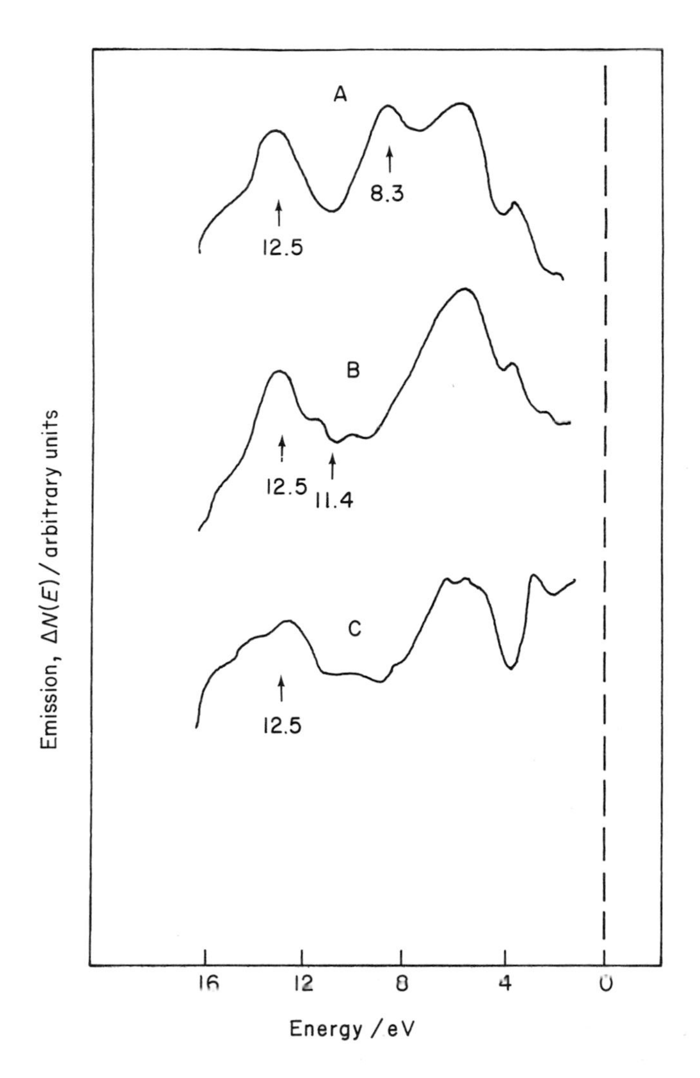

FIG. 22.6. Difference spectra of carbon monoxide chemisorbed on W(100); $E_p = 21{\cdot}2$ eV. Curve A, CO chemisorbed on clean W(100) at 300 K; curve B, after heating A to 700 K; curve C, after heating A to 1100 K.

22.7.3. Assignment of peaks to molecular orbitals

As noted above, there is agreement between the experimental results obtained by different workers for many CO/metal systems, but the assignment of the peaks to particular molecular orbitals is subject to controversy.

The two main peaks denoting binding energies of around 14 and 17 eV were originally assigned [1, 2] to the 5σ and 1π levels of gaseous carbon monoxide. The absence of a peak attributable to its 4σ orbital [a result later confirmed in the He(II) spectrum] was, however, unexplained. Theoretical calculations of the structure of metal carbonyls in which the C atom is bonded to a metal atom, however, suggested that this assignment should be inverted [26]. Furthermore, an experimental determination of the relaxation energy of the re-assigned CO 1π orbital agreed with an approximate theoretical value of 2 to 3 eV; whereas, for the original assignment, there would be virtually no relaxation of either the 5σ or the 1π levels [27–29], i.e. a relaxation contribution of the correct magnitude can only be obtained by a reversal of the original assignment. It is, therefore, probable that the lower energy band at 14 eV arises from a mixture of 5σ and 1π orbitals and that the higher spectral peak should be assigned to a 4σ orbital. An alternative assignment of a 1π orbital and a mixture of 4σ and 5σ orbitals, respectively, is less likely from consideration of the relative intensities of the experimental maxima. It is evident that definite assignments must await a more refined treatment of photoeffects in adsorbed layers.

Attention is, therefore, being increasingly directed to the interactions of large adspecies with metal surfaces; the finger-printing of surface complexes and the identification of the orbitals involved in the chemisorption bond have already been accomplished for some adsorption systems.

22.7.4. Identification of adspecies

An example is the dissociative chemisorption of methanol on W(100) at room temperature [30]. By exposure of the clean surface to gaseous CO and then to hydrogen, an UPS spectrum is obtained that is virtually identical with that produced by the chemisorption of gaseous methanol, and as expected, on heating to 700 K, hydrogen is evolved and the UPS spectrum of β CO develops (Fig. 22.7). Similarly, spectral peaks of nitrogen and hydrogen adatoms can be identified when small amounts of NH_3 are chemisorbed on a clean W(100) surface; features due to molecular ammonia complexes are only visible after increasing exposure to NH_3.

22.7.5. Bonding-orbital identification

An example of bonding-orbital identification has been provided in a study of the chemisorption of benzene on Ni(111) [31, 32]. At low temperatures, the UPS spectrum of physically adsorbed multilayers of benzene comprises five main peaks, all of which can be aligned in position and relative intensity

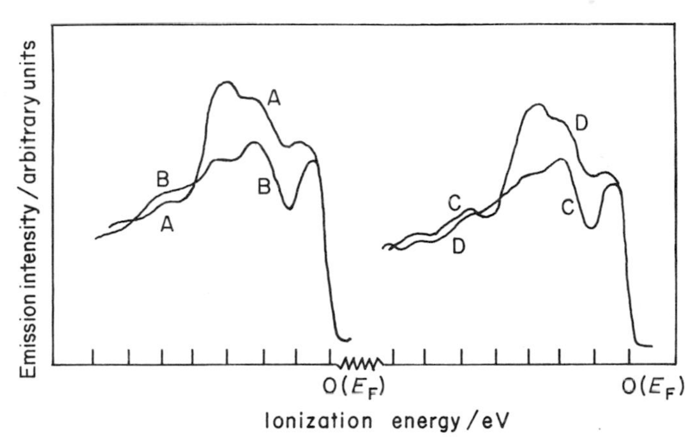

FIG. 22.7. Photoelectron spectra of methanol on W(100). Curve A, 0·3 L CH_3OH exposure to clean W(100) at room temperature; curve B, on heating A to 700 K to effect desorption of hydrogen; curve C, 0·21 L CO exposure to clean W(100) at room temperature; curve D, exposure of C to 0·9 L H_2 to reproduce the spectrum of methanol. Incident radiation $h\nu = 21{\cdot}2$ eV; reference level $E_F = 0$.

with the corresponding peaks in the gaseous benzene spectrum by inclusion of a constant relaxation energy. At room temperature, benzene is chemisorbed but the σ-orbital peaks can still be aligned with those in the gaseous spectrum by including a somewhat greater relaxation term. It is evident that many of the structural features of the gaseous molecule are retained in the molecularly chemisorbed molecule. The observation of greatest interest is that the energy of the highest doubly-degenerate π orbital of the free molecule is increased on chemisorption; this result denotes that the surface bonding involves only a π–d bond (Fig. 22.8).

Similarly, molecular chemisorption of ethylene on Ni(111) and W(100) surfaces takes place at 100 K [33], since the three main spectral peaks at 10·9, 12·7 and 14·1 eV can be aligned with the corresponding energy maxima in the spectrum of gaseous ethylene using a relaxation energy of 1·4 eV. A larger energy displacement of the fourth π peak is again consistent with π–d bonding of the surface complex.

On raising the temperature to 230 K, the original three peaks disappear and two new peaks at 9·1 and 11·1 eV develop. By including a relaxation energy of 3 eV, these peaks can be aligned with the two higher σ levels of acetylene, which has, therefore, been formed following dissociation of the chemisorbed ethylene (Fig. 22.9). The larger shift of the π-orbital peak shows that π–d bonding (of somewhat greater strength) is still involved.

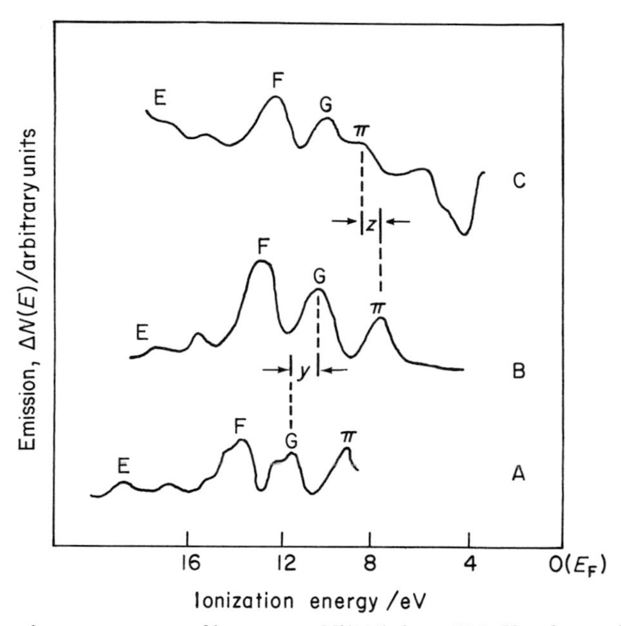

FIG. 22.8. Photoelectron spectrum of benzene on Ni(111). $h\nu = 21{\cdot}2$ eV; reference level $E_F = 0$. Curve A, spectrum of C_6H_6(g) referenced to (vacuum level − E_F); curve B, spectrum of multilayers of condensed benzene reference to $E_F = 0$; curve C, spectrum of chemisorbed benzene ($T = 300$ K) referenced to $E_F = 0$. E, F, G, positions of three of the main corresponding σ-orbital peaks; y = relaxation shift of σ-orbital energy levels; z = chemical bonding shift of π orbital.

Finally, on heating the C_2H_4/W(100) system to between 500 and 600 K, further dissociation ensues to give the spectrum of C_2 fragments, and at 1000 K, that of single C adatoms.

22.7.6. Identification of site geometry

The chemisorption of vinyl halides on different crystal planes of platinum provides an example of the specific effect of surface geometry [34]. On Pt(111), dissociative chemisorption persists up to monolayer coverages, but on

Pt(100) it is restricted to low surface concentrations above which only associative chemisorption takes place. Moreover, by presorption of small amounts of CO, the dissociative process on Pt(100) can be completely inhibited. Since the CO adspecies is believed to bond preferentially to the four-fold bridging sites on this plane, dissociative chemisorption apparently solely occurs at localities with this specific geometry.

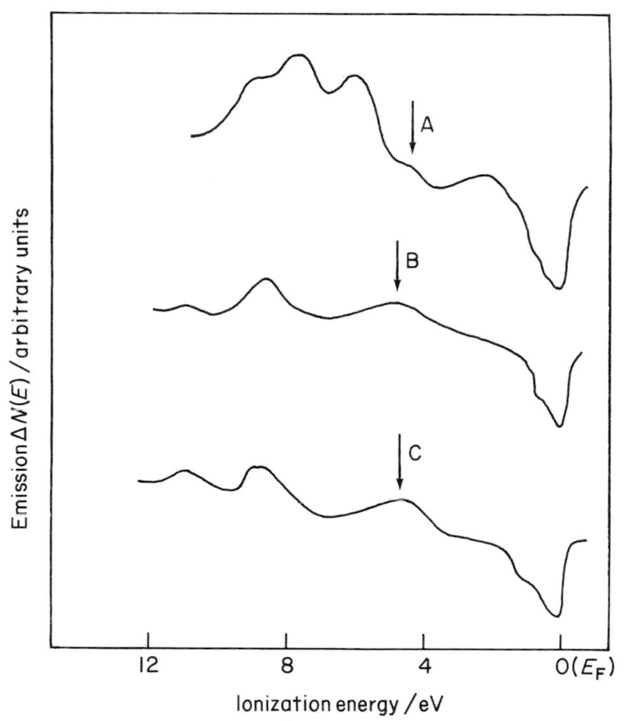

FIG. 22.9. Photoelectron spectrum of ethylene and acetylene on Ni(111). Incident radiation $h\nu = 21{\cdot}2$ eV. Curve A, absorption of C_2H_4 at 100 K; curve B, chemisorbed layer (A) heated to 230 K; curve C, adsorption of C_2H_2. Arrows denote the position of the π-bonding orbital. Reference level, $E_F = 0$.

22.7.7. Oxidation of metals

The various stages in the oxidation of metals can be partially delineated by observing the changes in the UPS spectrum [35–38]. Thus, on nickel [3], the initial process of chemisorption of oxygen is accompanied by a decrease in the intensity of the *d*-band emission from nickel, and an increase of the oxygen-induced level at ~6 eV as more oxygen is admitted (see Fig. 22.10).

For larger uptakes of oxygen, a peak *ca* 2 eV associated with the *d* orbitals of the oxide, together with broad peaks in the range 2·5 to 11 eV arising from the two *p* orbitals of the oxygen, grow in intensity, thereby indicating the formation of the oxide.

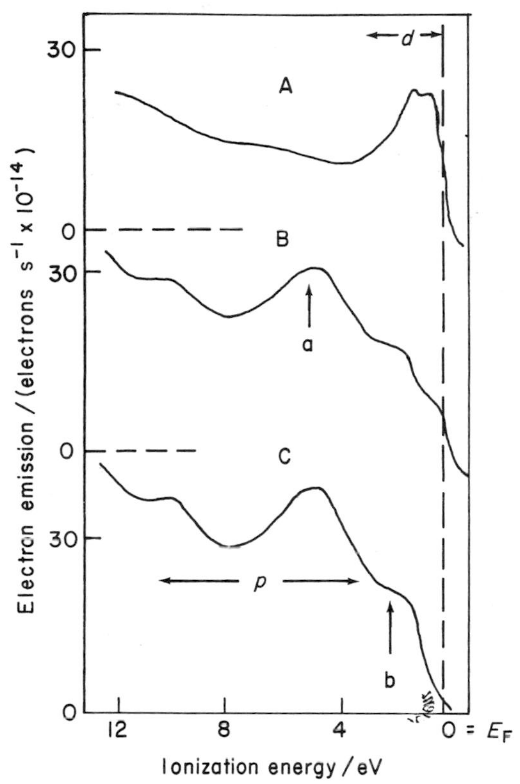

FIG. 22.10. Photoelectron spectra of clean nickel (curve A), after exposure to ~25 L of oxygen (curve B) and after exposure to ~60 L of this gas (curve C). The *d*-band emission (*d*) decreases with uptake of oxygen, and the intensity of the oxygen-induced level at ~6 eV increases. Oxidation is denoted in curve C by the development of a *d*-orbital peak of the oxide at b, and a broad 2*p* band at p.

22.7.8. Surface states

A surface state on a clean W(100) plane giving a peak at 0·4 eV below the Fermi level has been detected [39, 40] (see Fig. 22.5). As expected, the band rapidly decays when hydrogen (or carbon monoxide) is chemisorbed. The amount of chemisorbed hydrogen necessary to effect complete removal of the peak is consistent with the formation of a surface complex of empirical formula W_4H, assuming that each hydrogen adatom removes one surface state.

REFERENCES

1. D. E. Eastman and J. K. Cashion, *Phys. Rev. Letters*, 1971, **27**, 1520.
2. D. E. Eastman, *Phys. Rev. B*, 1971, **3**, 1976.
3. D. W. Turner, C. Baker, A. D. Baker and C. R. Brundle, "Molecular Photoelectron Spectroscopy", Wiley, New York, 1970.
4. J. K. Cashion, J. L. Mees, D. E. Eastman, J. A. Simpson and E. C. Kuyatt, *Rev. Scient. Instrum.*, 1971, **42**, 1670.
5. D. E. Eastman, W. D. Grobman, J. L. Freeouf and M. Erbudak, *Phys. Rev. B.*, 1974, **9**, 3473 (synchroton radiation source).
6. C. R. Brundle, *J. Vac. Sci. Tech.*, 1974, **11**, 212 (summary of experimental values of escape depth).
7. J. W. Gadzuk, *Phys. Rev. B*, 1974, **10**, 5030.
8. J. W. Gadzuk, *J. Vac. Sci. Tech.*, 1974, **11**, 275.
9. A. Liebsch and F. W. Plummer, *Faraday Disc. Chem. Soc.*, 1974, **58**, 19.
10. T. Koopmans, *Physica*, 1933, **1**, 104.
11. E. W. Plummer, *Topics Appl. Phys.*, 1975, **4**, 143.
12. J. R. Smith, S. C. Ying and W. Kohn, *Phys. Rev. Letters*, 1973, **30**, 610.
13. J. E. Demuth and D. E. Eastman, *Phys. Rev. Letters*, 1974, **32**, 1123.
14. C. R. Brundle, *Surface Sci.*, 1975, **48**, 99.
15. R. W. Joyner and M. W. Roberts, *J. Chem. Soc., Faraday Trans. I*, 1974, **70**, 891.
16. S. K. Atkinson, C. R. Brundle and M. W. Roberts, *Chem. Phys. Letters*, 1974, **24**, 175.
17. P. Biloen and A. A. Holster, *Faraday Disc. Chem. Soc.*, 1974, **58**, 106.
18. P. J. Page and P. M. Williams, *Faraday Disc. Chem. Soc.*, 1975, **58**, 80.
19. H. Conrad, G. Ertl, J. Kuppers and E. E. Latta, *Faraday Disc. Chem. Soc.*, 1974, **58**, 116.
20. A. M. Bradshaw, D. Menzel and M. Steinkilbert, *Chem. Phys. Letters*, 1974, **28**, 516.
21. A. M. Bradshaw and D. Menzel, *Ber. Bunsenges Phys. Chem.*, 1974, **78**, 1140.
22. D. E. Eastman, *Solid State Comm.*, 1972, **10**, 933.
23. W. C. Price, A. W. Potts and D. G. Streets, *in* "Electron Spectroscopy" (D. A. Shirley, ed.), North Holland Publ. Co., New York, 1972.
24. J. M. Baker and D. E. Eastman, *J. Vac. Sci. Tech.*, 1973, **10**, 233.
25. E. W. Plummer, B. Waclawski and T. V. Vorburger, *Chem. Phys. Letters*, 1974, **28**, 510.
26. D. R. Lloyd, *Faraday Disc. Chem. Soc.*, 1974, **58**, 136.
27. C. R. Brundle, *Faraday Disc. Chem. Soc.*, 1975, **58**, 125 *et seq.*
28. G. Blyholder, *J. Vac. Sci. Tech.*, 1974, **11**, 865.
29. I. P. Batra and O. Robaux, *J. Vac. Sci. Tech.*, 1975, **12**, 242.
30. W. E. Egelhoff, J. W. Linnett and D. L. Perry, *Faraday Disc. Chem. Soc.*, 1974, **58**, 35.
31. J. E. Demuth and D. E. Eastman, *Phys. Rev. Letters*, 1974, **32**, 1123.
32. J. E. Demuth and D. E. Eastman, *Jap. J. Appl. Phys.*, Suppl. 2, Part 2, 1974, 827.

33. E. W. Plummer, B. J. Waclawski and J. V. Vorburger, *Chem. Phys. Letters*, 1974, **28**, 510.
34. T. A. Clarke, I. D. Gay, B. Law and R. Mason, *Faraday Disc. Chem. Soc.*, 1975, **60**, 119.
35. A. M. Bradshaw, D. Menzel and M. Steinkilbert, *Jap. J. Appl. Phys.*, Suppl. 2, Part 2, 1974, 841.
36. B. J. Waclawski and E. W. Plummer, *J. Vac. Sci., Tech.*, 1973, **10**, 292.
37. A. M. Bradshaw and D. Menzel, *Ber. Bunsenges Phys. Chem.*, 1974, **78**, 1140.
38. J. M. Baker and D. E. Eastman, *J. Vac. Sci. Tech.*, 1973, **10**, 223.
39. B. J. Waclawski and E. W. Plummer, *Phys. Rev. Letters*, 1972, **29**, 783.
40. B. Feuerbacher and B. Fitton, *Phys. Rev. Letters*, 1972, **29**, 786.

23

Ion Neutralization Spectroscopy

With this particular technique, there seems little doubt that only the electron interactions at, and immediately above, the surface, are being examined. Unfortunately, the apparatus is extremely complicated and the mathematical procedures involved in transforming the experimental data to provide the local density of states are quite complex. In consequence, only a limited amount of work has been reported.

23.1. GENERAL PRINCIPLES

Excitation is brought about by bombardment of the surface with a monoenergetic beam of inert-gas ions [1–4]. The ion most frequently used is He^+, but Ar^+ and Ne^+ beams have also been employed. The process may be described as a two-electron Auger transition, involving one electron of energy $\delta - \Delta$, and the other of lower energy $\delta + \Delta$, these levels being symmetrically placed with respect to an energy level δ in the valence band (see Fig. 23.1). On impact of the He^+ ion with the surface, the lower-energy electron tunnels through to the vacant level in the inert gas ion, thereby neutralizing its charge. An amount of energy $I - \Phi - (\delta + \Delta)$ eV is released, I being the effective ionization potential of the He atom at the metal surface, and Φ the work function of the metal with respect to the vacuum level. This energy release causes the ejection of the higher-energy electron $(\delta - \Delta)$ out of the metal, the energy required being $\Phi + (\delta - \Delta)$ eV. This emitted Auger electron has kinetic energy equal to $I - 2(\Phi + \delta)$ eV, the magnitude of which is derived from an energy analysis similar to that employed in

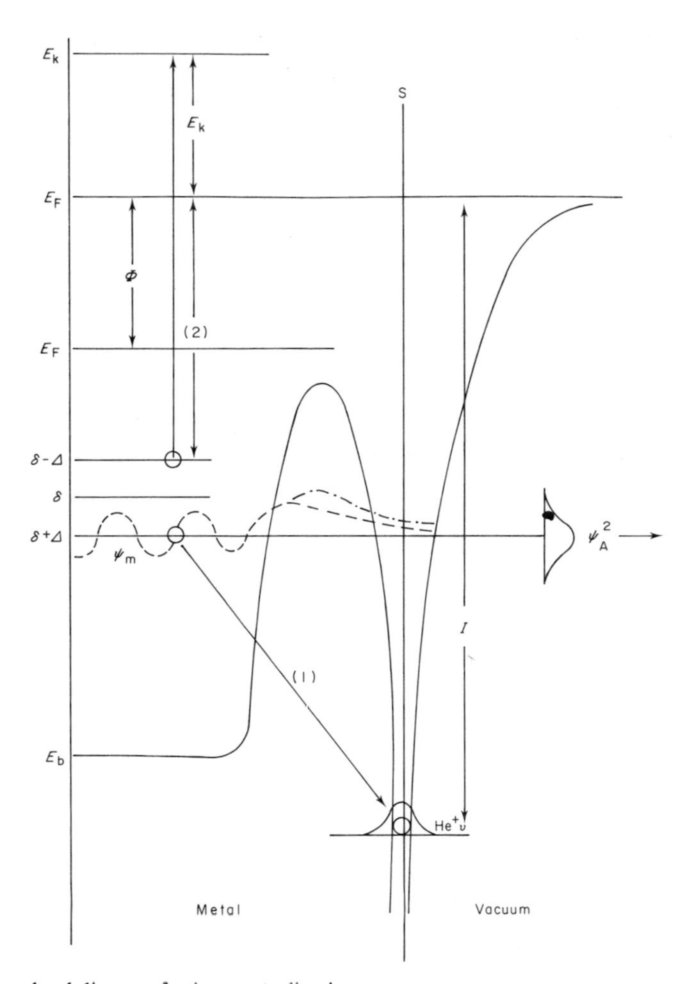

FIG. 23.1. Energy level diagram for ion neutralization spectroscopy.

E_b, lowest energy level in conduction band of metal; E_F, Fermi level; E_v, vacuum level; E_k, kinetic energy of ejected (Auger) electron; I, ionization potential of He atom at metal surface S; Φ, work function of metal. δ is energy level of valence electron; $\delta \pm \Delta$, valence energy levels symmetrically placed with respect to the δ energy level.

Mechanism of ejection of Auger electron: (1) neutralization of He^+ by an electron from the $\delta + \Delta$ level; energy release is $I - \Phi - (\delta + \Delta)$; (2) this energy is utilized to eject an electron from the $\delta - \Delta$ level with kinetic energy E_k. Hence,

$$I - \Phi - (\delta + \Delta) = E_k - (2\Phi + \delta)$$

or

$$E_k = I - 2(\Phi + \delta)$$

The wave function in the metal, ψ_m, is enhanced (·—·—) in amplitude within the resonance energy range of the adsorbed atom; ψ_A is the wave function of the emitted electrons.

electron impact Auger spectroscopy. The valence band can, therefore, be explored to a depth of $I - 2(\Phi + \delta)$, i.e. *ca* 10 eV below the Fermi level.

Since the first electronic transition is to a He^+ ion at the metal surface, the density of states under examination are described by wave functions that are immediately outside and at the surface itself. Thus, when a chemisorbed layer is present, the contribution of the topmost atomic layer of the solid is negligible. Moreover, the depth of electron penetration of the solid is less than the mean free path for electron scattering; consequently, few scattered electrons are emitted and a very low-intensity background spectrum is observed. The probability of ejection of an Auger electron per incident inert-gas is 1/10; it is, therefore, a highly efficient process and strong peaks are produced. The resolution is better than 0·2 eV, and multipeak structures are observable. Moreover, since the work function of a metal is characteristic of a specific crystal face, different bonding structures of simple adsorbates on the same metal may be distinquished by the different surface density of states. Such observations, together with the existence of multiple peaks, are consistent with the concept of a localized surface compound having specific molecular orbitals with discrete energy levels. The spectral peaks arise from orbital resonances of electrons in the chemisorption bond and, hence, provide information about the structure and local symmetry of the surface molecule.

23.2. CONVERSION OF ENERGY-ANALYSIS PLOT TO LOCAL DENSITY OF STATES

The experimental energy-analysis plot gives the transition density $U(\zeta)$, i.e. the product of the local density of states and the related transition probabilities. Since these latter values are determined in large part by the magnitude of the local wave functions, the structure of the $U(\zeta)$ plot is similar to that of the local density of states [or $D(\zeta)$] plot. The magnitude of $D(\zeta)$ is crudely given by the product of the average square of the wave function of energy $U(\zeta)$ in the vicinity of the chemisorbed atom A and the density of states $N(\zeta)$, where ζ is the energy below and with respect to the Fermi level, i.e.

$$D(\zeta) = [U(\zeta)]_A^2 N(\zeta). \tag{23.1}$$

When chemisorption takes place, $D(\zeta)$ approximates to the sum of the states of the metal substrate M and those of the surface molecule S, i.e.

$$D(\zeta) = \{[U_M(\zeta)]^2 + [U_S(\zeta)]^2\} N(\zeta). \tag{23.2}$$

Here $N(\zeta)$ is the density of states of the combined system.

In formation of the chemisorption bond, charge transfer from the metal takes place in order to equalize the chemical potential of electrons in the free adsorbate to that of the electrons at the Fermi level. In consequence, there is a shift and broadening of levels and resonance or virtual energy states are created. In other words, the valence electrons of the free adsorbate are distributed amongst the orbitals in the surface molecule and those wave functions in the vicinity of the adspecies are enhanced within the range of surface orbital resonances. Outside this range, there is no enhancement and the wave functions are essentially those of the clean metal surface.

23.3. APPLICATIONS TO CHEMISORPTION

Some results for the oxygen/Ni(111) system illustrate the type of information which has been extracted using ion neutralization spectroscopy. The oxygen atom has four p electrons in an energy well ζ_A and has an ionization energy of $\zeta_A + \Phi_A$, where Φ_A is the work function of the metal when its surface is covered with O-adatoms. These electrons are distributed amongst three bonding orbitals of different energies ζ_{s1}, ζ_{s2}, ζ_{s3}. The orbitals have different ionization energies and, therefore, three different spectra peaks are obtained, i.e. the three surface orbitals have different ionicity and bonding. The orbital resonance, and therefore the peak intensity, are strongest for bonding orbitals pointing outwards from the surface. The peaks are distinct from the original peak nearer the Fermi level arising from the Ni d band, since this band is outside the range of resonance levels created by the chemisorbed layer and its peak intensity is greatly reduced as the amount of chemisorbed oxygen is increased.

Most of the experimental data has been concerned with the chemisorption of oxygen, sulphur and selenium on the (111) plane of nickel (see Fig. 23.2). The intepretations of the results are tentative, but indicate the type of information that might more definitely emerge after further developments. For example, the spectra obtained from $c(2 \times 2)$O and p(2×2)O structures are virtually the same, apart from differences in peak intensities, and might indicate that the same local bonding is involved. Only a single broad peak is visible, possibly due to strong interactions between the three bonding orbitals. The energy displacement of the peak from the energy of oxygen p orbitals is larger than that for the corresponding systems when oxygen is replaced by S and Se, and shows that the greatest transfer of negative charge is to the oxygen atom in the adsorbed state. There is also some evidence of surface reconstruction.

The orbital spectra of c(2×2)S and p(2×2)S show large differences

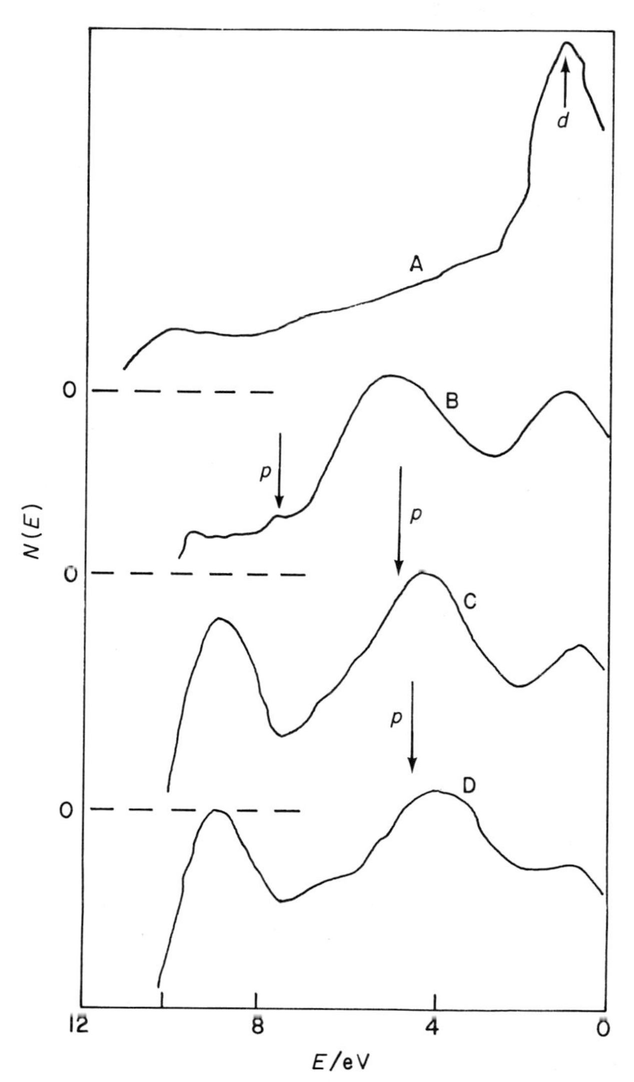

FIG. 23.2. Transition density functions ($U(\delta)$ for clean Ni(111) (curve A); after oxygen chemisorption (curve B); after sulphur chemisorption (curve C); and after selenium chemisorption (curve D). *d*, *d*-band peak at *ca* -1 eV is decreased by adsorption; *p*, energy of *p* orbitals of free oxygen atom, free sulphur atom and free selenium atom in curves B. C and D, respectively.

[as do c(2 × 2)Se and p(2 × 2)Se] amongst themselves (and from the corresponding oxygen structures) and it is evident that different local bondings are present in these surface complexes. Three peaks arising from the inequivalent orbitals of S and Se appear in c(2 × 2) spectra; and by analogy with the C_{2v} symmetry of H_2S (and H_2Se) a bridged structure Ni_2X across the diagonal of the Ni(1 × 1) surface mesh has been postulated. Similarly, the p(2 × 2) structure may be assigned to pyramidal Ni_2X complex having C_{4v} symmetry.

Ion neutralization spectroscopy investigations have so far been confined almost entirely to the Bell Telephone Laboratories, principally because of the experimental complexity and the theoretical difficulties in extracting the local density of electronic states from the results.

REFERENCES

1. H. D. Hagstrum, *Phys. Rev.*, 1966, **150**, 495.
2. H. D. Hagstrum and G. E. Becker, *Phys. Rev.*, 1967, **159**, 572.
3. H. D. Hagstrum and G. E. Becker, *Phys. Rev. Letters*, 1969, **22**, 1054.
4. H. D. Hagstrum and G. E. Becker, *Phys. Rev. B*, 1971, **4**, 4187.
5. G. E. Becker and H. D. Hagstrum, *J. Chem. Phys.*, 1971, **54**, 1015.
6. G. E. Becker and H. D. Hagstrum, *Surface Sci.*, 1972, **30**, 505.
7. H. D. Hagstrum, *J. Vac. Sci. Tech.*, 1973, **10**, 31.

24

Electron Energy-loss Spectroscopy

24.1. ORIGIN OF SPECTRAL LOSS PEAKS

When the surface of a metal is bombarded by a primary beam of electrons of energy E_p, some of the electrons are inelastically scattered in the outermost surface layers and suffer characteristic energy losses E_L in the range 0 to 40 eV. Secondary loss peaks, therefore, appear in the energy distribution spectrum of back-scattered electrons from metal surfaces and, when very thin metal foils are employed, in the energy spectrum of transmitted electrons [1–3]. The peak energy is given by $E_p - E_L$. Its position on the low-energy side of the energy distribution can, therefore, be altered by varying the energy E_p of the incident electrons. Loss peaks are also visible in the higher-energy range of Auger spectra, but Auger positions are independent of the primary beam energy, although the band shapes are modified. Consequently, distinction between Auger and loss peaks can be easily accomplished.

Energy losses result from four excitation processes.

(i) Intraband transitions of electrons from lower occupied to higher unoccupied energy levels in the same band, and interband transitions from one band to unoccupied levels of another band.

(ii) Volume plasmon excitation which occurs when the conduction electrons of the metal are subjected to an external perturbation, as, e.g., by irradiation with ultra-violet light of an appropriate frequency. In this case, the electro-magnetic field sets up a collective harmonic oscillation of the free electrons by displacing them with respect to the ion-cores and so opposes the attractive Coulombic force between the ion-cores and the electrons. The angular frequency of the volume electron-density oscillation is given by

$$\omega_v = (ne^2/\varepsilon_0 m_0)^{\frac{1}{2}}, \tag{24.1}$$

where e is the electronic change, m_0 the rest mass of the electron, ε_0 the vacuum dielectric constant, and n is the volume electron density. For transition metals, the d electrons are substantially localized and simultaneous interband transitions are induced by the primary excitation. The dispersion relationship for volume plasma is then given by

$$\frac{e^2}{\varepsilon_0 m_0} \sum \frac{f_{no}}{\omega_b^2 - \omega_{no}^2} = 1, \tag{24.2}$$

where f_{no} is the oscillation strength of a d-electron transition of frequency ω_{no} from its ground state to the nth excited state. The plasmon frequency ω_b is increased when $\omega_{no} > \omega_b$ and decreased when $\omega_{no} < \omega_b$.

(iii) Surface plasmon excitation [4] is due to the effect of the boundary conditions imposed at the surface; it represents a longitudinal oscillation of the surface free-electron density with a frequency ω_s, where, for a planar metal/vacuum interface,

$$\omega_s = \omega_v/\sqrt{2}. \tag{24.3}$$

In energy-loss spectroscopy, the inelastic scattering of the incident electrons is responsible for the creation of surface plasma; and electron penetration into the metal sets up localized Coulombic fields and generates volume plasma [5].

(iv) Excitation of lattice vibrations of the surface metal atoms: the phonon energies are very small ($<0{\cdot}1$ eV), and discrete peaks are not usually observed although the loss peaks arising from excitations [(i) to (iii)] are broadened.

24.2. THE ENERGY-LOSS SPECTROMETER

The spectrometer is essentially that used in Auger spectroscopy. Excitation is brought about by bombardment with a well-focused low-energy beam of electrons. The energy spread within the beam should be as narrow as possible in order to minimize the width of the energy-loss peak. Resolution of $\sim 0{\cdot}05$ eV is possible with a well-monochromatized primary beam of low energy (5 to 10 eV). The energy distribution curve is a plot of $N(E_k)$ as a function of E_k, where $E_k = E_p - E_L$; it comprises energy-loss peaks arising from valence electron transitions, surface plasmon and, to a lesser extent, volume plasmon oscillations, and also from Auger transitions. Separation of losses due to surface plasma oscillations and electron transitions can be, at least partly, separated by comparison with optical reflectance data.

The energy analysis of the scattered electrons is accomplished using an Auger analyser and collector; the electronic differentiation equipment gives directly the first derivative plot. The peak intensity compared with the total integrated back-scattered current is about the same as that obtained in Auger spectroscopy; angular distribution measurements can be made by incorporating a rotating Faraday-cup collector.

24.3. APPLICATIONS TO CHEMISORPTION

24.3.1. Vibration frequencies of chemisorbed species [6] (0–0·8 eV region)

A low incident energy (~5 eV), highly monochromatized to achieve good resolution (~0·05 eV), is required. The stretching frequency of the C—O bond of CO chemisorbed on W(100) plane has been evaluated from the loss peak in good agreement with the infra-red value [6]. For hydrogen and nitrogen, loss peaks corresponding to the stretching frequencies of the chemisorbed molecules were absent, indicating that dissociative chemisorption had occurred. Low frequency vibrations were, however, visible; these could be reasonably attributed to the vibrations of H adatom–W and N adatom–W bonds. Similarly, a loss peak corresponding to a W—O adatom stretching vibration was observed for water chemisorbed on W(100) at low coverages; the O—H vibrations of the water molecule were visible at higher coverages at which hydrogen bonding to the O adatoms in the primary layer would be expected.

24.3.2. Adsorbent and chemisorption energy levels

Some information has been obtained from the effect of chemisorption on the energy-loss peaks displayed by the clean metal. The primary beam energies are usually greater than 10 eV but less than 100 eV. Valence electron transitions and surface plasmon (and sometimes volume plasmon) oscillations occur, and separation is uncertain. Surface plasmon excitation can, in principle, be extracted from the angular dependence of the scattered intensity, by the attenuation of transmission through very thin metal foils by increasing their thickness, and by variation of the penetration depth of incident electrons by varying the primary beam energy. Interband transitions can also be identified, since their loss peaks, both in intensity and energy, are unaffected by chemisorption. However, there is little agreement in interpretation of results. Thus, a loss peak at 4·2 eV arising from a clean Ag(111) surface has been attributed to surface plasmon excitation, to a combination of surface and volume plasma, and to an interband transition of a 4*d* electron to an energy state near the vacuum level [7–9]. A loss at 8·2 eV has been interpreted [10] as a volume plasmon excitation, and two peaks at 12·1 and 17·5 eV are attributed to interband transitions [11].

For a chemisorbed layer of carbon monoxide on single planes of copper, silver, gold, platinum, palladium and nickel, a common peak at *ca* 13·5 eV is observed [12–15] (cf. Fig. 24.1). On the (100) plane of nickel, a second

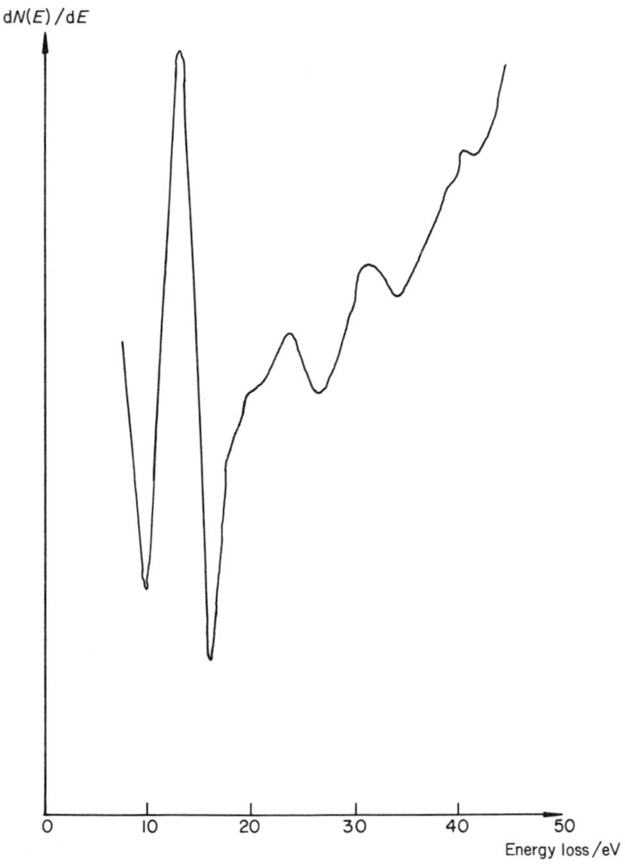

FIG 24. 1. Energy-loss (first-derivative) spectrum for a carbon monoxide overlayer on Au(100) at 81·5 K. E_p = 75 eV at normal incidence.

peak appears at 5·5 eV; this has been assigned to an electron transition from the Fermi level to an energy state 5·5 eV above E_F on the grounds that the density of unoccupied states attains a maximum at this level. The 13·5 eV peak has, therefore, been attributed to an electron transition from a level 8 eV below E_F to the 5·5 eV level of minimum occupancy [16] (Fig. 24.2).

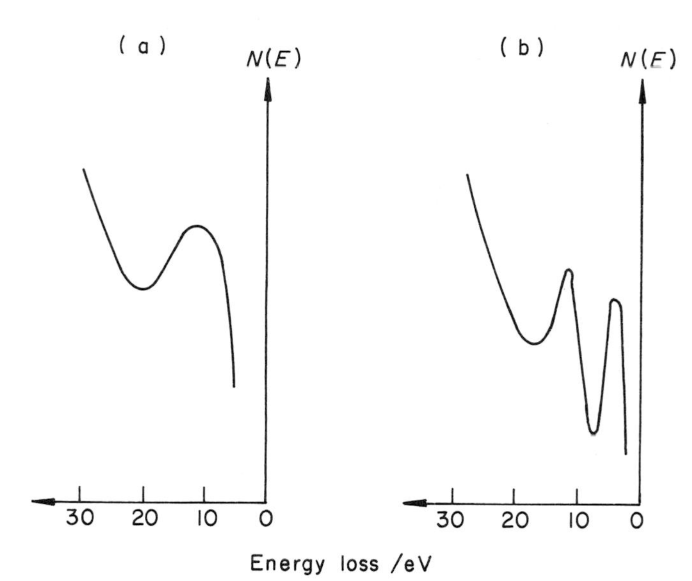

FIG. 24.2. Energy-loss spectrum of (a) a clean Ni(110) surface, and (b) after chemisorption of carbon monoxide (~monolayer). E_p = 62 eV; chemisorption peaks at 5·5 and 13·5 eV.

24.3.3. Charge transfer

The (100) plane of molybdenum [17] on exposure to an electron beam of energy 250 eV displays a broad peak at 23 eV and a narrow one (~1 eV width) at 11·6 eV. The intensity of this latter peak is decreased on chemisorption of carbon monoxide but its peak energy is increased to 12·2 eV. It was suggested that charge transfer accounted for this 0·6-eV increase in energy and that some of the localized electrons of the surface metal atoms were, thereby, freed to the extent of 0·3 electrons per Mo atom and these then contributed to the surface plasmon oscillation [17].

Electron energy-loss spectroscopy is in an early state of application and the interpretation of experimental results is, as yet, speculative. Nevertheless, with further progress, the method may have considerable potential value in future investigations of clean surfaces and, because of its high sensitivity to the presence of adsorbates, of chemisorption of gases by metals.

24.4. IONIZATION SPECTROSCOPY [18, 19]

An electron beam of high energy E_p excites inner-shell electrons of bonding energies E_B to unoccupied states E_u above the Fermi level. For back-scattered electrons with kinetic energy E_k the energy loss is $E_L = E_p - E_k = E_B + E_u$ with a minimum value at $E_u = E_F$. The fine structure of loss peaks is related to the density distributions of unoccupied states above E_F. From the energy of the loss peak (E_k) and the value of E_p, the binding energy E_B can be evaluated and used in the identification of surface impurities. The spectra are simpler than their Auger counterparts, since only two electronic states are involved, but the intensity of the peaks is much smaller. There have been few applications of this technique to chemisorption investigations.

REFERENCES

1. D. Bohm and D. Pines, *Phys. Rev.*, 1952, **85**, 338; 1953, **92**, 609.
2. H. Raether, *Surface Sci.*, 1967, **8**, 233.
3. A. J. Bennett, *Phys. Rev. B*, 1970, **1**, 203.
4. R. H. Ritchie, *Phys. Rev.*, 1959, **113**, 1254.
5. R. A. Ferrell, *Phys. Rev.*, 1958, **111**, 1214.
6. F. M. Propst and T. C. Piper, *J. Vac. Sci. Tech.* 1966, **4**, 53.
7. M. P. Seah, *Surface Sci.*, 1971, **24**, 357.
8. J. Daniels, *Z. Naturforsch.*, 1966, **21a**, 662.
9. B. R. Cooper, E. L. Kreiger and B. Segal, *Phys. Rev. B*, 1971, **4**, 1734.
10. E. B. Pattinson and P. R. Harris, *J. Phys. D*, 1972, **5**, L 59.
11. G. McElhiney, H. Papp and J. Pritchard, *Surface Sci.*, 1976, **54**, 617.
12. H. Papp and J. Pritchard, *Surface Sci.*, 1975, **53**, 371.
13. F. P. Netzer and J. A. D. Matthews, *Surface Sci.*, 1975, **51**, 352.
14. G. Doyen and G. Ertl, *Surface Sci.*, 1974, **43**, 197.
15. G. McElhiney and J. Pritchard, *Surface Sci.*, 1976, **54**, 592.
16. J. Küppers, *Surface Sci.*, 1973, **36**, 53.
17. J. Lecante, *in* "Adsorption–Desorption Phenomena" (F. Ricca, ed.), Academic Press, New York and London, 1972, p. 369.
18. R. L. Gerlach, J. Houston and R. L. Park, *Appl. Phys. Letters*, 1970, **16**, 179.
19. R. Haydock, V. Heine and M. J. Kelley, *J. Phys. C*, 1972, **5**, 2845.

25

Appearance Potential Spectroscopy

Soft X-ray spectroscopy was first used to determine the electronic band structure of bulk metals, but surface properties can be investigated by employing incident electron beams of low-energy (< 2000 eV) [1, 2]. Investigations have been largely restricted to clean metal surfaces and only a limited amount of data has been acquired about chemisorbed layers.

25.1. GENERAL PRINCIPLES

Emission of soft X-rays from a metal is brought about by bombarding its surface with low-energy electrons. The spectrum is generated by recording the total X-ray yield as a function of the energy of the primary beam; it consists of a series of very small peaks (~1 % of the total integrated intensity) on a rapidly-rising high-intensity background (see Fig. 25.1). The position and number of peaks are characteristic of the electronic configuration of the metal. The energy of a peak with respect to that of the incident beam is the threshold energy to excite an inner-shell electron of the metal atom to the Fermi level. The excited state comprises the residual core vacancy, an excited core electron and a scattered primary electron. The de-excitation of the core vacancy is brought about either by a band-to-band transition and emission of X-rays with energy characteristic of the band gap, or by an Auger transition followed by excitation and emission and ejection of an Auger electron (Fig. 25. 2).

These de-excitation processes compete with each other. Should the excited core electron originate from an inner-shell level corresponding to a binding energy of less than 1000 eV, the fluorescence intensity is very small, since the lifetime of the core vacancy is predominantly determined by the Auger transition rate. For a Coster–Kronig transition the lifetime is of particularly short duration, and the width of the spectral peak is large because of the operation of the uncertainty principle; its peak height, or measured intensity, is, therefore, abnormally small. An additional factor effects further reduction in intensity. The primary electron loses part of its energy at the threshold value in producing the core vacancy; consequently, the back-ground intensity is diminished at the peak position, and the corresponding characteristic peak is less prominent.

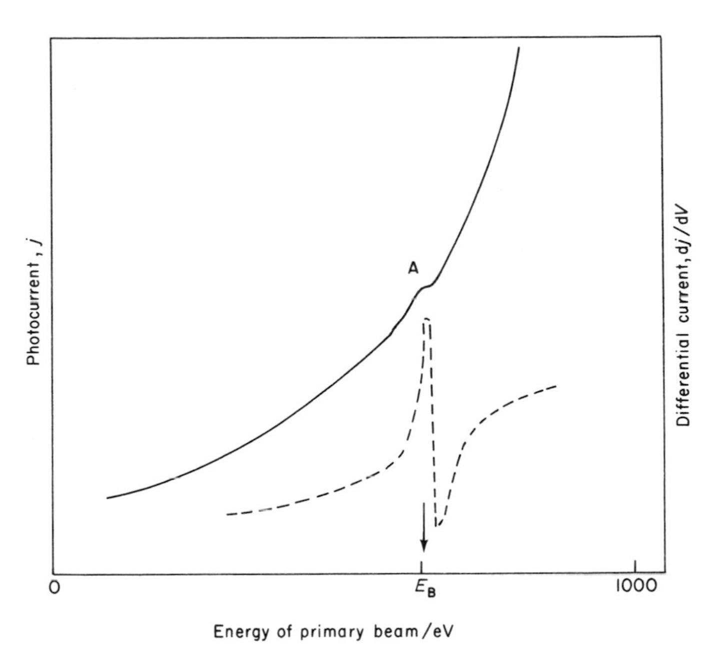

FIG. 25.1. Schematic representation of the total X-ray yield recorded as photocurrent from the photocathode as a function of the energy of the primary electrons. Increase of emission at A indicates the threshold energy, or binding energy E_B of an excited core electron. The differential current, or first derivative, plot is indicated by a dashed line.

The X-ray emission is from a surface region of depth approximately equal to the elastic mean free path of the primary electrons, and the peak intensities preferentially reflect the properties of the outermost layer of atoms, the depth being 5 to 20 Å in the range 5 to 1500 eV. The surface sensitivity is, therefore, similar to that of Auger spectroscopy.

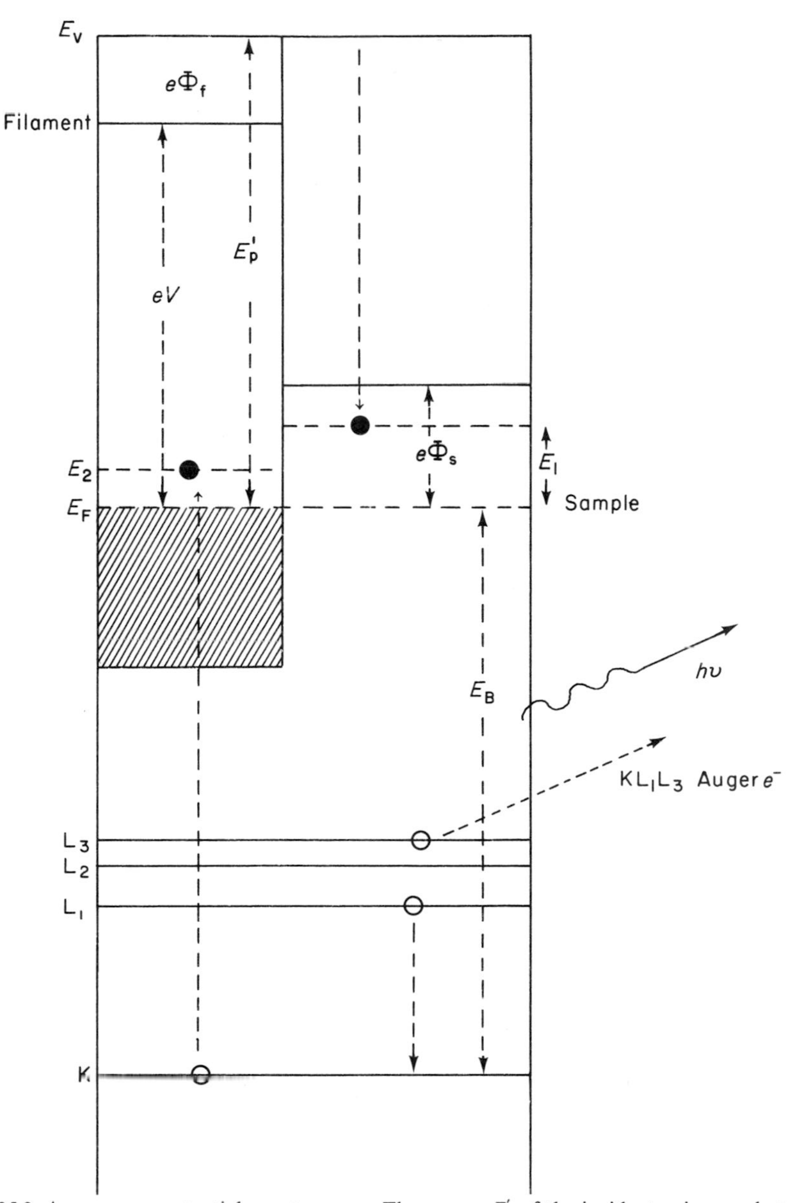

FIG. 25.2. Appearance potential spectroscopy. The energy E_P' of the incident primary electron is $eV + e\Phi_F$, where Φ_f is the work function of the filament. It is captured in an energy state E_1 above the Fermi level E_F and excites a core electron from the K state of the sample to a state E_2 above E_F, so that $E_2 + E_B = eV + e\Phi_F - E_1$, where E_B is the binding energy of the K electron with respect to the Fermi level. The minimum, or threshold, incident energy for excitation is $eV_{min} + e\Phi_F = E_B$. The core-electron excitations give rise to increased emission of characteristic X-rays in the total X-ray yield.

25.2. THE APPEARANCE POTENTIAL SPECTROMETER [3–5]

The primary electron beam is generated by electrical heating of either a clean, or a thoria-coated, tungsten filament (see Fig. 25.3). Since the coated filament has a lower work function, higher electron currents (1 to 10 mA) are obtained at much lower filament temperatures, and the heating of the sample is considerably less than that by the high temperatures necessary with uncoated filaments. Accelerating voltages of 0 to 1000 eV are applied to the

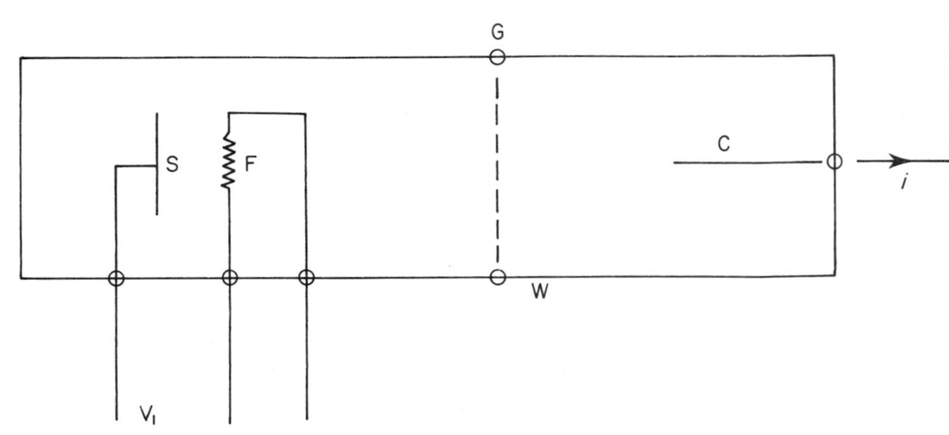

FIG. 25.3. Schematic representation of spectrometer. S, sample; F, heated filament (electron source); V_1, variable accelerating potential to accelerate electrons from filament to sample; G, grid at negative potential to screen out secondary electrons; W, chamber wall functioning as photo-cathode to convert X-rays to secondary electrons; C, central collector electrode wire at positive potential; *i*, measured photocurrent.

incident electrons across a gap of about 1 cm between the sample and the filament. Emission of soft X-rays and production of secondary electrons take place, but the latter are screened out by means of a grid maintained at negative potential. After passing through the grid, the X-rays, over a wide acceptance angle ($\sim 130°$), enter the collector, the walls of which function as a photocathode. The secondary electrons that are formed are collected by a central wire maintained at a positive potential to give a photocurrent which increases, approximately linearly, with increase of the primary electron flux. The signal-to-noise ratio is small, but, since the noise level increases as the square root of the primary flux, it can be made more favourable by using larger primary currents.

25.3. DERIVATION OF THE DENSITY OF STATES [6]

The intensity of the X-ray peaks is predominantly determined by the product of the density of states $[N(E_B)]$ having energies in the close vicinity of the binding energy E_B of the inner-shell electron that is being excited, and the density of unoccupied states $N(E_1)$ above the Fermi level. The energy of the excited state comprises the sum of the energy E_2 of the level to which an electron is ejected and the energy E_1 of the captured primary electron; and energy conservation requires (cf. Fig. 25.2) that

$$E_1 + E_2 = E'_P - E_B, \tag{25.1}$$

where E'_P is the energy of the incident electron, referenced to the Fermi level of the filament. Consequently, the density of unoccupied states is approximately given by the integral

$$\int N(E'_P - E_B - E_2)\, N(E_2)\, dE_2. \tag{25.2}$$

The lower limit of integration is $E_F = 0$ and refers to the condition that the energy of the captured electron referenced to the Fermi level is zero, i.e. $E_2 = E'_P - E_B$ from Eq. (25.1). The upper limit corresponds to the situation with the excited core electron at E_F, i.e. $E_1 = E_P - E_B$. The integration, therefore, includes all possible combinations of the energies E_1 and E_2 and is directly proportional to the density of the final states. The rate of excitation, which is roughly linearly related to the peak intensity, is, therefore, proportional to

$$N_i(E_B) \int N(E_P - E_B - E_2) N(E_2)\, dE_2, \tag{25.3}$$

i.e. its value is determined by the local density of states in the vicinity of the nucleus with energies close to E_B and the density of unoccupied states at and above the Fermi level. An implicit assumption is that the transition probabilities of the two excited electrons are the same, i.e. independent of the energy of their final state and that of the primary beam.

In the X-ray spectrum, the total X-ray intensity increases as the square of the incident electron energy; consequently, its first derivative is linearly related to the accelerating potential applied to the primary electron beam. Differentiation of Eq. (25.3) then indicates that the intensity, or the height of the characteristic peak, is approximately proportional to the square of the density of the unoccupied states at and above the Fermi level. The results are, therefore, complementary to those obtained in photoelectron spectroscopy which provide information on the density of the occupied states near the Fermi level.

25.4. APPLICATION TO METAL SURFACES AND CHEMISORBED OVERLAYERS

25.4.1. Binding energy of core electrons and surface analysis

From the appearance potentials or the threshold energies of the X-ray peak, binding energies E_B of inner shell electrons can be evaluated. The energy E_e of the primary electron referenced to the Fermi level of the filament is the sum of the work function of the filament $e\Phi_f$ and the kinetic energy E_P which is determined from the accelerating voltage applied across the gap between the filament and the sample; this voltage also determines the potential difference between the Fermi levels of the sample and the filament. The threshold energy E_T is, therefore, given by

$$E_T = E_B - e\Phi_f. \quad (25.4)$$

The exact location of the threshold energy is often uncertain; it is usually evaluated from the first peak in the second derivative of the plot of the X-ray yield against incident energy, since this peak locates the inflexion point of the first derivative curve.

The value of E_B equals the energy difference between the initial and final states of the system; the orbital energy is obtained by including a correction for relaxation effects. Since APS is concerned with two electrons of low energy and a core vacancy, whereas only one electron of higher energy and a core vacancy is involved in XPS, the two relaxation energies are not the same; as a consequence, the APS binding energies are generally lower than those evaluated from XPS data. APS can also be used for surface analysis, but the sensitivity compared with AES is very low and many elements cannot be detected.

25.4.2. Band structure of metals [6]

A result of interest to surface chemists is that the band-width of the lowest-energy peaks in the APS spectra of a series of 3*d* transition metal has been found to approximate to the width of the empty *d* band that is theoretically derived assuming a rigid *d* band. With alloys, the APS results are in good accord with the coherent potential approximation of the minimum polarity theory in which the two metal components have separated *d* bands.

25.4.3. Chemical shifts [6–7]

The appearance potential peaks of some 3*d* metals are shifted to higher values (0·5 to 2·0 eV) when oxygen is absorbed. These higher binding energies originate from excited states in the valence band and hence provide direct information about the chemisorption bond. The chemical shifts are different from those observed in AES and XPS, where the change in the binding energies of the inner-shell electrons arises from the different chemical environment at the surface created by the chemisorbed species.

With further development, appearance potential spectroscopy may well become important in chemisorption studies. The technique has experimental convenience and provides chemical information on valence electrons, but surface damage is comparatively high and the interpretation of the results is, as yet, only semi-quantitative.

REFERENCES

1. R. L. Park, J. E. Houston and D. G. Schriener, *Rev. Scient. Instrum.*, 1970, **41**, 1810.
2. R. L. Park and J. E. Houston, *J. Vac. Sci. Tech.*, 1971, **8**, 91.
3. R. L. Park and J. E. Houston, *Surface Sci.*, 1971, **26**, 664.
4. A. M. Bradshaw and D. Menzel, *Phys. Stat. Solid. B*, 1973, **56**, 135.
5. R. G. Musket and S. W. Taatjes, *J. Vac. Sci. Tech.*, 1972, **9**, 1041.
6. J. E. Houston and R. L. Park, *J. Chem. Phys.* 1971, **55**, 4601.
7. R. L. Park and J. E. Houston, *Phys. Rev. B*, 1972, **6**, 1073.

26

Electronic Properties of Adsorbed Layers†

26.1. EXPERIMENTAL ASPECTS OF ELECTRON SPECTROSCOPIC INVESTIGATIONS

Electron spectroscopy investigations are primarily concerned with the characterization of the energy spectra of electrons taking part in bond formation between an adsorbate and an adsorbent, i.e. the aim is to acquire information about the surface electronic structure and nature of the adspecies–metal surface bond, and to relate the measured quantities to those derived from quantum-mechanical calculations.

Common features of the techniques are: (i) the primary process is the excitation of electrons of the surface metal atoms (and of chemisorbed overlayers), either by bombardment with electrons under controlled acceleration conditions, or by electromagnetic radiation, such as X-rays or ultraviolet light of known photon energy;‡ (ii) the measured quantity is the distribution of kinetic energies of electrons ejected from, or scattered by, the surface under examination; and (iii) the depth examined in most cases is restricted to the few outermost layers of the surface. The escape depth of the emitted electron is limited because the cross-section of inelastic collision of electrons of energy < 2 keV is large.§

† See Plummer [1] and Gomer [2].

‡ The use of synchrotron radiation as a source of high-intensity and continuous energy distribution is now possible [3, 4].

§ Escape-depth data as a function of the kinetic energy of an emitted electron [5].

Ion-neutralization (INS) and field-emission (FES) spectroscopy provide data on wave functions at, or immediately above, the surface. The escape depth in ultra-violet photon spectroscopy (UPS), at glancing angle incidence, is about 5 Å, and in Auger (AES) and X-ray photoelectron (XPS) spectroscopy around 10 Å.

The range of energy states accessible to measurement also varies. In FES it is less than 2 eV below E_F, and limited to 10 eV in INS. In UPS it is about 30 eV, and extends to 2 keV and higher for AES, XPS and electron energy loss spectroscopy (EELS). Two categories of energy states are examined: (i) the density of occupied states in the valence band by UPS, INS, FES (and of core states in XES and XPS); and (ii) the density of unoccupied states at or above E_F, in the vicinity of the atom the core of which is being excited by EELS and APS (and XPS). The former group is used to investigate surface energy states, and the latter group is more concerned with energy states in the bulk.

An experimental limitation in electron spectroscopy is the signal-to-noise ratio, which should be as high as possible. The peak intensity depends primarily on the rate of intial ionization. X-ray excitation has a larger ionization cross-section than that of an electron beam of the same energy, and the secondary electron production responsible for the spectral background is less. However, high-intensity electron beams are easily generated, and separation of the background by electronic differentiation is very effective. The maximum ionization rate for core electrons is for a primary energy/electron binding energy ratio of 2 to 3, and the rate increases roughly as the fourth power of the atomic number of the atom. Ion neutralization spectroscopy is unique in that the penetration depth is less than the mean free path of the electrons; consequently, only a few scattered electrons contribute to give a low-intensity background.

A major disadvantage of all electron spectroscopic techniques is the surface damage brought about by the primary beam; in this respect, X-ray excitation is less harmful. The damage is severe with high-energy electron beams, but low-energy electrons are also responsible for electron-impact desorption and dissociation of chemisorbed layers, not only by the incident beam but also by the low-energy secondary electrons that are produced. Favourable experimental conditions are low-energy excitation, small beam currents, and the use of pulse-counting techniques. In field-emission spectroscopy, complications arise from the perturbation of the electron energy levels of the surface complex by the high externally-applied electric field, but there is no surface damage.

Information about the electronic structure of the adsorbate–metal surface molecule is derived: (i) from the energies of spectral peaks which are closely related to the electron binding energies; and (ii) from the chemical shifts of peaks when inner-shell electrons are excited. These strongly-bound electrons take no direct part in chemisorption bond formation and the chemical shift reflects in a complex manner only the change of valence-electron environment. The theoretical analysis of such shifts is very complicated and, as yet, little knowledge of bond formation has emerged. The data are particularly non-informative in AES in which three electronic transitions are involved; but XPS, with highly monochromatized X-ray beams and greatly improved analysers may well become increasingly useful, since only one electron-transition is involved [6].

The primary excitation is confined to valence electrons in INS, FES and UPS. In INS, an electron hole in the conduction band is created by neutralization of the surface inert-gas ion and the energy released is used to eject another electron by an Auger two-electron excitation process. The act of neutralization depends in detail on the nature of the surface potential and the theoretical interpretation of the spectrum is, therefore, uncertain. Coupled with its experimental complexity, this spectroscopy is not in general use, despite the fact that it could supply complementary information to that obtained by UPS [7]. Field-emission spectroscopy requires the measurement of the enhancement factor arising from resonance tunnelling through virtual levels that are formed when discrete levels of the free adsorbate have their energy counterparts in the continuum of metal states. The factor is related to the one-dimensional density of states at the surface; details of the electronic structure of clean metal surfaces, and of chemisorbed layers can be resolved [8].

Nevertheless, ultra-violet photoelectron spectroscopy is, at present, the most valuable source of information about surface electron properties, since comparison of the spectra of overlayers with those of free molecules allows an approximate assignment of bonding orbitals. However, transforming energies referenced to the Fermi level of the metal to those of free molecules referenced to the vacuum level is still an uncertain procedure. The addition of the effective work function to the former is customary and, preferably, the value relevant to the particular concentration of the adsorbate should be used. However, possible changes of the inner potential of the metal and modification of the surface dipole probably take place. Relaxation effects are also of uncertain magnitude; theoretical calculations for free molecules have been performed, but there is an additional unknown contribution due to the relaxation of the free electrons of the metal towards the positive hole in the final excited state. Experimental estimates have been made but the accuracy is not high.

Indeed, it is not clear whether the photo-emission from an adsorbed layer is the photo-ionization of the surface molecule or the ejection of an electron from the N-electron ground state of the metal–adsorbate–vacuum complex.

Theoretical analysis of photo-emission has, therefore, been mainly concentrated on one-electron spectroscopy, as in UPS. The use of difference spectra, obtained by subtraction of the spectrum of the clean surface from that of adsorbate-covered surface, is a possible means of eliminating the structure originating from the substrate (although it is not yet clear whether this procedure can be justified). The electronic structure of the adsorbate is then described in terms of density of states derived from the difference spectrum. It is probably a reasonably good representation when the adsorbate levels are below the conduction band, since the density of metal states with energies near the virtual levels of the adsorbate then make a neglible contribution. Thus, for CO chemisorbed on transition metals for which the adsorbate-induced levels are well separated from *d*-band energies of the metals, the photo-emission current is probably closely related to the density of states at the surface.

The most definitive information is obviously obtained from investigations using a single metal crystal plane as adsorbent. Effective cleaning procedures are available† and the freedom from contaminants can be effectively monitored by Auger spectroscopy. The determination of surface geometry by LEED is essential, since the structure of the chemisorbed overlayer is largely determined by that of the substrate, and the LEED data, reflecting the long-range periodicities of overlayers, provide valuable information complementary to those given by electron spectroscopy. In particular, with the accumulation of intensity data and the use of theoretical procedures that are now being refined, LEED data may soon supply reliable values of surface bond-lengths and the symmetry characteristics of the surface complex.

Most UPS measurements have been made with a primary beam of photon energy of 21·2 eV [He(I)] at a fixed incident angle and the measured electron flux is collected at a wide acceptance angle. A glancing incident beam ensures the maximum contribution of the outermost surface plane to the signal, but little work on the effect of varying the angle has been reported although depth profiling to explore attenuation effects and the variation of density of states within the first few layers would be valuable.

Other angle-dependent factors, however, are also important (see, for example, Gadzuk [10]). Thus, the description of the total energy distribution curve includes transition probabilities which vary not only with angle of incidence of the primary beam and its polarization, but also the orientation of the surface orbitals, e.g. for an orbital perpendicular to the surface the transition probability would be greatest when the electrical field vector of the primary beam is at normal incidence. Similarly, the effect of variation

† However, according to Bradshaw and Menzel [9], in order to avoid contamination, a scan time of only ~ 20 min is allowed in a background pressure as low as $\sim 10^{-9}$ Pa.

of the incident photon energy has not received much attention, although from existing data (see, for example, Feuerbacher and Adriaens [11] and Bradshaw *et al.* [12]) it would appear that a higher energy [40·8 He(II)] source might provide: (i) a simpler spectrum from the clean metal surface; (ii) an easier separation of bulk and surface emission with some gain in surface sensitivity; and (iii) values of the different relative peak intensities which, by comparison with those of a He(I) spectra, would assist in assignment of levels.

Recently, increasing interest has been shown in the theory and measurement of the anisotropic angular distribution of photoemitted electrons which arises from the specific orientation of orbitals to the surface. Maxima and minima appear at certain angles; their positions are related to the geometry of the metal surface, and their relative intensities to the geometry of adsorbate bonding. Thus, in the interaction of a p orbital of an adatom with the metal d orbitals, bridge bonding between two metal atoms and symmetrical bonding with four metal atoms should theoretically be distinguishable [13], although final-state scattering may make this distinction difficult.

It is emphasized that the concept of a localized surface molecule is implicit throughout these considerations; indeed, the position and extent of broadening of the virtual levels are the basic parameters that are used in the theoretical analysis of interactions in the chemisorbed state. However, the photoemission process is complex and such factors as the perturbation of the metal by the adsorbate, scattering of photoemitted electrons, interference effects between bulk and surface emission, relaxation effects, absence of an experimentally accessible reference energy state for adspecies, etc., hinder an accurate determination of orbital energies or of local density of states.

In circumstances where reasonable accuracy is sufficient, UPS has considerable value in finger-printing bonding orbitals and in identifying ad-

TABLE 25.1. Comparison of various techniques

	Experimental convenience (5)	Chemical information on inner shells (5)	Chemical information on valence electrons (5)
Auger electron spectroscopy	4·3	1·9	1·5
Appearance potential spectroscopy	3·7	2·3	1·0
Electron energy-loss spectroscopy	3·7	1·6	1·7
Ion neutralization spectroscopy	1·2	0·0	3·5
Ultra-violet photo-emission spectroscopy	3·1	0·2	4·1
X-ray photo-emission spectroscopy	3·2	4·3	2·7

species that are formed by surface dissociation or reaction of chemisorbed species.

A useful guide, representing an average assessment provided by workers in surface chemistry, for comparing the potentialities and usefulness of various electron spectroscopic techniques, has been recently published (Table 25.1) [14].

The main conclusions are: (i) AES ranks highest for sensitivity and speed of analysis for monitoring surface compositions and degrees of cleanliness; and (ii) UPS, in which photo-emission involves only one electronic level, is the most valuable technique for chemisorption investigations because the interpretations of UPS results is more likely to be rapidly developed in a quantitative manner than for the other techniques.

26.2. THEORETICAL APPROACHES TO CHEMISORPTION

Theoretical treatments of the electronic properties of clean metal surfaces have been outlined in Chapter 9; because of the ease of applying an extension of the bulk-lattice model to adatom/metal atom chemisorption bonds, this theoretical approach was also outlined. The recent advent of new electron spectroscopic techniques and, in particular, of field-emission resonance tunnelling and ultra-violet photo-emission studies, have stimulated further advances in the formulation of adatom/metal interactions, and calculations of theoretical parameters that are directly related to the experimental results obtained spectroscopically have been accomplished.

The chemisorption non-ionic surface bond involves electron sharing but it differs from normal chemical bond in that all the conduction electrons of the metal are available for participation in the formation of the adatom/

applied to surface studies [14]

Ease of interpretation of chemical information (5)	Lack of surface damage (5)	Ease of quantification (5)	Sensitivity and speed of analysis (5)	Total (Max. = 35)
1·8	2·1	2·9	4·8	19·3
1·5	0·9	1·4	2·5	13·3
1·7	2·2	1·3	2·4	14·6
1·9	3·7	1·4	2·1	13·8
2·7	4·4	1·8	3·1	19·4
3·4	4·3	3·4	3·1	24·4

metal complex. The characteristics of such quasi-localized bonding were first discussed by Gurney [15] in 1935, and his treatment still forms the basis of the model Hamiltonian approach.

The main purpose of this short review is to outline theoretical approaches of importance for the interpretation and analysis of electron spectroscopic data and for the evaluation of the density of states of one-electron energy levels at the adsorbate from which charge transfer and binding energies of adatom/metal systems can be predicted [10, 16–31].

26.2.1. Model Hamiltonian [10, 16–31]

When an isolated adsorbate atom approaches a metal surface, its discrete electron-occupied energy level ε_a associated with the wave function ϕ_a interacts with the continuum of metal states described by wave functions ϕ_m with overlap of the wave functions. The states of the adatom/metal system are then constructed from linear combinations of the unperturbed metal states $|\mathbf{k}\rangle$ with eigenvalues $\varepsilon_{\mathbf{k}}$ and the ground state of the isolated atom $|a\rangle$ of eigenvalue ε_a, where $\mathbf{k}$ is the quantum wave number of the metal states.

At the adatom/surface equilibrium distance, electron tunnelling through the potential barrier separating the metal and adatom, both from the metal to the adatom and by the adatom electron to the metal, takes place, with the consequence that a metal electron has a transient existence on the adatom. The characteristic tunnelling time is denoted by τ. In accord with the uncertainty principle, the original narrow electron level of the adatom is broadened to a half-width Δ_a, where $\Delta_a \simeq h/\tau$, and a virtual or resonance adsorbate state is formed. This broadened level provides the spectrum of the molecular orbitals of the surface complex.

The bonding interaction arising from orbital overlap is responsible for a downward shift of the energy peak of the virtual state to below the free-atom level. At the same time, the intra-atomic Coulomb repulsion U between two electrons of opposite spins transiently present at the adatom produces an upward shift. Another contribution must also be included because the ionization potential of the atoms is increased and its electron affinity is decreased by the attractive image potential ($V_{im} = e^2/4x$, at a distance x from the surface), thereby reducing the repulsion to $U' = U - 2V_{im}$. When U' is less than $\pi\Delta_a$, i.e. when the electron spends very little time at the adatom, the Hartree–Fock approximation remains valid and a non-magnetic solution of the Hamiltonian can be obtained. The net energy shift is referred to as the level shift function $\Lambda(\varepsilon)$.

The properties of the broadened virtual level can be expressed in terms of the local density of states ρ_a at the adsorbate, where $\rho_a(\varepsilon)$ is given by

$$\rho_a(\varepsilon) = \sum_{\mathbf{k}} \langle a|\mathbf{k}\rangle^2 \delta(\varepsilon - \varepsilon_{\mathbf{k}}). \tag{26.1}$$

In Eq. (26.1), $\varepsilon_{\mathbf{k}}$ is the eigenvalue of the eigenvector ($H|\mathbf{k} = \varepsilon_{\mathbf{k}}|\mathbf{k}\rangle$) and $\delta(\varepsilon - \varepsilon_{\mathbf{k}})$ is the density of states at the energy ε. Use is made of Green's function, $G = 1/(\varepsilon - H - i\alpha)$, where α is a small positive quantity; in particular,

$$G_{\mathbf{kk}} = 1/(\varepsilon - \varepsilon_{\mathbf{k}} - i\alpha), \tag{26.2}$$

in which the imaginary part is

$$\mathrm{Im}\, G_{\mathbf{kk}} = i\alpha/[(\varepsilon - \varepsilon_{\mathbf{k}})^2 + \alpha^2]. \tag{26.3}$$

The local density of states $\rho_a(\varepsilon)$ can then be written as

$$\begin{aligned} \rho_a &= (1/\pi)\, \mathrm{Im} \sum \langle a|\mathbf{k}\rangle \langle \mathbf{k}|G|\mathbf{k}\rangle \langle \mathbf{k}|a\rangle \\ &= (1/\pi)\, \mathrm{Im}\, G_{aa}; \end{aligned} \tag{26.4}$$

and Im G_{aa} is closely related to the electron spectroscopic results.

26.2.2. The Newns–Anderson approach

Newns [24, 25] has extended Anderson's treatment [29] of an impurity atom in a metal to an adatom at the surface of a metal; use is made of the restricted Hartree–Fock approach (the self-consistent-field molecular-orbital treatment) for the case where $U < \pi\Delta_a$, i.e. where the electron correlation contribution can be neglected. Up and down spins are balanced, i.e. the average charge density is then the sum of contributions from each of the occupied one-electron states.

The full Hamiltonian H is equal to the sum of spin-dependent Hamiltonians H^σ

$$H = \sum_{\sigma} H^\sigma, \tag{26.5}$$

and

$$H^\sigma = H_m + V^\sigma = H_m + \langle n_\sigma\rangle + V_{a\mathbf{k}}. \tag{26.6}$$

In this equation, H_m is the unperturbed metal Hamiltonian and V^σ the perturbation due to the presence of the adatom; $\langle n_\sigma\rangle$ is the average population of σ-spin electrons on the adatom.

All off-diagonal matrix elements of H are neglected except those coupling $|\mathbf{k}\rangle$ to $|a\rangle$, i.e. $H_{a\mathbf{k}} = V_{a\mathbf{k}}$. It then follows that

$$H^{\sigma}_{aa} = \varepsilon_a + \langle n_{\sigma}\rangle U_e + V_{im} \equiv \varepsilon_{a\sigma}, \tag{26.7}$$

where U_e is the effective Coulomb repulsion of the electrons obtained by correcting U for correlation (and screening) effects.

The adatom Green's function is

$$G^{\sigma}_{aa}(\varepsilon) = \left[\varepsilon - \varepsilon_{a\sigma} - \sum_{\mathbf{k}} \frac{|V_{a\mathbf{k}}|^2}{(\varepsilon - \varepsilon\mathbf{k} - i\alpha)}\right]^{-1}, \tag{26.8}$$

from which the density of states of spin σ at the adatom is

$$\rho^{\sigma}_{aa} = -(1/\pi)\,\mathrm{Im}\,G_{aa}(\varepsilon) = \frac{\Delta_a/\pi}{(\varepsilon - \varepsilon_{a\sigma} - \Lambda(\varepsilon))^2 + \Delta_a^2(\varepsilon)}. \tag{26.9}$$

The electron charge number of electrons with spin σ can then be evaluated using Eq. (26.10):

$$\langle n_{a\sigma}\rangle = \int_{-\infty}^{0} \rho_{aa}(\varepsilon)\,\mathrm{d}\varepsilon; \tag{26.10}$$

and

$$\Delta_a(\varepsilon) = \pi \sum_{\mathbf{k}} |V_{a\mathbf{k}}|^2 \delta(\varepsilon - \varepsilon_{\mathbf{k}}). \tag{26.11}$$

The possible energy levels of the adatom $\varepsilon_{a\sigma}$ are real roots of Eq. (26.12):

$$\varepsilon - \varepsilon_{a\sigma} - \Lambda_a(\varepsilon) = 0. \tag{26.12}$$

Two cases arise.

(i) $\Delta_{\alpha}(\varepsilon) = 0$ (as for strong chemisorption). Here ε is within the allowed energy band of the metal (cf. Eq. (26.11)), i.e. some values of $\varepsilon_{\mathbf{k}} = \varepsilon$; and the adatom level is a virtual or resonance state that has undergone a bonding–antibonding splitting with one empty state above the metal band and one filled state below it. Bonding with formation of a surface complex ensues, since there is a net lowering in the energy of all the filled states below E_F. A strong localized interaction can, therefore, give rise to different adsorption states on different crystal faces and also on the same face due to existence of these split-off states which have characteristic orbital symmetries and different bonding energies. For less strong interaction, however, a localized surface complex state at the lower energy with a virtual state at higher energy is formed (see Fig. 26.1).

(ii) $\Delta_a(\varepsilon) = 0$ (as for weak interaction). Here ε falls outside the metal band. When the original unperturbed level of the adatom ε_a is also outside the band, only one virtual state below (or above) the band is a bonding state; bur characteristic bonding is absent and the energy level spectrum is featureless.

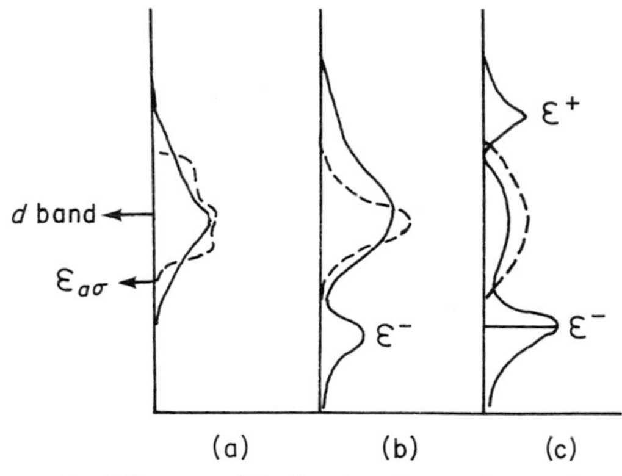

FIG. 26.1. Energy-level diagram of the density of states for an adatom chemisorbed on a transition metal: (a) no localized bonding; (b) formation of a higher-energy virtual state and a lower-energy surface molecule state (ε^-); (c) formation of two states ε^+, ε^- with energies outside the d band.

The chemisorption energy may also be evaluated as the energy difference between the initial E_i and final state from Eq. (26.13):

$$E_c = E_i - \sum_{\sigma} \int_{-\infty}^{0} \varepsilon \rho_{aa}^{\sigma}(\varepsilon)\, d\varepsilon + U_e \langle n \rangle^2, \qquad (26.13)$$

where E_i is the energy of the electrons involved in bonding with $V_{ak} = 0$.

26.2.2.1 *Group orbitals*

The adatom/metal bond energy can be evaluated from an approximate form of Eq. (26.11), i.e.

$$\Delta_a \approx \pi \rho(\varepsilon) \langle |V_{ak}|^2 \rangle_{(\text{average})}, \qquad (26.14)$$

where $\rho(\varepsilon_\sigma)$ is the density of metal states at energy ε; and ρ_{aa}^{σ}, given by Eq. (26.9), is then a Lorentzian defined by $\varepsilon_{a\sigma}$ and Δ_a. However, Δ_a is experimentally found to be $\lesssim 1$ eV and, from Eq. (26.14), V_{ak} is then $\lesssim 0{\cdot}75$ eV. But for strong chemisorption the bond energy is $\gtrsim 3$ eV.

This disagreement can, however, be resolved by introducing the concept of group orbitals. The molecular orbitals of the surface complex are formed from the valence atomic orbital of the adsorbate; a group of localized orbitals of a specific symmetry associated with a small set of metal atoms is chosen to correspond with the local symmetry of the (assumed) bond configuration. This concept leads to a localized bond of a character that is now consistent with the band theory of metals. The substrate orbitals are rep-represented by Bloch states (periodic atomic orbitals modulated by a plane-wave function) whose eigen states are given by

$$|\mathbf{k}\rangle = (1/\sqrt{N}) \sum e^{i\mathbf{k}.\mathbf{R}_i} \phi(\mathbf{r} - \mathbf{R}_i), \tag{26.15}$$

where $\mathbf{R}_i$ is the position of the ith ion-core; $\phi(\mathbf{r} - \mathbf{R}_i)$ is the atomic-like orbital on the ith ion-core, and N is the number of atoms comprising the metal. The group orbital of the correct symmetry for coupling with the adatom is written as

$$|g\rangle = (1/\sqrt{N_g}) \sum_i a_i |\phi_g(\mathbf{R}_i)\rangle, \tag{26.16}$$

where N_g is the number of centres in the orbital, ϕ_g is the metal orbital of symmetry g, and $a_i = \pm 1$ depending on the orbital symmetry. The level width is then given by

$$\Delta_g(\varepsilon) \approx \pi \sum_{\mathbf{k}} \langle a|H|g\rangle \langle g|\mathbf{k}\rangle^2 \delta(\varepsilon - \varepsilon_g(\mathbf{k})), \tag{26.17}$$

where $\Delta_g(\varepsilon)$ is the dominant term in the total $\Delta(\varepsilon)$, and $|a\rangle$ couples only with that part of $|\mathbf{k}\rangle$ which projects on to $g>$;

$$\langle a|H|g\rangle \langle g|\mathbf{k}\rangle = (1/N_g) \sum_{ij} a_i a_j \langle a|H|\phi_g(\mathbf{R}_i)\rangle \langle \phi_g(\mathbf{R}_j)|\mathbf{k}\rangle \tag{26.18}$$

describes the overlap of atomic orbital of the absorbate with one of the N_g equivalent metal atoms. The level width function $\Delta_g(\varepsilon)$ becomes

$$\Delta_g(\varepsilon) \equiv \pi |N_g \beta_g|^2 \rho_g(\varepsilon), \tag{26.19}$$

where

$$N_g \beta_g \equiv \sum_i a_i \langle |H|\phi_g(\mathbf{R}_i)\rangle \tag{26.20}$$

and a specific band structure $\varepsilon = \varepsilon_g(\mathbf{k})$ has been assumed. The surface density of states $\rho_g(\varepsilon)$ at ϕ_g is then given by

$$\rho_g(\varepsilon) = \sum_{\mathbf{k}} |\langle g|\mathbf{k}\rangle|^2 \, \delta(\varepsilon - \varepsilon_g(\mathbf{k})). \tag{26.21}$$

Strong interaction is represented by the overlap integrals between ϕ_a and the ϕ_g states, which must be in a tight-binding band such as the d band

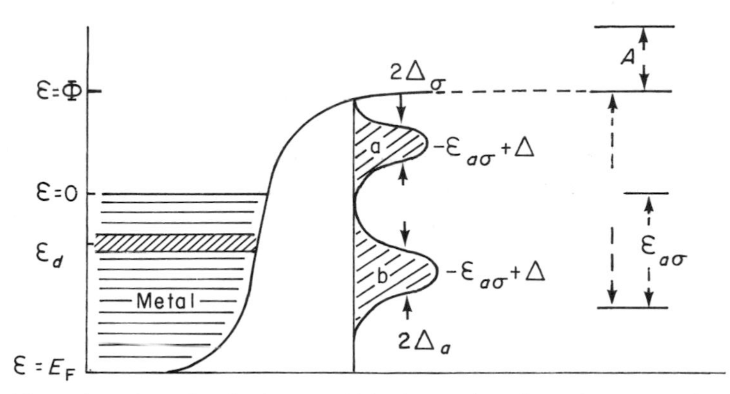

FIG. 26.2. Formation of two localized states; (a) by interaction of atomic state ε_a with d-band states (ε_d), followed by (b) weak interaction of molecular-orbital states ε^+, ε^- with the s-band states to form two virtual molecular states.

of transition metals, and the projection of the density of states of the metal on ϕ_g. Two localized states corresponding to discrete bonding and antibonding levels outside this narrow d band of energy peaked at ε_d are formed (see Fig. 26.2). These discrete virtual states are at energy ε^+ and ε^-, where

$$\varepsilon^{\pm} = \frac{\varepsilon_a + \varepsilon_d}{2} \pm \tfrac{1}{2}[(\varepsilon_a - \varepsilon_d)^2 + 4(N_g\beta_g)^2]^{1/2}. \tag{26.22}$$

The term $(N_g\beta_g)^2$ is evaluated from the experimental level shift using Eq. (26.23):

$$\Lambda(\varepsilon) = |N_g\beta_g|^2/(\varepsilon - \varepsilon_d). \tag{26.23}$$

Consequently, the binding energy results from the formation of the surface molecule from these states. The level width, however, is brought about by their additional weak interaction with the s-band continuum (see Fig. 26.3). The disagreement between the value of $V_{a\mathbf{k}}$ deduced from Λ_ε using Eq. (26.14) and the experimental binding energy is, therefore, resolved.

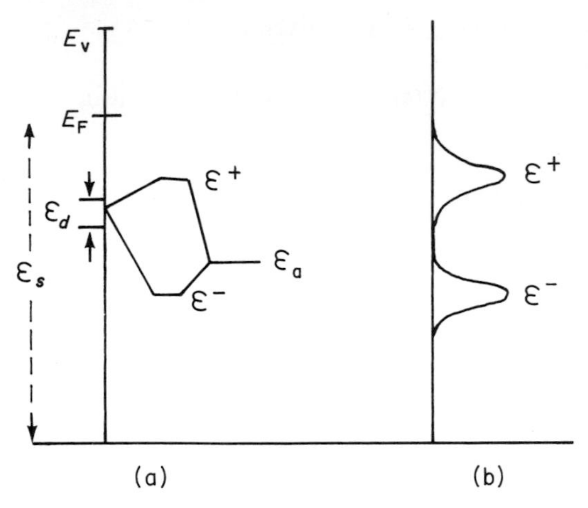

FIG. 26.3. Interaction of an adatom chemisorbed on a transition metal showing (a) ionic adspecies with valence level avove E_F largely unoccupied and (b) co-valent binding with adsorbate virtual state below E_F fully occupied.

26.2.2.2. *Comparison of some theoretical and experimental results*

Experimental data from field-emission resonance tunnelling and ultraviolet photoelectron spectra are available for the H adatom/W(100) system. Virtual levels appear at ~1 eV (below the Fermi level) with a half-maximum width 0·6 eV, and at ~5·4 eV with width 0·7 eV. The group orbital resonance is identified with the maximum in the density of states of the *d* band at ~2 eV as derived from UPS studies. For an (assumed) effective Coulomb repulsion U_e of ~5 cV, the "1-eV" resonance peak and the binding energy are evaluated using Eq. (26.12),

$$\varepsilon - \varepsilon_{a\sigma} - \Lambda(\varepsilon) = 0,$$

which may be transformed to

$$\varepsilon = \varepsilon_{a\sigma} + U_e\langle n\rangle/2, \qquad (26.24)$$

where $\langle n\rangle$ is the total spin population and $\langle n_\sigma\rangle = \langle n_{-\sigma}\rangle = \frac{1}{2}$ to give the non-magnetic solution. The binding energy is then ~3 eV, i.e. fairly close to the experimental value, and the resonance peak energy is roughly 1 eV.

There is also a weak *d* band photoemitted from the W(100) plane with a maximum at ~4·5 eV: from this value, bonding and antibonding levels at ~6 and ~1 eV below E_F can be estimated. From both calculations, the adatom carries an excess negative charge (as in some bulk hydrodes); this is also consistent with the positive change of work the function observed experimentally for this system.

26.2.2.3 *Adsorbate–adsorbate interactions* [23, 36, 37]
Repulsion by electrostatic dipole–dipole interactions of neighbouring chemisorbed atoms decreases rapidly with increasing distance between the adatoms. In addition, analogous to the presence of an impurity atom in a metal whereby local perturbations of electron density of a long range and oscillatory character are set up, adatoms can also interact indirectly through the metal. The effect is to lower the local potential field within the metal and wave functions of neighbouring adatoms have an oscillatory character. Thus, for the chemisorption of alkali metals on (110)W, a virtual level near the Fermi level is created. The perturbation by the adatom is found to decrease rapidly in the (110) direction but less so in all other directions; it is responsible for a change in occupancy $\Delta n(\mathrm{R})$ of an orbital level of a surface metal atom at a distance R from the adatom such that $\Delta n(\mathrm{R})$ is proportional to $(\sin 2{\cdot}8R)/R^x$ with $x = 3$ along the (110) direction and $x = 5$ in other directions. The perturbation is, therefore, oscillatory, anisotropic and long range.

The change of occupancy results from a splitting of the original virtual level of the isolated adatom into two virtual overlapping levels of different widths when a second adatom occupies a neighbouring position. The electron population of the levels is, thereby, changed; if the lower level is the narrower, its electron occupancy is increased, and the net interaction between the adatoms decreases as $1/R^2$. The effect is significant only when the virtual level of the isolated adatom is close to E_F; it is, therefore, negligible for most gaseous adsorbates. For hydrogen, there is a small repulsion between two adatoms on neighbouring sites (although the free atom 1s level is ~9 eV below E_F). Nevertheless, repulsion between two electrons on this level is very large (~15 eV), and is sufficient to decrease the energy of the virtual level to close to the Fermi level. A detailed treatment [36] of the theory of indirect interactions between H adatoms has successfully predicted adlayer structures at different coverages closely consistent with LEED observations.

26.2.3. The valence-bond (Schreiffer–Paulson–Gomer) theory [38, 39]

A major difficulty of the LCAO–MO method is to include correlation effects arising from electron tunnelling from adsorbate to metal and, thereby, to avoid an excessively large ionic contribution to the binding energy. In the valence-bond method, such contributions are ignored and electron correlations are included by assuming electron-pair bonding. It is then necessary to promote metal electrons from the Fermi level in order to create free spins for pairing with the adsorbate spin. The subsequent formation of the valence bond provides energy which overcompensates for that required for electron promotion. This theory correctly predicts that the binding energy of an adatom will increase as the *d* band narrows, i.e. that stronger chemisorption is observed on those transition metals having narrower *d* bands. However, this many-body approach makes quantitative analysis difficult and imposes severe restrictions on its application.

26.2.3. The linear response method

The basis of this method, in which the metal is treated as a jellium, has been given in Chapter 12. On minimization of the total energy of the adsorption system, electron-density distributions at energies near the Fermi level and work functions may be obtained in good agreement with experimental values. The model ignores the band structure of the metal and is, therefore, highly approximate for transition metals; but calculations of dipole moments, resonance levels and less so, of ionic adsorption energies are reasonably in accord with experimental data for a bare proton embedded in the electron gas at the metal surface. Extensions to other systems are being attempted, but calculations pose some considerable difficulties.

REFERENCES

1. E. W. Plummer, *Topics Appl. Phys.*, 1975, **4**, 153.
2. R. Gomer, *Adv. Chem. Phys.*, 1974, **27**, 211.
3. D. E. Eastmann, W. D. Grobman, J. L. Freeouf and M. Erbudak, *Phys. Rev. B*, 1974, **9**, 3473.
4. K. Codling, *Reports Prog. Phys.*, 1973, **36**, 541.
5. I. Lindau and W. E. Spicer, *J. Electron Spectr.*, 1974, **3**, 409.
6. J. T. Yates, T. E. Madey and N. E. Erickson, *Surface Sci.*, 1974, **43**, 257 and 526.
7. H. D. Hagstrum, *J. Vac. Sci. Tech.*, 1973, **10**, 264.
8. D. Penn and E. W. Plummer, *Phys. Rev. B*, 1974, **27**, 211
9. A. M. Bradshaw and D. Menzel, *Vakuum Technik*, 1975, **24**, 15.
10. J. W. Gadzuk, *Phys. Rev. B*, 1974, **10**, 5030.
11. B. Feuerbacher and M. R. Adriaens, *Surface Sci.*, 1974, **45**, 553.
12. A. M. Bradshaw, D. Menzel and M. Steinkilbert, *Faraday Disc. Chem. Soc.*, 1975, **58**.
13. A. Liebsch, *Phys. Rev. Letters*, 1974, **32**, 1203.
14. R. W. Joyner and M. W. Roberts, *in* "Surface and Defect Properties of Solids" (M. W. Roberts and J. M. Thomas, eds), The Chemical Society, London, 1975, Vol. 4, Chap. 3, p. 68.
15. R. W. Gurney, *Phys. Rev.*, 1935, **47**, 479.

16. J. W. Gadzuk, *Surface Sci.*, 1967, **6**, 133, 159.
17. J. W. Gadzuk, *J. Phys. Chem. Solids*, 1969, **30**, 2307.
18. J. W. Gadzuk, *Surface Sci.*, 1974, **43**, 44.
19. T. B. Grimley and S. M. Walker, *Surface Sci.*, 1969, **14**, 395.
20. T. B. Grimley, *J. Vac. Sci. Tech.*, 1971, **8**, 31.
21. T. B. Grimley, *in* "Adsorption–Desorption Phenomena" (F. Ricca, ed.), Academic Press, New York, 1972, p. 215.
22. T. B. Grimley and B. J. Thorpe, *J. Phys.*, 1971, F1, L14.
23. T. B. Grimley and M. Torrini, *J. Phys. C*, 1973, **6**, 868.
24. D. M. Newns, *Phys. Rev.*, 1969, **178**, 1123.
25. D. M. Newns, *J. Chem. Phys.*, 1969, **50**, 4572.
26. D. M. Newns, *Phys. Rev. B*, 1970, **1**, 3304.
27. D. M. Newns, *Phys. Letters*, 1972, **38A**, 341.
28. D. M. Newns, *Phys. Rev.*, 1970, **25**, 1575.
29. P. W. Anderson, *Phys. Rev.*, 1961, **124**, 41.
30. J. Hubbard, *Proc. Roy. Soc. A*, 1963, **276**, 238.
31. J. R. Schrieffer and D. C. Mattis, *Phys. Rev. A*, 1965, **140**, 1412.
32. B. J. Thorpe, *Surface Sci.*, 1972, **33**, 306.
33. J. W. Gadzuk and E. W. Plummer, *Rev. Mod. Phys.*, 1973, **45**, 487.
34. B. J. Waclawki and E. W. Plummer, *J. Vac. Sci Tech.*, 1973, **10**, 292.
35. B. Feuerbacher and B. Fitton, *Phys. Rev. B*, 1973, **8**, 4890.
36. R. L. Einstein and J. R. Schrieffer, *Phys. Rev. B*, 1973, **7**, 3629.
37. D. L. Adams, *Surface Sci.*, 1974, **42**, 12.
38. J. R. Schrieffer and R. Gomer, *Surface Sci.*, 1971, **25**, 315.
39. R. H. Paulson and J. R. Schrieffer, *Surface Sci.*, 1975, **48**, 329.
40. N. D. Lang and W. Kohn, *Phys. Rev. B*, 1971, **3**, 1215.
41. N. D. Lang, *Solid State Phys.*, 1973, **28**, 225.
42. J. R. Smith, S. C. Ying and W. Kohn, *Phys. Rev. Letters*, 1973, **30**, 610.
43. P. Hohenberg and W. Kohn, *Phys. Rev. B*, 1964, **136**, 864.
44. W. Kohn and L. J. Sham, *Phys. Rev. A*, 1965, **140**, 1133.

27

Review of Some Experimental Data

The main objectives of chemisorption studies are: (i) to interpret the nature and magnitude of the chemisorption bond in terms of the geometric, electronic and vibrational properties of the adsorbate–adsorbent system; and (ii) to explain the kinetics and mechanisms of adsorption and desorption rate processes. An assessment of the present progress towards these aims is now attempted. It is restricted to the simple adsorbate molecules, hydrogen, carbon monoxide and nitrogen and, for the most part, to tungsten as adsorbent, since these systems, being the most extensively and intensively investigated, provide the most definitive and detailed information. The extensive data on the chemisorption of oxygen by transition metals are much more difficult to interpret because of complications of incorporation, surface reconstruction and oxidation.

27.1. PROGRESS TOWARDS THE IDEAL SYSTEM

Experimental results reported prior to 1950 have little fundamental interest because of the substantial contamination of the metal surfaces and their unknown surface structures. These difficulties have been virtually eliminated by use of effective cleaning procedures and ultra-high vacuum techniques, and the availability of spectroscopically ultra-pure metals. Contamination in an amount equivalent to $\sim 1\%$ monolayer capacity can be detected (and its nature identified) by Auger spectroscopy, but this sensitivity has only been achieved in the last decade as a result of the introduction of electronic differentiation, improvements in the signal-to-noise ratio and higher instrumentation resolution. It has now become evident that a clean surface may be subsequently contaminated by slow diffusion of trace impurities from the bulk of a high-purity metal to its surface [1], and that surface structures different from the bulk may possibly be stabilized by this impurity.

Investigations with clean surfaces were first performed using polycrystalline filaments of tungsten and evaporated films. Chemisorption on many transition metals proved to be a non-activated process with a consequence that residual pressures of 10^{-7} to 10^{-8} Pa must be maintained to avoid substantial contamination within the time-period of experimentation. Considerable improvements in the sensitivity of ionization gauges opened the way to measurements of the initial rates of adsorption (and sticking probabilities) and desorption, and meaningful (mean) values of equilibrium thermodynamic properties were recorded.

The recognition that chemisorption parameters are dependent on the surface crystallography of the metal adsorbent awaited the advent of field-emission microscopy in 1950 and of field-ion microscopy ten years later; rapid developments, both experimentally and in the theoretical interpretation of the results of low-energy electron diffraction measurements, then provided quantitative information on the geometry of individual crystal planes and their influence on the long-range periodicity and dimensions of the unit cell of chemisorbed layers.

Finally, the application of various electron spectroscopic techniques, which had already yielded magnitudes of valence-orbital energies of free molecules, to chemisorbed layers during the last 5 or 6 years, has provided the means to investigate the electronic properties of chemisorbed species.

27.2. EXPERIMENTAL APPROACHES

27.2.1. Crystallographic studies

Detailed knowledge of the surface crystallography of the metal adsorbent is an essential requirement before an investigation of the properties of a chemisorbed layer is commenced. The use of ion bombardment and subsequent thermal annealing has greatly extended the number of clean metal surfaces that can be examined, particularly since the efficacy of the cleaning process can be sensitively monitored by Auger spectroscopy. Long-range order and symmetry characteristics are determined by LEED, and the location of individual atoms and the presence of disorder can be examined by field-ion microscopy. Non-ideality (see Clarke *et al.* [2]) of the surface structure (i.e. different from the bulk structure) often occurs. Thus, the low-index planes of b.c.c. transition metals (e.g. W, Mo) normally have the ideal (1 × 1) structure but some of the (111) planes are slightly disordered. The (111) planes of the f.c.c. metals (e.g. Pt, Ni) are invariably ideal but the (100)Pt face has a (5 × 1) symmetry due to an hexagonal outermost layer of atoms. The presence of non-ideality is often responsible for "abnormal" properties of some overlayers. In general, planes of high symmetry and low surface energy are most likely to be ideal and to have a minimal defect concentration.

The information derived from LEED patterns of chemisorbed layers is limited; the dimensions and orientation of the unit cell can be unambiguously derived, but the location of an adatom within the cell is unknown or, at best, uncertain. In principle, this knowledge should be obtainable from spot intensity measurements, but the theoretical computations involve parameters the magnitudes of which are imprecisely known. Similarly, the presence of multiplet states, and the assessment of degree of order in submonolayer coverages, cannot be directly inferred from the LEED pattern.

Nevertheless, other qualitative conclusions can be drawn. Thus, the frequent observation of a (2 × 2) structure confirms the existence of repulsive interactions between the adsorbed molecules, since the occupation of adjacent sites is energetically unfavourable. Such interactions are attributed *inter alia* to dipole–dipole repulsion and "inter-penetration" of valence electron shells at saturation coverage. However, additional effects result from indirect overlap of wave functions because the free electrons of the metal become involved in the partial delocalization of the adsorbate–adsorbent bond. This interaction is oscillatory, in that it may be repulsive or attractive depending on the distance of the adspecies from the "localized" adsorption site [3–5]. The attractive interactions are responsible for island formation which may also involves *s–d* hybridization in the surface plane. Its occurrence can be inferred from the change of intensity of half-order beams in the LEED pattern, since the intensity increases approximately linearly with coverage when islands are formed and are distinguished from random occupation of unoccupied sites for which the intensity increases as the square of the coverage [6, 7].

For the CO/Ni system, the fairly abrupt decrease of adsorption heat at ~50% coverage when the (2 × 2) structure is complete [8] suggests that random filling of sites takes place at higher coverages, but this conclusion is not supported by the change of the LEED pattern, which indicates the formation of a close-packed layer, i.e. the adspecies are displaced from the original high-symmetry adsorption sites and, at saturation coverage, a compressed layer is formed [9].

27.2.2. Kinetic studies

Detailed studies of the kinetics of desorption, initially using flash-desorption and later temperature-programmed procedures, revealed the existence of multiplet states of a single adsorbed species. From the thermal desorption spectra, the number of states, their binding energies and their population can be evaluated and, from the kinetic order of the evolution rate, a dissociative chemisorption process may be distinguished. Quantitative analysis of the desorption profiles involves simplifying assumptions. Departures arise from intrinsic and induced heterogeneity, interconversion between states, co-operative and other interaction effects. The use of single crystal planes in place of polycrystalline filaments simplifies the interpretation of the results, but whether two sub-states have distinct bonding structures or reflect the effect of interactions within a single sub-state is often uncertain [8]. Nevertheless, particularly when these results are complemented by those from LEED and other techniques, it remains the most effective and informative method for multiplet-state investigations.

Measurements of rates of adsorption have provided magnitudes of sticking probabilities, evidence for the existence of precursor states (and consequently the need to reformulate the kinetics), and have shown that a low population of a particular state or sub-state may be kinetically restricted by a very small sticking probability and not by its thermodynamic properties.

27.2.3. Equilibrium studies

Isosteric heats of chemisorption on specific crystallographic planes have recently been determined for the first time, for CO on Ni(100), Pd(100) and on Cu(100), using the change of work function to monitor the surface concentration [9–12].

Work function measurements, however, have been made with most gas/metal systems. The dipole character of individual states and sub-states can often be evaluated and, because of the difference in magnitude (and sometimes of sign), the change of work function with coverage provides evidence for sequential population (or otherwise) of some sub-states. Theoretical treatments of chemisorption phenomena involve a knowledge of the extent of charge transfer, the magnitude of which can be semi-quantitatively inferred from the surface potential of the adspecies. Unfortunately, the dipole values are not sufficiently precise because the bond length is uncertain and has often to be assumed to be equal to the covalent radius of the adatom.

In the next section, the present status of our knowledge of chemisorption phenomena, as inferred from the published results obtained from various experimental procedures, is outlined for the CO/W, N_2/W and H_2/W systems.

27.3. SOME GAS/TUNGSTEN SYSTEMS

27.3.1. Carbon monoxide

This system has been one of the most extensively investigated [13] because: (i) clean W surfaces can be produced simply by high-temperature flashing; (ii) field- and ion-emission microscope tips of W can be easily fabricated; and (iii) CO is molecularly chemisorbed (although this conclusion may be questioned) at room temperature and below.

Evidence for (iii) comes from various sources: (i) prolonged exposure of a clean microscope tip to 10^{-6} Pa pressure of CO at room temperature does not give any indication of the highly characteristic C adatom pattern [14, 15]; (ii) repeated adsorption and flash desorption of CO from a polycrystalline W filament does not modify the adsorption isotherm of CO on the clean metal (both of these tests are very sensitive to trace contamination by C adatoms); and (iii) the kinetics of desorption are first-order [16].

Nevertheless, bombardment of a CO overlayer by low-energy ($\sim$12 eV) electrons (as in electron-stimulated desorption) effects dissociation into C and O adatoms.† Moreover, incipient dissociation of CO on W(110) is indicated in the UPS spectrum, in that its spectral features can be simulated by a linear summation of the individual spectra of C and O adatoms.‡ Rupture of the C—O bond, and formation of this bond by combination of C and O adatoms separately adsorbed, are both thermodynamically feasible [16, 17]; and a second-order desorption process is experimentally observed from one of the multiplet states [16]. This latter, however, might be due to desorption of pairs of CO molecules as was postulated to account for the isotopic mixing [18] of $^{12}C^{18}O$ and $^{13}C^{16}O$ via an intermediate complex

```
C—O
|  |
O—C
```

bonded to an array of four W atoms by alternate C and O atoms.

The probable explanation is a two-site adsorption of a CO molecule lying flat on the W surface with the C and O atoms bonded to two adjacent W atoms. The strong perturbation arising from such bonding may: (i) greatly weaken, but not rupture, the C—O bond (incipient dissociation); or (ii) break the bond while the complex retains a pseudo-molecular identity due to the strong localized interaction between the neighbouring unlike adatoms. At the temperature of desorption, these adatoms combine and molecular CO is evolved [19].

† See Chapter 4.
‡ See Chapter 22.

A two-site adsorption state can also be formed by a C atom bond to two adjacent W atoms as has been assigned to infra-red bands below 2000 cm^{-1} for CO chemisorbed on many transition metals. However, the two typical infra-red bands prominently displayed for such systems are absent, or at best only faintly visible, with W as adsorbent, although they have been observed after evaporation of tungsten in the presence of a low pressure of CO, or after its co-evaporation with CaF_2 and exposure to this gas.† The low-intensity, or absence, of the bands is consistent with a flat two-site structure, rather than a co-bridging state.

Either type of structure is consistent with: (i) a c(2 × 2) LEED pattern at low coverages; (ii) a ratio greater than unity for the number of H adatoms to CO admolecules in saturated "monolayers" of these gases; and (iii), the variation of chemisorption heats with crystal plane because of different distances apart of adjacent W atoms.

Two main chemisorbed states can also be distinguished in UPS,‡ viz.: (i) a β state formed at lower coverages; and (ii) an α state which largely obscures the β state at higher concentrations and gives two peak energies similar in magnitude to those obtained on other transition metals; it probably has a linear one-site bonding structure of lower desorption energy.

These two states have been previously [20, 21] distinguished in thermal desorption data from polycrystalline filaments, as (i) a low-temperature α state having peak temperatures between 200 and 500 K with binding energies of 80 to 120 kJ mol^{-1}, and (ii) a β state desorbed between 1100 and 1850 K with energies of 200 to 420 kJ mol^{-1}. In addition, sub-states (α_1, α_2(?)) and three β substates with binding energies of 220 to 250 kJ mol^{-1} (β_1), 285 to 325 kJ mol^{-1} (β_2) and 310 to 420 kJ mol^{-1} (β_3) were later identified. The range of values probably results from different concentrations of the two states on the various low-index planes of the polycrystalline filament.

Some results for individual crystal planes are now available [16, 22]. The three β states on W(100) have binding energies of 240 kJ mol^{-1} (β_1), 270 kJ mol^{-1} (β_2) and 315 kJ mol^{-1} (β_3), and populate *ca* 60% of the sites, the contributions from the β_1 and β_2 states being only *ca* 20%. The β_3 state is absent on W(110), but the binding energies of β_1 (210 kJ mol^{-1}) and β_2 (275 kJ mol^{-1}) are similar to those on the (100) plane, and have a population of about 50% of the total number of adsorption sites. Work functions increase for the β_1 state on W(100) and W(110), but decrease for β_2 and β_3 (indicating a positive pole away from the surface). On both planes, two α substates can be distinguished with energies of 100 to 125 kJ mol^{-1} (100), and 63 to 84 kJ mol^{-1} (110) and a total population of about 50%. The general

† See Chapter 16.

‡ See Chapter 22.

picture even on single planes is still complex and detailed interpretation of the various binding energies on different planes and of the different signs of the dipoles is still uncertain.

The detection of a third, or virgin, state by field-emission microscopy [15, 23, 24] exemplifies the value of examination of adsorbate–adsorbent systems by various techniques; it is formed by adsorption of CO on the W tip at 50 K or below. The emission pattern of the clean tip is unchanged but the work function increases by 0·8 eV, thereby indicating chemisorption; this decreases due to physisorption on this primary layer at higher coverages. The virgin state probably comprises two CO molecules attached to one W atom in a disordered layer. Above 200 K, sufficient thermal energy is available to transform this state to the stable α and β states with preferential desorption of the α state. On cooling to 50 K, additional adsorption into the α state is effected, with no increase in concentration of the virgin state, as denoted by the change of work function, since the dipole moment of the α state is positive away from the surface whereas that of the virgin state is negative. Although this latter state is present on all crystal faces after initial adsorption at 77 K, the transformation to α and β states at higher temperatures is markedly dependent on the surface crystallography of the various planes.

The LEED pattern of the chemisorbed overlayer [25, 26] formed at room temperature indicates an immobile layer without long-range order. At a temperature sufficiently high for the onset of mobility, a c(2×2) pattern is developed. The work function decreases during this transformation and may result from one or more of the following processes: (i) a disorder–order conversion [27], (ii) transformation from a linear one-site to a two-site bonding [28]; and (iii) some reconstruction of the surface structure.

Similar states are formed on molybdenum, e.g. 3β and 1, or 2, α states on Mo(100) [29]. The population ratios differ to some extent and, in particular, the α state is substantially less populated.

27.3.2. Nitrogen

Three main states α, β, γ are observed in the thermal desorption spectrum [30, 31]. The peak temperatures and corresponding binding energies are: γ(150 K, $\sim$40 kJ mol^{-1}), α($\sim$500 K, 75 to 100 kJ mol^{-1}) and β(1200 K, 300 kJ mol^{-1}). The γ state has been found on the (100), (110) and (111) planes, and is undoubtedly present with about the same binding energy on higher-index planes. It is a weak molecularly (physisorbed) bound state which is desorbed with first-order kinetics and does not undergo isotopic mixing. It comprises three substates: γ [possibly atomic on W(110)] and γ^+ are electropositive species; and γ^- [on W(100)] is electronegative and probably recessed into the surface [32].

The α state is formed on the (111) plane but not on the (100) and (110) planes; the kinetic order of desorption is unity and the state is molecular. Probably the α state is present only when sites of fourfold symmetry are available [33, 34]. It may be converted to the β state by a low-energy activated process, since, with increasing nitrogen pressure, its surface concentration initially increases to a maximum before decreasing to zero at saturation coverage.

The β state has been investigated in detail [32–35]. The evolution process is second-order indicating an atomic adspecies; the binding energy is virtually the same on all planes that have been studied, and is independent of coverage on the (001), (012) and (013) planes. The sticking probabilities are initially 0·25 to 0·4 and are independent of coverage, thereby indicating that a precursor precedes the final chemisorption. The sticking probabilities on the (011), (111) and (211) planes are very low (0·04 to 0·004) and vary with coverage; in particular, the decrease with coverage varies as $(1 - \theta)^2$ on the (110) plane, indicating a two-site adsorption state. The LEED pattern on W(100) is a (2×2) structure resulting from occupation of alternate sites by N adatoms (34, 36).

The β state displays two sub-peaks β_1 and β_2 with similar binding energies; the former appears to be molecular, since it is desorbed with first-order kinetics, but a non-integral higher order is found for the β_2 state. However, both states undergo complete isotopic mixing [37, 38] as expected for atomic adspecies, but the dipole moments are opposite in sign. One explanation [39] of these unusual results is that an equilibrium of a molecular complex with adatoms is set up, and first-order kinetics arise from desorption of adatom pairs. However, the interpretation of much of the data is still unsatisfactory and uncertain.

Some results on nickel films and silica-supported metal are of interest [40]. Three states, labelled γ_1, γ_2, γ_3 with peak temperatures of 100, 130 and 195 K, are observed and may arise from the presence of (111), (100) and (110) planes, respectively. On the (110) plane, the work function is increased, but decreased on the other two planes. The infra-red spectroscopic results are [41, 42] consistent with a molecularly adsorbed species and identification of the band-peak with the N–N stretching frequency with the nitrogen molecule perpendicular to the surface. Tentative assignments are [40] $(Ni{-}N{=}N)^-$ to γ_3 and the molecular ion structure $(N{=}N)^+$ to the γ_1 and γ_2 states.

27.3.3. Hydrogen

A γ state, probably a physisorbed molecular layer over the primary chemisorbed β state, with a binding energy of 5 to 10 kJ mol^{-1}, is formed on all planes; it is desorbed with first-order kinetics at $\sim$190 K. Definite evidence

[43] for an α state desorbing between 200 and 300 K is lacking. In an earlier field-emission microscopic study [44], four main states were reported: A, (211), electronegative, 85 to 195 kJ mol^{-1}; B, (411), electropositive, 65 to 85 kJ mol^{-1}; C, (411), electronegative, 35 to 65 kJ mol^{-1}; D, (411), (211), electropositive, 25 to 45 kJ mol^{-1}. (In the above list, the plane with the highest population, the sign of the dipole, and the binding energy range, are given in that order.) Confirmatory evidence, however, is lacking.

The most detailed knowledge is available about the β states [45–47]. The kinetic order of evolution is two, consistent with an atomic state. Thermal desorption spectra from four single planes have been determined in good agreement in various laboratories. Two sub-states β_1, β_2 are formed on the (100), (110), (211) and (111) planes, and additional β_3, β_4 states on the (111) plane. For the first three planes, the β_2 state is more strongly bound [(211), 160 kJ mol^{-1}; (110), 140 kJ mol^{-1}; (100), 135 kJ mol^{-1}] than the β_1 state [(110), 115 kJ mol^{-1}; (100), 110 kJ mol^{-1}; (211), 85 kJ mol^{-1}]. The population ratio (β_1/β_2) is unity for the (110) and (211) planes, and two for the (100) plane.

The results on W(100) are the most extensive. The LEED pattern is initially c(2 × 2) with a maximum intensity at ~0·2, and changes to the (1 × 1) structure at saturation coverage [48]. The saturation population has been reported as $1{\cdot}5 \times 10^{15}$ to $2{\cdot}0 \times 10^{15}$ adatoms cm^{-2}, the highest value having been determined from a molecular-beam study [49]. At $\theta = 0{\cdot}25$, the 5×10^{14} H adatoms cm^{-2} are associated with 1×10^{15} atoms cm^{-2} as expected from the c(2 × 2) structure; this β_2 state, therefore, probably bridges two adjacent W atoms. Above $\theta \sim 0{\cdot}2$, the β_1 state apparently sequentially populates the remaining unoccupied sites. To accord with a population ratio (β_1/β_2) of two, β_1 might occupy fourfold symmetry sites, but the maximum intensity would then occur at $\theta = 0{\cdot}33$ and a (1 × 1) structure would not occur at saturation [45, 48].

In contrast, the data provided by field-emission elastic resonance and inelastic tunnelling† indicate that β_1 and β_2 are not separate binding states. On W(100), the elastic scattering enhancement at 0·4 eV (below E_F) from the clean metal corresponding to a surface state [50] disappears on adsorption of hydrogen up to $\theta \sim 0{\cdot}2$, and two new levels at 0·9 and 1·1 eV appear [51]. These two β_2 peaks decay and a broad level at ~2·5 eV develops when θ is increased to its saturation value. The interpretation of these results is that: (i) the β_1 and β_2 peaks in the thermal desorption spectra are not associated with two distinct binding states; (ii) the β_1 state does not sequentially populate sites after the saturation value of the β_2 state has been attained at $\theta = 0{\cdot}2$; and (iii) above $\theta \sim 0{\cdot}2$, each additional H-atom adsorption converts one (or more) of the β_2 adatoms to a new state, i.e. a coverage-dependent change

† See Chapter 17.

in binding energy brought about by adsorbate–adsorbate interactions takes place.

The results obtained from inelastic tunnelling emission experiments [51] support these conclusions. For hydrogen adsorption on W(100) at 77 K, up to $\theta \sim 0{\cdot}2$, two vibrational modes at 0·14 and 0·07 eV are observed, and, moreover, are displaced in accord with the mass ratio when deuterium replaces hydrogen. (These vibrational modes had previously been observed by electron energy-loss spectroscopy [52].) When the coverage is increased to saturation, both peaks disappear, probably as a result of a change in the vibrational and electronic structure of the chemisorbed layer.

The choice between the two models: (i) two distinct binding sites β_1, β_2, and (ii) one β_2 state that undergoes a binding transition arising from lateral interactions within the adsorbed layer, cannot be made with any certainty.

In contrast, the adsorbed layer at low coverage on the W(110) plane has a p(2 × 1) structure with β_2 occupying alternate sites along one axis. A sequential filling by the β_1 state to form a saturated (1 × 1) structure with one H adatom per W site is here in accord with the unit population ratio of the two states, and also with the change of work function with coverage the plot of which displays two linear sections with a change of slope at $\theta = {\sim}0{\cdot}5$. On W(100), however, monotonic linear increase of the work function supports the concept of a single β_2 state on this plane.

27.4. STATUS OF THE PRESENT SITUATION

For the chemisorption of simple gases on single crystal planes of metals, results of the characterization and measurement of electronic and geometric properties of overlayers are now available with fairly high precision. Determination of the long-range order and the dimensions of the unit cell has become a routine exercise, and the advance in the theoretical treatments of multiple scattering processes provide hope that the exact location of an adatom in the unit cell and the accurate evaluation of chemisorption bond length will be derived from intensity measurements. The full potential of electron spectroscopic techniques has not yet been exploited but they are already providing valuable information on the local density of occupied levels, the nature of bonding orbitals, transition probabilities and other parameters essential for the quantification of theoretical treatments of the problem of chemisorption at the equilibrium configuration. At present, despite the complexities of the simplest adatom/metal system, quantitative and reproducible experimental results are available, but further advance in theory is necessary before unambiguous interpretations of the number of multiplet states, the nature of the bonds and binding energies can emerge.

REFERENCES

1. R. W. Joyner, J. Rickman and M. W. Roberts, *Surface Sci.*, 1973, **39**, 445.
2. T. A. Clarke, R. Mason and M. Tescari, *Surface Sci.*, 1973, **40**, 1.
3. T. B. Grimley, *Proc. Phys. Soc.*, 1967, **90**, 751; **92**, 776.
4. T. B. Grimley, *Surface Sci.*, 1969, **14**, 395.
5. T. L. Einstein and J. R. Schrieffer, *Phys. Rev. B*, 1973, **7**, 3629.
6. P. J. Estrup and E. G. McRae, *Surface Sci.*, 1971, **25**, 1.
7. J. E. Houston and R. L. Park, *Surface Sci.*, 1971, **26**, 286.
8. D. L. Adams, *Surface Sci.*, 1974, **42**, 12.
9. J. C. Tracey, *J. Chem. Phys.*, 1972, **56**, 2736, 2748.
10. J. C. Tracy and P. W. Palmberg, *Surface Sci.*, 1969, **14**, 274.
11. A. J. Pritchard, *J. Vac. Sci. Tech.*, 1972, **9**, 895.
12. H. Conrad, G. Ertl, J. Koch and E. E. Latta, *Surface Sci.*, 1974, **43**, 462.
13. R. R. Ford, *Adv. Catalysis*, 1970, **21**, 51.
14. R. Klein, *J. Chem. Phys.*, 1958, **31**, 1306.
15. L. Swanson and R. Gomer, *J. Chem. Phys.*, 1963, **39**, 2813.
16. L. R. Clavenna and L. D. Schmidt, *Surface Sci.*, 1972, **33**, 11.
17. C. G. Goymour and D. A. King, *J. Chem. Soc., Faraday Trans. I*, 1973, **69**, 736, 749.
18. T. E. Madey and J. T. Yates, *J. Chem. Phys.*, 1965, **44**, 1675.
19. Y. Viswanath and L. D. Schmidt, *J. Chem. Phys.*, 1973, **59**, 4184.
20. P. A. Redhead, *Trans. Faraday Soc.*, 1961, **57**, 641.
21. J. T. Yates and D. A. King, *Surface Sci.*, 1972, **32**, 479; 1973, **36**, 739; 1973, **38**, 114.
22. C. Kohrt and R. Gomer, *Surface Sci.*, 1971, **24**, 77; 1973, **40**, 71.
23. A. E. Bell and R. Gomer, *J. Chem. Phys.*, 1966, **44**, 1065.
24. T. Engel and R. Gomer, *J. Chem. Phys.* 1970, **52**, 1832.
25. J. W. May and L. H. Germer, *J. Chem. Phys.*, 1966, **44**, 2895.
26. J. Anderson and P. J. Estrup, *J. Chem. Phys.*, 1967, **46**, 563.
27. D. L. Adams and L. H. Germer, *Surface Sci.*, 1971, **27**, 21; 1972, **32**, 205.
28. R. A. Armstrong, *Can. J. Phys.*, 1968, **46**, 949.
29. L. D. Schmidt, *in* "Adsorption–Desorption Phenomena" (F. Ricca, ed.), Academic Press, New York, 1972, p. 391.
30. T. W. Hickmott and G. Ehrlich, *J. Phys. Chem. Solids*, 1958, **5**, 47.
31. G. Ehrlich, *J. Chem. Phys.*, 1961, **34**, 29.
32. T. A. Delchar and G. Ehrlich, *J. Chem. Phys.*, 1965, **42**, 2686.
33. P. J. Estrup and J. Anderson, *J. Chem. Phys.*, 1967, **46**, 567.
34. D. L. Adams and L. H. Germer, *Surface Sci.*, 1971, **26**, 109; 1971, **27**, 21.
35. P. W. Tamm and L. D. Schmidt, *Surface Sci.*, 1971, **26**, 286.
36. L. R. Clavenna and L. D. Schmidt, *Surface Sci.*, 1970, **22**, 365.
37. J. T. Yates and T. E. Madey, *J. Chem. Phys.*, 1965, **43**, 1055.
38. L. J. Rigby, *Can. J. Phys.*, 1965, **43**, 532.
39. J. L. Robins, W. K. Wasburton and T. N. Rhodin, *J. Chem. Phys.*, 1967, **46**, 665.
40. D. A. King, *Surface Sci.*, 1968, **9**, 375.
41. R. P. Eischens and J. Jacknow, *in* "Proceedings of the 3rd International Congress on Catalysis", North Holland Publ. Co., Amsterdam and New York, 1965, p. 627.
42. R. van Hardeveld and A. van Montfoort, *Surface Sci.*, 1966, **4**, 396.
43. T. W. Hickmott, *J. Chem. Phys.*, 1960, **32**, 810.
44. W. J. M. Rootsaert, L. L. van Reijen and W. H. M. Sachtler, *J. Catalysis*, 1962, **1**, 416.

45. P. W. Tamm and L. D. Schmidt, *J. Chem. Phys.*, 1969, **51**, 5352; 1971, **54**, 47.
46. D. L. Adams and L. H. Germer, *Surface Sci.*, 1972, **32**, 205.
47. B. D. Barford and R. R. Rye, *J. Chem. Phys.*, 1974, **60**, 1046.
48. P. J. Estrup and J. Anderson, *J. Chem. Phys.*, 1966, **45**, 2254.
49. T. E. Madey, *Surface Sci.*, 1972, **32**, 355; 1973, **36**, 281.
50. B. J. Waclawski and E. W. Plummer, *Phys. Rev. Letters*, 1972, **29**, 786.
51. E. W. Plummer and A. E. Bell, *J. Vac. Sci. Tech.*, 1972, **9**, 583.
52. F. M. Probst and T. C. Piper, *J. Vac. Sci. Tech.*, 1967, **4**, 53.
53. B. D. Barford and R. R. Rye, *J. Chem. Phys.*, 1974, **60**, 1046.

Author Index

Numbers in parentheses are reference numbers and are included to assist in locating references in the text where the authors' names are not mentioned. Numbers in *italics* indicate the pages on which references are listed in full.

H

I

J

K

N

O

P

Q

R

Subject Index

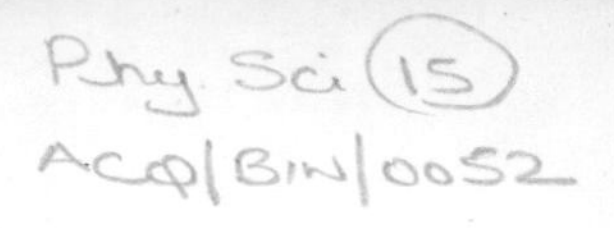
Phy. Sci (15)
ACQ/BIN/0052